REPROGRAMMABLE RHETORIC

REPROGRAMMABLE RHETORIC

Critical Making Theories and Methods in Rhetoric and Composition

EDITED BY
MICHAEL J. FARIS AND STEVE HOLMES

UTAH STATE UNIVERSITY PRESS
Logan

© 2022 by University Press of Colorado

Published by Utah State University Press
An imprint of University Press of Colorado
245 Century Circle, Suite 202
Louisville, Colorado 80027

All rights reserved

 The University Press of Colorado is a proud member of the Association of University Presses.

The University Press of Colorado is a cooperative publishing enterprise supported, in part, by Adams State University, Colorado State University, Fort Lewis College, Metropolitan State University of Denver, University of Alaska Fairbanks, University of Colorado, University of Denver, University of Northern Colorado, University of Wyoming, Utah State University, and Western Colorado University.

ISBN: 978-1-64642-257-9 (paperback)
ISBN: 978-1-64642-258-6 (ebook)
https://doi.org/10.7330/9781646422586

Library of Congress Cataloging-in-Publication Data

Names: Faris, Michael J., editor. | Holmes, Steve, 1983– editor.
Title: Reprogrammable rhetoric : critical making theories and methods in rhetoric and composition / edited by Michael J. Faris and Steve Holmes.
Description: Logan : Utah State University Press, [2022] | Includes bibliographical references and index.
Identifiers: LCCN 2022017922 (print) | LCCN 2022017923 (ebook) | ISBN 9781646422579 (paperback) | ISBN 9781646422586 (ebook)
Subjects: LCSH: English language—Rhetoric—Study and teaching (Higher)—United States. | English language—Composition and exercises—Study and teaching (Higher)—United States. | Critical pedagogy—United States. | Multimedia systems—Study and teaching (Higher)—United States.
Classification: LCC PE1405.U6 R475 2022 (print) | LCC PE1405.U6 (ebook) | DDC 808/.042071173—dc23/eng/20220520
LC record available at https://lccn.loc.gov/2022017922
LC ebook record available at https://lccn.loc.gov/2022017923

Cover illustration © Sashkin/Shutterstock

CONTENTS

Introduction
    Michael J. Faris and Steve Holmes   3

SECTION 1: FRAMING CRITICAL MAKING

1. Noise Composition: A Story of Co-Design and Relationality
   Steven Hammer   29

2. The Circulation of Touch: Very Simple Machines for Creating Tactile Textual Experiences
   David M. Sheridan   45

SECTION 2: TEXT MINING AS CRITICAL MAKING

3. The Woman Who Tricked the Machine: Challenging the Neutrality of Defaults and Building Coalitions for Marginalized Scholars
   Cana Uluak Itchuaqiyaq   75

4. Critical Text Mining: Ethical Paradigms for Determining Emoji Frequency in #blacklivesmatter
   Kellie M. Gray and Steve Holmes   92

5. Reprogramming the Faciloscope: A Software Development Story
   Ryan Omizo   108

6. Big Data, Tiny Computers: Making Data-Driven Methods Accessible with a Raspberry Pi
   Aaron Beveridge and Nicholas Van Horn   129

SECTION 3: EVERSION AND CRITICAL MAKING

7. Touch-Interactive Rhetorics: Exploring Our "First Sense" as a Rhetorical Act of Eversion
   Matthew Halm and David M. Rieder   145

8. What the Computer Said: Poetic Machines, Rhetorical Adjuncts, and the Circuits of Eloquence
   Andrew Pilsch   165

9. Actionable Monuments: Making Critical Augmented Reality Activism
   Sean Morey and M. Bawar Khan   180

SECTION 4: CRITICAL PLAY AS CRITICAL MAKING

10. Reparative Making: Re-Orienting Critical Making for Queer Worldmaking
    Michael J. Faris   201

11. Developing *A Strong Fire*: Bridging Critical Making, Participatory Design, and Game Design
    Wendi Sierra   222

12. Twisted Together: Twine Games as Solidarity Machines
    Kendall Gerdes   242

SECTION 5: CRITICAL MAKING AS INSTRUCTIONAL DESIGN

13. Cultivating Critical Makers: Crafting with Paper-Electronic Circuits in an Online First Year Composition Course
    Bree McGregor   265

14. Crafting in the Classroom: Carpentry and Pedagogy in Rhetoric and Composition
    John Jones   286

Index   301

REPROGRAMMABLE RHETORIC

INTRODUCTION

Michael J. Faris and Steve Holmes
Texas Tech University

reprogramme | *reprogram*, v.
　"Transitive. *To programme differently or again; to supply with a new programme.*"

(OED 2020b)

Reprogramming describes the activity of revising or rewriting an existing program. The definition of reprogramming contains within itself the presupposition that a current program exists and functions in some way. To be sure, the act of reprogramming does not necessarily entail a positive or negative connotation. Even the most popular software can benefit from periodic upgrades to layer new functionality over an older infrastructure. At other times, however, entire programs or parts of programs need to be rewritten and reconceived entirely because they are not functioning well.

　This collection addresses a specific program: rhetoric and composition scholars' past and present engagements with critical making and maker cultures. While there is no single overarching program that can characterize the diversity of this work, our field's early engagements have nevertheless settled into some familiar subroutines. In turn, these subroutines necessitate a sustained and dedicated act of reprogramming, which *Reprogrammable Rhetoric* seeks to address: First, an all too familiar lack of *critical* in critical making and, second, a related need to rethink how we employ critical making to negotiate the theory and practice divide in rhetoric and composition studies. We will unpack these claims in this introduction as the effort to reprogram these two subroutines offers a rationale for why we have set up this edited collection in the way that we have.

MAKING AND CRITICAL MAKING OUT-
SIDE OF RHETORIC AND COMPOSITION

The term "critical making" is understood differently throughout academic and nonacademic contexts. It generally describes a wide range of practices, theories, and methods that emphasize the potential of making, hacking, and remaking to effect some sort of social or political change—that is, to do rhetoric. Matt Ratto's (2011) theories and practices of critical making remain an ongoing conceptual touchstone for many makers in different disciplinary and practitioner audiences. Critical making "signals a desire to theoretically and pragmatically connect two modes of engagement with the world that are often held separate—critical thinking, typically understood as conceptually and linguistically based, and physical 'making,' goal-based material work" (Ratto 2011, 253). Rhetoric and composition scholars have had many debates over the past decades regarding the relationship between theory and practice. Thus, Ratto's articulation of critical making is appealing because it allows theory (concepts, analysis, critique) to connect to material and practical forms of enactment and composition. Furthermore, Ratto allows making practices themselves to be a starting place through which to build reflective theoretical arguments (see also Ratto and Hockema 2009).

Extending Ratto's early work, Ratto and Megan Boler (2014) published an edited collection titled *DIY Citizenship: Critical Making and Social Media*. They note that critical making tends to focus on digital media, which is unsurprising given the complex and massive ways in which contemporary technologies structure and mediate identity in the present. However, they explain that DIY citizenship goes far beyond traditional craft-making and digital media considerations to examine how hybrid material compositions like yarn-bombing activism function as a form of public or counterpublic sphere participation.

Beyond Ratto's work, media artist-theorist Garnet Hertz (2012) has published an influential collection of critical making manifestos in zine form. He also runs a critical making lab at Emily Carr University in Vancouver, Canada. Similar to the idea of DIY citizenship, Hertz's work emphasizes critical making's genealogy in civil disobedience, which was even more clearly underlined in another zine manifesto collected by Hertz (2016a) called *Disobedient Electronics*. In both sets of manifestos, critical making emerges as the natural and necessary outgrowth of the tactical media famously theorized and practiced by the Critical Art Ensemble (CAE). As Hertz (2016b) argues elsewhere, the potential for critical making lies in its potential to reintroduce criticality into making and maker cultures that have become depoliticized, creating opportunities for the making of "built and functional devices" that "materially articulate particular stances and ideas" and "enable individuals to reflect on the personal and social impact of new technologies."

As this reference to the CAE highlights, it is important to observe that critical making practices draw on a number of historical lineages, including tactical media, "hacktivism," and, more recently, the emergence of digital humanities research and pedagogy and the popular cultural "makers" movement. The makers movement includes a number of humanities-based "makers' labs" that have been started in universities throughout the United States and Canada. Art schools, from Rhode Island School of Design (RISD) to the University of California, Berkeley, also offer courses in critical making. Nearly a decade ago, RISD published an edited collection titled *The Art of Critical Making* (Somerson and Hermano 2013), which described its critical-making philosophies.

CRITICAL MAKING IN RHETORIC AND COMPOSITION STUDIES

This conceptual shift from the act or mentality of making itself (Hertz 2012, 2016b) to the potential for making things to do something offers points of overlap with rhetoric and composition studies. A great deal of previous work on multimodal composition and material rhetoric is and was already a form of critical making in all but name. Literal reprogramming that engages physical computing, critical maktivism, circuit programming, and related phenomena such as rhetorical processes and compositions has been one common approach that our field has imported from making discourses. Helen J. Burgess and David M. Rieder's (2015) special issue of *Hyperrhiz* ("Kits, Plans, Schematics") offered a landmark engagement with critical making and composition, which was followed by Rieder's (2017) book *Suasive Iterations: Rhetoric, Writing, and Physical Computing*. Most recently, Burgess and Roger Whitson (2019) published a special issue of the online journal *Enculturation* that was devoted to executable approaches to critical making: building kits and schematics that enable readers/viewers to reprogram existing digital programs and physical objects. Digital rhetoric and composition scholars have also explored software and coding (Brooke 2009; Brown 2015; Brock 2012; Jones and Hirsu 2020; Vee 2017) and digital humanities (Ridolfo and Hart-Davidson 2015).

Echoing positions like Ratto's refusal to divide theory from practice, a number of multimodal composition scholars have challenged the reduction of writing to the print-based analytical essay. Alongside exploring digital forms of writing, Jody Shipka (2011) added that considerations of multimodality should include *all* modalities: "Texts that explore how print, speech, still images, videos, sounds, scents, live performance, textures (for example, glass, cloth, paper affixed to plastic), and other three-dimensional objects come together, intersect, or overlap in innovative and compelling ways" (8). Other scholars have more directly pushed the notion of multimodality toward the playful and critical experimentation with the materiality of digital and nondigital objects through engagements with maker cultures in general. David M. Sheridan (2010) articulated an early argument for rhetoric and composition researchers to explore how to compose material objects through 3D printers in the writing classroom. Importantly, Sheridan productively suggested that the field's reasoning for including visual and digital forms would also lend itself to supporting the use of digital technologies to fabricate physical objects in the spirit of many still-popular maker practices and technologies.

While such arguments are aimed at researchers in the present, it is important to note that materiality has always had a role—if an unacknowledged one—in rhetorical practice. Early examples include Demosthenes's embodied embrace of the canon of delivery by shouting at waves to train his speaking voice as well as the use of the Greek *pynx*, or a small hill, to amplify an oral speaker's voice (Morey 2015). To study and enact rhetoric has always been a study of multimodality even if the modes privileged or studied in a given historical moment have been limited (McCorkle 2012). Even prior to rhetoric and composition studies' embrace of digital technologies, Gregory Ulmer (1994) noted that *chora*, as discussed in Plato's *Timaeus* dialogue, functions as a similar space of cognitive, embodied, and material potentiality for invention before rhetoric actualizes as practice (see also Rickert 2013).

While there are a number of excellent reasons that scholars have offered for studying and composing through critical engagements with rhetoric and technology's material character, David M. Sheridan, Jim Ridolfo, and Anthony J. Michel's (2013) discussion of the role of multimodality as a kairotic mode of discovery for public rhetoric pedagogy remains particularly compelling. Rhetoric in the Greco-Roman tradition historically seeks to prepare students for participation in civic life. Insofar as more forms of public interaction are occurring in digital and material ways, Sheridan et al. argue that writing teachers foreclose in advance their students' abilities to participate in these spaces if they do not explore how rhetoric works through different mediums. Furthermore, they suggest that to teach multimodality is to teach many of the same principles that we privilege for print-based writing students. To offer a recent illustration, Michael J. Faris et al. (2018) demonstrated how to teach a graduate new media rhetoric course through littleBits, a set of electronic building blocks, as an important part of learning about the risks of composition, experimentation, and failure. For this reason, the authors declare, "material composition is within the disciplinary purview of rhetoric and composition" (Faris et al. 2018, "Situating"); or, as Sheridan (2010, 257) powerfully states, "It's ours."

To sum up, maker cultures and critical making are not something new to be added on to rhetoric and composition studies. Rather, these conversations and practices can help us to continue to explore in new ways our historic interests in materiality and multimodality. In his webtext, "A Maker Mentality Toward Writing," Sheridan (2016) offers a cogent illustration of what writing studies can gain through examining and integrating the design techniques of "makerspaces." Once strictly the domain of engineers, art and design students, and computer scientists, now a growing number in digital rhetoric are starting to experiment with the vocabularies, tools, and design techniques of physical computing, coding, and related practices. For example, Steven Hammer and Aimée Knight (2015) have advocated tinkering with circuit-bending as a way to privilege invention and acts of discovery. As part of their exigency for their special issue, Burgess and Whitson (2019) noted a desire to continue the work of the original *Hyperrhiz* special issue on "executable culture." Whether theoretical or practically inclined, making things should equip readers or viewers to make physical end products themselves. In a definition we will revisit again in the next section, Burgess and Whitson (2019) situated critical making as a process and series of relations that is not reducible to the production of an end product or technical knowledge alone. Rather, critical making as part of an ethic of executable culture involves "a special focus on sharing and the various processes involved in the construction of objects and knowledge" (Burgess and Whitson 2019). In addition to some of the individuals we have already mentioned (Faris, Rieder, Burgess, and Sheridan), Burgess and Whitson (2019) also pointed to a panoply of related concepts like "tactical media" (Raley 2009), "speculative design" (Dunne and Raby 2013), "prototyping" (Sayers 2015), and "adversarial design" (DiSalvo 2012) that are increasingly part of the vocabulary of digital rhetoric and composition scholars and teachers.

CULTURAL CODING ERRORS IN COMPOSITION'S CRITICAL MAKING PROGRAM

While critical making is still emerging as a subfield in rhetoric and composition, there is already enough work in our field and in critical making discourses outside of our field to call for some acts of reprogramming. In part, one of our challenges for *Reprogrammable Rhetoric* was not just in identifying some of the problems within how critical making has been defined and continues to be defined but also in determining to what extent these problems are reprogrammable. Clearly, we believe that some of these problems are indeed reprogrammable, and we hope that our contributors' efforts will help in these activities of reprogramming.

As a call to action, Joyce Locke Carter's (2016) chair's address to the Conference on College Communication and Composition membership called for the field to adopt a maker mentality. This approach went beyond mere technological training (or tool learning) to also include and yet conceive differently our traditional focus on ethics, audience, situation, and motive. In support of Carter's call to think about ethics, the continued insistence or implicit default to apolitical conceptions of the technical and material remains a broader problem that faces our field's past and present engagements with maker cultures.

As a powerful example, Burgess and Whitson (2019) pointed to Hertz's (2018) effort to reprogram *Make* magazine's highly influential "Maker's Bill of Rights." Early conceptions of maker culture were populated by technical concerns such as, "Components, not entire subassemblies, shall be replaceable" (Jalopy, Torrone, and Hill 2006). By comparison, Hertz's (2018) update refused to divorce technical concerns from political ones. He offered new ethical axioms such as, "If women don't have a pivotal voice at an event, panel or exhibition, I'm not participating." To put it simply, engaging critical making and maker cultures must be accompanied by a firm ethical commitment or, at least, an ethical commitment needs to emerge out of whatever it is we are making. Context matters. Even if the purpose or subject matter a given making activity is not explicitly political in orientation, to be an actualized thing in this world is to be already shot through with political structure, power, and other forms of relationality. As a case in point, Burgess and Whitson (2019) observed that far from politically neutral areas of rhetorical invention, makerspaces on college campuses all too often function as low- or unpaid development labs for large technical companies.

An additional problem with viewing technical skills—digital and nondigital—as apolitical lies in contributing to a historic and ongoing white masculinist bias within making discourses. Christina Dunbar-Hester (2014) noted that nonwhite activists who seek to embrace critical making have struggled with "inscribed historical patterns of inclusion and exclusion, as electronics tinkering has long been associated with white masculinity" (76). Many treatments of critical making may pay lip service to inclusivity, but nevertheless continue to embody a particular set of tools that often require advanced expertise in forms of knowledge and practices from white male-dominated fields. Examples include

Raspberry Pis, physical circuit building, data visualization and analysis, and Arduino microcontrollers (Gollihue 2019). In the introduction to *Making Things and Drawing Boundaries*, Jentery Sayers (2017) affirmed that critical making still needed to critique "the normative assumptions and effects of popular maker cultures—usually white, cisgender, straight, male, and able-bodied" (7).

In the history of composition studies and new media studies, the theories informing such criticality have perhaps been too narrowly limited in their genealogies. Malea Powell (2016) and Angela Haas (2007) have both suggested that making is not a newly theorized practice but has intellectual roots in Indigenous practices—practices that also theorize rhetoric and composition. Following from their arguments, we should understand practice as operative theory. Even calling to overturn or deconstruct a theory–practice divide by engaging critical making can still turn on other unacknowledged divisions like the ongoing colonization of knowledge in our field and in maker discourses.

In recognition of this problem, some of our chapters in this edited collection trace excluded alternative genealogies of making. Steven Hammer's chapter explores how non-Western maker ontologies predate many of the Western thinkers' interests in nonhuman agency such as "rhetorical carpentry" (Brown and Rivers 2013), Wendi Sierra's chapter explores Indigenous game design and play as a form of critical making, and, finally, Cana Uluak Itchuaqiyaq's chapter explores critical text mining in the context of studying academic citation practices through text mining and decolonial theory. Kellie M. Gray and Steve Holmes also recontextualize text mining into webscraping and analyzing tweets for #blacklivesmatter to participate in data curation as a form of activist engagement for racial equality.

These are the types of approaches that we hope this edited collection features in order to help rhetoric and composition researchers, teachers, and makers reprogram our early efforts to engage critical making discourses and practices. To be sure, more work than what this edited collection represents is needed. In reprogramming the "Maker's Bill of Rights," Hertz (2018) also acknowledged that the broad claim that technology can solve society's issues often obscure the fact that these same invoked technologies are still causing many of them. Other historically excluded or unrecognized makers have specifically challenged this issue, such as the work that occurs in feminist hacker spaces. As the title of Amy Burek et al.'s (2017) zine chapter is titled, "Feminist Hackerspaces: Hacking Culture, Not Devices." Other hackathons or makerspaces outside of institutional spaces have specifically endeavored to support people of color, women, and gender minorities. Echoing Burgess and Whitson's (2019) emphasis on processes (executability) and not just end products in critical making, Krystin Nicole Gollihue (2019) noted that the Tuscone Women Techmakers Hackathon makes ethical dispositions like civility ("Be excellent to one another" [quoted in Gollihue 2019, 15]) as a fundamental element of critical making. For these and other excluded makers, to ask about critical making is not about tool use or technological mastery per se, but about what type of ethical community of relations and identities a given making activity supports or sustains.

Similarly, other maker collectives like Machine Room, Noisebridge, Seattle Attic, and Double Union also situate their work in a comprehensive ethical code as an alternative to mainstream maker cultures. Patrick Jagoda's (Bennett et al. 2018; Ehrenberg, Jagoda, and Gilliam 2018; Jagoda et al. 2015) work with economically disadvantaged youth in Chicago with game building offers another alternative history alongside queer game designers who explore how the materiality of controllers can shape perception (Pozo 2018). Countless examples exist. We have to participate (as some already have) more broadly in recognizing these efforts. Furthermore, part of our reprogramming activities must include understanding the mechanisms in our field and making discourses outside our field that continue to prevent us from engaging them. These structures run deep in critical making, but also through digital rhetoric and digital humanities scholarship. In a review of the role of "critique" in critical making and digital humanities scholarship, Jagoda (2017) complained, "When we engage and test the ideas of [notable critical makers], however, it is through methods of bricolage, remixing, modding, and design" (361). In other words, all too often scholars continue to revert to the cultural programming routines of the old "Maker's Bill of Rights" instead of using Hertz's (2018) reprogrammed one or reprogramming new ones on their own.

This tendency is one that cuts to the central problem with answering this question about the extent to which critical making is reprogrammable. As a case in point, consider the turn to "object-oriented rhetoric" or "things" (Barnett and Boyle 2016), which undergird certain approaches to critical making such as rhetorical carpentry (Bogost 2012; Brown and Rivers 2013). In Andrea Riley-Mukavetz et al.'s (2016) Cultural Rhetorics Conference panel (titled "Three Queer/Feminist/Indigenist Rants and a Critique of Heteropatriarchal Colonialism in Object-Oriented Theory"), they argued that object-oriented rhetoric (OOR) and object-oriented ontology (OOO) reinscribed colonial relations (see also Powell 2016). While OOR and OOO are arguably more complex than some of their critics have allowed, it is undeniable that OOO, by metaphysical design, offers no answers for ethics and politics beyond a Heideggerian quietism and the perpetual claim that objects' realities are deeper than our knowledge of them. OOO and OOR do not shift easily from an ontological "is" to an ethical "ought." Furthermore, there is an undeniable ethnocentrism in OOO. Graham Harman (2011) began with Heidegger and other mostly European phenomenologists as the basis for OOO rather than considering or even attempting to acknowledge prior Indigenous epistemic traditions. From the perspective of making critical-making discourses, it is difficult to overlook these historical contexts because no philosophical form or theory ever fully lifts free from contingencies of culture and language (Felski 2015). Metaphysical claims in print about the ontological nature of reality are material instantiations and shot through with the contingencies of history.

To sum up, the *critical* part lies in connecting any discussion or use of technologies and making to their ethical and political contexts. If rhetoric and composition scholars only privilege, for example, 3D printers and even text mining or coding, we may unwittingly reproduce a colonial mentality ("can," Burgess and Whitson [2019] argue, is a privileged form of making, whereas not all college classrooms will have a research university's corporate-funded makerspaces). For a similar reason in a different context, Rick Wysocki and co-authors argue in their manifesto "On Multimodality" that "practices of making and critical activity must be rendered mutually supportive" (2019, 21). That is, criticality and composition are not separate activities, but constantly in conversation—in conjunction—with each other. Rhetoric and composition scholars arguably need our own paradigm since our field approaches writing and rhetoric through a generative sense not seen in other fields. Politics in rhetoric and composition's interests in critical making, Burgess and Whitson (2019) clarify, is at once about activism for a particular cause like Black Lives Matter but also an interconnected and broader sense of what it means to be human in a collective society that is built on reciprocity and relationality. Thus, reprogramming critical making has to connect making culture to the material conditions that produce kairotic opportunities for interventions of all types.

Alternative genealogies of critical making should seek to trace how institutional *and* noninstitutional forms of genealogies of making work. We need to ask questions as a field such as "Do our histories of making include how women, BIPOC, queer, and working class peoples collaborate, make, tactically appropriate and critical engage with technology?" In other words, we cannot simply make things and build our theories as we make them without acknowledging the existence of other cultural binaries and forms of epistemic and material colonization that structure sites, spaces, materials, and access to making in particular spaces and places. Thus, theories enacted by and produced by practices of making should draw from a variety of intellectual traditions. As Wysocki et al. (2019, 21) argue, "We must negotiate and continuously reorient ourselves across a spectrum of theoretical framing and practical doing."

REPROGRAMMING CRITICAL MAKING IN COMPOSITION

So far, we have laid out some of the functioning programs for critical making in order to suggest some methods for reprogramming them. Reprogramm*able* is the adjective form: "Capable of being reprogrammed" (OED 2020a). As this edited collection will hopefully testify to, there are some productive ways to reprogram some of these issues. While we have talked about programming as a verb, the noun "program" possesses some useful etymological resonances along these lines:

> **program** (n.) 1630s, "public notice," from Late Latin programma "proclamation, edict," from Greek *programma* "a written public notice," from stem of *prographein* "to write publicly," from *pro* "forth" (see pro-) + *graphein* "to write" (see -*graphy*). General sense of "a definite plan or scheme" is recorded from 1837. (Harper 2020)

One needn't be a full-fledged Derridean to appreciate this "always already" connection between the public and the activity of programming in itself. This etymological connotation is yet another reason why we argue that this particular word—reprogrammable—is worth emphasizing for this collection. To reprogram always constitutes the possibility of change and productive (or unproductive) deviation for reaching the Other. To program and to reprogram are to admit these sometimes-neglected sites of construction and relationality. It is to admit that any program is part of a complex and emerging nature–culture assemblage. To reprogram is also to presuppose that something functioning can be done differently for ourselves and the others who use it, even if our goal is to negate, bracket, or ignore that Other. Someone or some*thing* (human and nonhuman) is required to run the program. Programs emerge socially and materially; they enable and disable.

Taken together, our chapters in this collection constitute and initiate some new programs of action for critical making in composition studies. We aim to offer expanded discussions of ethics and politics aimed at guiding the *critical* part of "critical making." We also hope to foster a more traditional engagement with critical making among rhetoric and composition studies teachers who may still be reluctant to view spending thirty dollars on a Makey Makey kit or an Arduino as equivalent to assigning a textbook on digital writing or multimodal composition. By design, we have asked a number of contributors to feature the actual assignments and technological instructions that they taught in actual college courses. Many of the authors have generously agreed to maintain the source code or instruction sets at a durable online location or to otherwise make their programming tutorials accessible to readers of this edited collection and the general public as part of the executable or kit-generating part of critical making. We aim to equip readers both to think about making and to make things themselves.

Some of our chapters are more theoretical and some are more practice-based. However, we want to affirm that we do not see the theory versus practice distinction as a useful one. Lisa Ede (2004, 119–29) and Bruce Horner (2019, xii–xiii), among others, have argued that theory is a form of material, social, and situated practice. And further, practice can be a way to make theory. Ede (2004) writes that we can "use practice as a means of thinking *through* complex scholarly and professional issues" (16), a sentiment that resonates with critical making, which Ratto and others identify with "doing theory" in a way that "entails moving beyond shallow critical reflection" and "attempt[s] to reconcile a schism between those who purportedly create" and those who critique and theorize (Resch et al. 2017, 152). Consequently, we understand practices of critical making as an opportunity to reinvent theory. As Horner (2019) writes, rhetoric and composition classes are sites where students can be invited to "theorize . . . differently": "To theorize is to reinvent, and reinvention requires theorizing" (xiii). Erin Manning (2016), in a discussion of her work at the SenseLab in Montreal, also argues that *making* can open up new avenues of knowledge. She asks, "How does practice that involves making open the way for a different idea of what can be termed knowledge?" (11). Part of this reinvention of practice (making) and theory/knowledge entails, we suggest, identifying conditions that enable and disable certain practices, which remains a valuable part of any critical making project for rhetoric and composition. One only has to look at how some critical making practices have elided intersectional and decolonial considerations to realize that more theoretical or, certainly, ethical discussions are essential. It does our field little good to *do things* and *make things* if such conversations are not accompanied by robust and rigorous political and ethical frameworks to differentiate which forms of critical making help us to build a better and more equitable or just community.

By exploring these themes, *Reprogrammable Rhetorics* explores ways to approach several overlapping questions that we believe our edited collection engages. First, *what ethical and political theories are important for our field to explore in relation to critical making?* Clearly, not all forms of political making fit into left-leaning or progressive social justice or public rhetoric scholarship. In other words, while we do not wish to limit critical making's spirit of experimentation, we and, indeed, our chapters in this collection, strive for more than "making for the sake of making." Second, *what additional intersections between critical making scholarship and digital rhetoric and writing studies can help to extend both the multimodal scholarship and material rhetoric scholarship that our field already explores?* *Reprogrammable Rhetoric* as a whole engages how an explicit engagement with critical making scholarship and practices can extend the various material, embodied, affective, and political dimensions that our field has already enacted while trying to offer new (or neglected) political directions and making practices to explore. Third, *what does our field offer critical making scholarship that it does not necessary attend to as strongly or as explicitly?* For example, does our historic attention to issues of audience or public rhetoric help offer more systemized methods of theorizing the activity of critical making itself? Could critical making scholarship and practice learn from exploring our scholarship on the history of invention or delivery (including decolonial, queer, and feminist interrogations of these histories)? Does our work on intersectional concerns in writing and social justice lend alternative forms of methodological, conceptual, or practical extension to areas and objects of concern for critical making? For example, our chapters on "critical text mining" in section 2 offer examples of how some of our contributors have reconceptualized data collection and analysis methods—methods that may still be treated as an apolitical technology by many. In a comment we in no way mean as critical or presumptuous, perhaps our disciplinary interest in very expansive definitions of materiality and technology might be useful to help critical making discourses shake up some of the "terministic screens" that may have started to settle into place as this research and making area has stabilized around some common objects of interest. By keeping these broad tensions in mind as a primary exigency, it is our hope that *Reprogrammable Rhetoric* offers a new inroad for both critical making and rhetoric and composition audiences.

Our edited collection is divided into five categories, which reflect these ends.

SECTION 1. FRAMING CRITICAL MAKING

Chapters in this section frame critical making as part of rhetoric and composition and specifically as a political practice. Steven Hammer's essay, "Post-Noise: A Story of Co-Design and Relationality," directly engages ethical issues with regard to making but from the standpoint of Indigenous making practices and accessibility on behalf of disabled users. Hammer's essay offers additional relevance for the field's interests in points of overlap and departure between material and cultural rhetorics (object-oriented rhetoric, new materialist rhetorics, etc.). By appealing to the need to ground critical making in prior marginalized non-Western relational ontologies, Hammer importantly situates a turn toward critical making not as a "new turn" but as a need to highlight the prior work with object rhetorics from oppressed populations and counter-histories of rhetorical materialism. He grounds this discussion in a study of accessibility and the "intention of helping deviant bodies perform traditional tasks on passive instruments." In this regard, he offers a specific manifestation of the purpose of this collection by reprogramming some of the ontological and epistemological divisions that the field has yet to fully engage in critical making scholarship.

Similarly, David M. Sheridan's essay reminds audiences that "critical making" is a term that has a genealogical and political history that is built into the idea of reprogramming itself. For example, this collection has a goal, which echoes the concerns of other critical makers, of creating a shared repository to enable executability. This goal has precedents, including the Creative Commons' web sharing, but, importantly, through artistic models such as postcards as in Craig Saper's discussion of "networked composition." In "The Circulation of Touch: Very Simple Machines for Creating Tactile Textual Experiences," Sheridan offers a theory of invention for critical making grounded in what he calls a "metaphorical reinterpretation" of the concept of "reprogrammable circuits." He explains this concept by describing many of his practical illustrations for teaching critical making through the creation of paper writing devices (PWDs), which can load text in analogous ways to physical computing circuits to create compelling interactive experiences for users. Like Hammer, Sheridan's essay importantly frames rhetoric and composition's interest in critical making as one that must interrogate the theory/practice divide rather than necessarily settle on one side or another.

SECTION 2. TEXT MINING RESEARCH METHODS AS CRITICAL MAKING

This section represents an attempt to reprogram the data visualization component of traditional practices of critical making. For clarification, David Staley (2017) declared that "the 'maker turn' expands the range of objects humanists might construct" to include "non-textual and perhaps even non-discursive objects" (37). He argues that "making, designing, and experiencing these visual, tactile, and material objects are hermeneutic acts, which afford the kind of inquiry expected in the humanities" (33). As this section will argue, Staley's point is perhaps taken to highlight a need to connect any performances of the screen to their public rhetoric instantiates beyond academia or making for making's sake. There are also concerns among critical makers (Sayers 2017) about the explicit need to make scholarly explorations of making as data analysis or data generation an ethical practice that is distinct from technological entrepreneurism or the "cathedral of computation" (Bogost 2015). After all, coding and programming is both a material and a political practice. In response, this section is designed to align some of our field's interest in critical making with our ongoing interest in text mining, digital humanities, linguistic methods, and descriptive statistics with a particular eye toward both enabling new text mining practices and, specifically, thinking about politics: How can we fashion or employ text mining tools to locate new data points to then form new topological models for how the field functions?

In her chapter, "The Woman Who Tricked the Machine," Cana Uluak Itchuaqiyaq retells a narrative of feminist and Indigenous critical making practices through an ethical application of text mining. Itchuaqiyaq employs quantitative methods with decolonial theory to study the status of citation practices of historically marginalized individuals in technical communication scholarship. While some continue to draw lines between the neutral use of quantification methods and algorithmic critique, Itchuaqiyaq productively documents how the two can be linked into what we might call "critical text mining."

The specific term "critical text mining" is a merger of critical making discussions with text mining methods. This term is the specific subject of Kellie M. Gray and Steve Holmes's essay, "Critical Text Mining: Ethical Paradigms for Determining Emoji Frequency in #blacklivesmatter." Their approach reflects especially Hammer's framing efforts to show that the idea of reprogramming is historical and culturally specific. As they explain, text mining is often critiqued as an object of neoliberal oppression or circumscribed as a narrow empirical method. Yet Gray and Holmes resituate critical text mining as a middle position that combines ethics with text mining methods to create or maintain datasets in public ways. They demonstrate critical text mining through a tutorial hosted on this collection's companion website that uses the data science programming language R and the twitteR package to webscrape tweets to produce activist datasets on #blacklivesmatter.

In a similar collaborative vein, Ratto (2011) argues that critical making projects could include "the act of shared construction itself as an activity and a site for enhancing and extending conceptual understandings of critical sociotechnical issues" (254). As an additional example, Ryan M. Omizo's approach to text mining is more methodological and instruction-based than either Gray and Holmes's or Itchuaqiyaq's respective essays. However, Omizo's essay, titled "Reprogramming the Faciloscope: An Experience Report on the Search for Genre," offers one of the best performances in this collection of what reprogrammable means since he is describing his methods and technical executions for revising an existing digital humanities project: the Faciloscope with Bill Hart-Davidson. The updated Faciloscope 2.0 offers an interactive interface that allows users to upload text in order to have software locate genre signals (Miller 1984). Omizo also includes instructions for using it as well as code for participating in the ongoing evolution of the Faciloscope 2.0. In relationship to critical making discourses, Omizo's essay fulfills the spirit of executability and contains an open invitation to readers to participate in the ongoing development and refinement of the Faciloscope 2.0 as a possible conduit to enable critical text mining activities.

Aaron Beveridge and Nicholas Van Horn's chapter, "Big Data, Tiny Computers: Making Data-Driven Methods Accessible with a Raspberry Pi," similarly offers code and tutorials for a different and more traditional form of critical making—programming the reprogrammable circuit board, the Raspberry Pi—but through a nontraditional purpose: text mining and data archiving. While Gray and Holmes engage text mining, Beveridge and Van Horn explore "big data" in relation to critical making. Big data has more resonances with datasets assembled and analyzed by largescale commercial software or proprietary knowledge sets that may require substantial amounts of technical know-how, financial capabilities, and processing power to engage. Yet Beveridge and Van Horn argue that everyday researchers and makers already have access to DIY tools of building their own large datasets through purchasing a relatively inexpensive Raspberry Pi computer (roughly $35), which, they contend, can address key hardware and workflow issues for long-term data collection projects. While the bulk of their chapter is instructional in helping readers connect a Raspberry Pi to their open-source MassMine webscraping tool suite, they do offer specific connections to how data curation can function as a form of critical making.

SECTION 3. EVERSION AND CRITICAL MAKING

One way to reprogram an existing program lies in using neglected theoretical concepts to help shed light on the deficiencies or limitations of prior ones. With respect to the theory and practice division, the concept of eversion is useful, which comes from William S. Gibson's (2007) *Spook Country*, which describes the "eversion of cyberspace." What he meant by this term is a situation through which the virtual or immaterial inverts itself and leaks out into the physical world. Work on wearable rhetorics and the Internet of Things are perhaps familiar illustrations of eversion. In a traditional critical making approach, eversion captures this sense of working with haptic wearable media or physical computing, which Rieder (2017) identifies as providing possibilities "to combine the virtual with the real in new ways, and to creatively bend the conventional experience of reality toward some suasive end, by folding into it some of the affordances of the virtual" (5). As Steven E. Jones (2018) notes in the context of discussing eversion and the digital humanities, *eversion* is also a call to *ground* any form of making back its material, social, cultural, and material contexts, which is one of the exigencies for this collection.

The first chapter in this section demonstrates how eversion applies to multiple chapters in this edited collection and not just in this particular section. In "Touch-Interactive Rhetorics: Exploring Our 'First Sense' as a Rhetorical Act of Eversion," Matthew Halm and David M. Rieder draw on eversion to highlight the need to develop new vocabularies and concepts for the field to be able to explore political interactions through touch and haptic design. For example, they rightly ask what it means for rhetoric and composition scholars to compose through forms of physical computing when wearable sensors can collect data unseen and largely unregulated from the haptic interactions of embodied users. Such new forms of theory and data collection can lead toward a reciprocal feedback loop of composing new compositions that make users more aware of how forms of physical computing monitor and control their agency: "With that transduced data in hand, a digital rhetor can generate multimodal feedback directed at their audience that leads to a stylized experience."

The Gibsonian spirit of merging and yet reconceiving differently two historically distinct knowledge domains was also reflected in another chapter by Andrew Pilsch, titled "What the Computer Said: Poetic Machines, Rhetorical Adjuncts, and the Circuits of Eloquence." He explains how the rhetorical concept of "eloquence" can be coupled with computer-generated compositions. Using a computer to compose means that both the human programmer and the computer come together to create an emergent product that neither could have produced without the other. He declares, "By imagining the computer as an eloquence adjunct, we can further think through digital rhetoric as a product of intimate relationships with our devices." Pilsch, who is well known in our field for creating bots, such as @InfiniteQuintilian on Twitter, uses this theoretical framework to offer some critical self-reflection on how his approach to building bots evolved, including specific discussions of the type of code and datasets that he used.

The concept of eversion also has alternative rhetorical histories in all but name. Sean Morey and M.Bawar Khan, in "Actionable Monuments: Making Critical Augmented Reality Activism," combine a theory of rhetorical/invention and making (Ulmer's "electronic monuments" or a "Memorial") with coding an augmented reality (AR) technology in an AR application devoted to bearing witness to the neglected animal costs of the Louisiana Gulf Coast BP oil spill of 2010. This application, which is available in the essay as well as their code available on this collection's companion website, is coded with Unity and the Vuforia AR SDK, corresponding image and video assets, and C# scripts. These scripts will link this app with the user's automobile via a Bluetooth onboard diagnostic scanner to provide real-time data on the user's relationship to the petrol economy and the nonhuman animal sacrifices that result from this participation. Here, Morey and Kahn illustrate how programming, AR interfaces, and theorists of electrate invention can intersect with critical making, eversion, and public rhetoric issues to address ongoing kairotic exigencies.

SECTION 4. CRITICAL PLAY AS CRITICAL MAKING

Critical making discourses can and should involve more than just Arduinos. In this section, contributors investigated material making through play as a form of ethical practice. Getting straight to the issue of eliding theory and practice, Michael J. Faris, in "Reparative Making: Re-Orienting Critical Making for Queer Worldmaking," repurposes Ratto's concept of critical making to theorize what he calls, drawing on Eve Kosofsky Sedgwick's work, *reparative making*. As he explains, reparative making involves "restorative making for individuals and groups that can assist in queer worldmaking practices, opening up the world to new ways of thinking and being." Faris turns to queer indie video game scenes, explaining how queer and transgender individuals make indie games for survival and reparative purposes; how these games often challenge or queer normative assumptions about gameplay, gender, sexuality, and disability; and, finally, how physical interfaces for video games and maker movements themselves can be queered as reparative modes of making.

Wendi Sierra, in "Developing *A Strong Fire*: Bridging Critical Making, Participatory Design, and Game Design," describes the ethics of invention through participatory critical making in the context of video game design. Part reflective-narrative and part theory-building, Sierra's chapter defines and draws on Elizabeth LaPensée's concept of Indigenously determined game design to describe how she and other actually affected Indigenous stakeholders have worked together on *A Strong Fire*, which is an Oneida culture and language game developed with the Oneida Nation of Wisconsin. In this regard, Sierra also reminds readers that our epistemic starting places for critical making matter in terms of whether we theorize and enact from Western or Indigenous (decolonial) starting places.

Many of us may not have the knowledge or classroom time to teach video game programming. However, critical play can be taught with a variety of tools, as Kendall Gerdes demonstrates in "Twisted Together: Twine Games as Solidarity Machines." Gerdes offers a rhetorical theoretical approach to the critical making practices seen in the Twine game *Depression Quest*, by Zoë Quinn. *Depression Quest* became the lightning rod for the 2014 GamerGate controversy. Since it defies so many conventional expectations of mainstream video game narrative, gameplay, and character archetypes, Gerdes uses Ratto's idea that critical making can "parse ... a world that exceeds language's meaning-making powers" to offer a productive rereading of feminist critical making approaches with regard to game design. Far from just a description, Gerdes, who draws on Avital Ronell's and Diane Davis's work, also uses a performative second-person address to reframe author–audience perspectives in the parallel way to how *Depression Quest* operates.

SECTION 5. CRITICAL MAKING AS INSTRUCTIONAL DESIGN

While many chapters touch on pedagogical implications, two of our chapters place the college classroom as a specific focal lens for critical making discourse and practice. In "Cultivating Critical Makers: Crafting with Paper-Electronic Circuits in an Online First Year Composition Course," Bree McGregor discusses how physical computing can be enacted through accessible nondigital means such as poster boards and graphite pencils to draw physical circuits can be taught in an undergraduate classroom. She examines the use of critical making assignments—paper-electronic circuits—to help students engage in tactile, reflective practices and rethink how they reach audiences and achieve rhetorical goals. McGregor includes an examination of the pedagogical design for an online, maker-themed course and discusses resources and support—including a website she built specifically for this purpose—for both instructors and students as they engage in new modes and mediums of composing.

Similarly, John Jones's chapter, "Crafting in the Classroom: Carpentry and Pedagogy in Rhetoric and Composition," offers another conceptual and practical application directly aimed at the undergraduate classroom. His chapter explores the idea of carpentry—making things that do rhetorical work (Bogost 2015; Brown and Rivers 2013)—and its relation to rhetoric and composition pedagogy. Using the lens of an upper-level digital media course that focused on making with the Arduino platform, Jones addresses how the creation and function of a course can be considered a form of carpentry and suggest how this framing can benefit both instructors who wish to integrate reprogrammable circuits into their teaching as well as those who do not. He examines the syllabus, a document that is primarily textual, but in its function exhibits tool-like qualities, exploring it in turn as a program, an object, and a platform and examining what is at stake in each of these framings. Speaking to the political issues of OOR and OOO, the chapter also addresses recent concerns in rhetoric and cultural studies related to these terms and discusses how instructors can integrate diverse voices into making-centered courses.

CONCLUSION

We (re)wrote and revised this introduction during the 2020–2021 global coronavirus pandemic (and many authors revised their chapters heavily during this pandemic as well). It's amazing how many people across the world turned to making during the early months of the pandemic. For instance, flour became a scarce commodity in many markets because so many US residents turned to baking bread as a pastime as they stayed at home to help suppress the spread of the novel coronavirus. Others turned to crafting and making practices like knitting, cooking, and sewing (especially of masks) and even more elaborate engineering feats like YouTuber and engineer Mark Rober's (2020) squirrel maze he built in his backyard to pass time and test how quickly squirrels could figure out his maze and successfully earn their treats.

But to state that so many turned to making during the pandemic is also to point out the privileged ways we often understand making. While many turned to baking bread or feats of whimsical engineering to pass time, others were still working full-time in service occupations, risking their health to continue to earn a living. Others weren't afforded the luxury of leisurely making as education moved online in spring 2021 and childcare become unavailable. This disparity was particularly gendered, as women academics, for example, became more responsible for stay-at-home childcare and at-home online education while their male counterparts (in comparison) continued publishing at rates higher than women (Flaherty 2020; Viglione 2020).

The coronavirus pandemic is not the only context for this collection, of course. But this kairotic exigency along with pressing issues such as Black Lives Matter and structural racism (Kendi 2019) highlights the ongoing need to situate critical making as a context of "both/and" instead of "either/or" when it comes to refining a flexible operating program for rhetoric and composition. The very selection of materials on offer or the ability to engage in making are themselves shot through with political relations and hierarchy. Admitting this should not be interpreted as suggesting a barrier to overcome or call to avoid experimentation and tinkering. Instead, the present era requires that we strive to connect making to all of its forms of affectivity and political inscription. At the minimum, we hope that this collection helps makers inside and outside of our field grapple productively with this set of tensions.

We have titled this collection *Reprogrammable Rhetoric: Critical Making Theories and Methods in Rhetoric and Composition*. We hope that the chapters in this collection provide models for reprogramming rhetoric, for thinking and enacting rhetoric anew as a way to engage in the world. We like to think of *rhetoric as making*—a way to critically engage in the world through enacting new practices that open up possibilities for world-making. This collection, we hope, joins critical making scholars who see critical making as a politicized endeavor in which "making becomes a means of not only designing a more just social future but composing a transformative tomorrow" (Wargo and Morales 2021, 137).

REFERENCES

Barnett, Scot, and Casey Boyle. 2016. "Introduction: Rhetorical Ontology, or, How to Do Things with Things." In *Rhetoric, Through Everyday Things*, edited by Scot Barnett and Casey Boyle, 1–16. Tuscaloosa: University of Alabama Press.

Bennett, Ireashia, Jennifer Brier, Patrick Jagoda, Gary Kafer, Márquez Rhyne, and Chelsea Ridley. 2018. "Transmedia Collage." *Thresholds*, no. 3. http://openthresholds.org/3/transmediacollage.

Bogost, Ian. 2012. *Alien Phenomenology, or What It's Like to Be a Thing*. Minneapolis: University of Minnesota Press.

Bogost, Ian. 2015. "The Cathedral of Computation." *The Atlantic*, January 15, 2015. https://www.theatlantic.com/technology/archive/2015/01/the-cathedral-of-computation/384300/.

Brock, Kevin. 2012. "One Hundred Thousand Billion Processes: Oulipian Computation and the Composition of Digital Cybertexts." *Technoculture: An Online Journal of Technology in Society* 2. https://tcjournal.org/vol2/brock.

Brooke, Collin Gifford. 2009. *Lingua Fracta: Toward a Rhetoric of New Media*. New York: Hampton Press.

Brown, James J., Jr. 2015. *Ethical Programs: Hospitality and the Rhetorics of Software*. Ann Arbor: University of Michigan Press.

Brown, James J., Jr., and Nathaniel Rivers. 2013. "Composing the Carpenter's Workshop." *O-Zone: A Journal of Object-Oriented Studies* 1 (1): 27–36. https://www.academia.edu/18103756/Composing_the_Carpenter_s_Workshop.

Burek, Amy, Emily Alden Foster, Sarah Fox, and Daniela K. Rosner. 2017. "Feminist Hackerspaces: Hacking Culture, Not Devices (the Zine!)." In *Making Things and Drawing Boundaries: Experiments in the Digital Humanities*, edited by Jentery Sayers, 221–31. Minneapolis: University of Minnesota Press.

Burgess, Helen J., and David M. Rieder, eds. 2015. "Kits, Plans, Schematics." Special issue, *Hyperrhiz* 13. http://hyperrhiz.io/hyperrhiz13/.

Burgess, Helen J., and Roger Whiston. 2019. "Introduction: Critical Making and Executable Kits." *Enculturation: A Journal of Rhetoric, Writing, and Culture* 29. http://enculturation.net/critical-making-and-executable-kits.

Carter, Joyce Locke. 2016. "2016 CCCC Chair's Address: Making, Disrupting, Innovating." *College Composition and Communication* 68 (2): 378–408.

DiSalvo, Carl. 2012. *Adversarial Design*. Cambridge, MA: MIT Press.

Dunbar-Hester, Christina. 2014. "Radical Inclusion? Locating Accountability in Technical DIY." In *DIY Citizenship: Critical Making and Social Media*, edited by Matt Ratto and Megan Boler, 75–88. Cambridge, MA: MIT Press.

Dunne, Anthony, and Fiona Raby. 2013. *Speculative Everything: Design, Fiction, and Social Dreaming*. Cambridge, MA: MIT Press.

Ede, Lisa. 2004. *Situating Composition: Composition Studies and the Politics of Location*. Carbondale: Southern Illinois University Press.

Ehrenberg, Philip, Patrick Jagoda, and Melissa Gilliam. 2018. "S.E.E.D.: Creating and Implementing an Alternate Reality." *Kairos: A Journal of Rhetoric, Technology, and Pedagogy* 22 (2). http://kairos.technorhetoric.net/22.2/praxis/ehrenberg-et-al/index.

Faris, Michael J., Andrew M. Blick, Jack T. Labriola, Leslie Hankey, Jamie May, and Richard T. Mangum. 2018. "Building Rhetoric One Bit at a Time: A Case of Maker Rhetoric with littleBits." *Kairos: A Journal of Rhetoric, Technology, and Pedagogy* 22 (2). http://kairos.technorhetoric.net/22.2/praxis/faris-et-al/index.html.

Felski, Rita. 2015. *The Limits of Critique*. Chicago, IL: University of Chicago Press.

Flaherty, Colleen. 2020. "Something's Got to Give." *Insider Higher Ed*, August 20, 2020. https://www.insidehighered.com/news/2020/08/20/womens-journal-submission-rates-continue-fall.

Gibson, William. 2007. *Spook Country*. New York: Putnam's.

Gollihue, Krystin Nicole. 2019. *"Re-making the Makerspace: Bodies, Power, and Identity in Critical Making Practices."* PhD diss., North Carolina State University. http://www.lib.ncsu.edu/resolver/1840.20/36743.

Haas, Angela. 2007. "Wampum as Hypertext: An American Indian Intellectual Tradition of Multimedia Theory and Practice." *Studies in American Indian Literatures* 19 (4): 77–100. doi:10.1353/ail.2008.0005.

Hammer, Steven, and Aimée Knight. 2015. "Crafting Malfunction: Rhetoric and Circuit Bending." *Harlot: A Revealing Look at the Arts of Persuasion* 14. https://web.archive.org/web/20170425034058/http://harlotofthearts.org/index.php/harlot/article/view/261/173.

Harman, Graham. 2011. *Tool-Being: Heidegger and the Metaphysics of Objects*. Chicago, IL: Open Court.

Harper, Douglas. 2020. "Program." *Online Etymology Dictionary*. https://www.etymonline.com/word/program.

Hertz, Garnet, ed. 2012. *Critical Making*. http://www.conceptlab.com/criticalmaking/.

Hertz, Garnet, ed. 2016a. *Disobedient Electronics*. http://www.disobedientelectronics.com/.

Hertz, Garnet. 2016b. "What Is Critical Making?" *Current* 7. http://current.ecuad.ca/what-is-critical-making.

Hertz, Garnet. 2018. "We Need Something Better Than the Maker Movement." *Neural*, no. 60, summer 2018.

Horner, Bruce. 2019. "Foreword." In *Reinventing (with) Theory in Rhetoric and Writing Studies: Essays in Honor of Sharon Crowley*, edited by Andrea Alden, Kendall Gerdes, Judy Holiday, and Ryan Skinnell, xi–xv. Logan: Utah State University Press.

Jagoda, Patrick. 2017. "Critique and Critical Making." *PMLA* 132 (2): 356–63. doi:10.1632/pmla.2017.132.2.356.

Jagoda, Patrick, Melissa Gilliam, Peter McDonald, and Christopher Russel. 2015. "Worlding through Play: Alternate Reality Games, Large-Scale Learning, and *The Source*." *American Journal of Play* 8 (1): 74–100. https://www.journalofplay.org/issues/8/1.

Jalopy, Mister, Phillip Torrone, and Simon Hill. (2006). "The Maker's Bill of Rights." *Make*. https://cdn.makezine.com/make/MAKERS_RIGHTS.pdf.

Jones, John, and Livinia Hirsu, eds. 2020. *Rhetorical Machines: Writing, Code, and Computational Ethics*. Tuscaloosa: University of Alabama Press.

Jones, Steven E. 2018. "Turning Practice Inside Out: Digital Humanities and the Eversion." In *The Routledge Companion to Media Studies and Digital Humanities*, edited by Jentery Sayers, 267–73. New York: Routledge.

Kendi, Ibram X. 2019. *How To Be an Antiracist*. New York: One World Publishing.

Manning, Erin. 2016. *The Minor Gesture*. Durham, NC: Duke University Press.

McCorkle, Ben. 2012. *Rhetorical Delivery as Technological Discourse: A Cross-Historical Study*. Carbondale: Southern Illinois University Press.

Miller, Carolyn. 1984. "Genre as Social Action." *Quarterly Journal of Speech* 70 (2): 151–67. doi:10.1080/00335638409383686.

Morey, Sean. 2015. *Rhetorical Delivery and Digital Technologies: Networks, Affect, Electracy*. New York: Routledge.

Oxford English Dictionary. 2020a. "Reprogrammable." *Oxford English Dictionary*. https://oed.com/view/Entry/275282.

Oxford English Dictionary. 2020b. "Reprogramme." *Oxford English Dictionary*. https://oed.com/view/Entry/163110.

Powell, Malea. 2016. "I Got Your Maker Space Right Here: Objects, Things, Making and Relations (Seems Like I've Heard This Song Before)." Presentation at the Cultural Rhetorics Conference, September 30–October 2, 2016, East Lansing, MI.

Pozo, Teddy. 2018. "Queer Games After Empathy: Feminism and Haptic Game Design Aesthetics from Consent to Cuteness to the Radically Soft." *Game Studies: The International Journal of Computer Game Research* 18 (3). http://gamestudies.org/1803/articles/pozo.

Raley, Rita. 2009. *Tactical Media*. Minneapolis: University of Minnesota Press.

Ratto, Matt. 2011. "Critical Making: Conceptual and Material Studies in Technology and Social Life." *The Information Society: An International Journal* 27 (4): 252–60. doi:10.1080/01972243.2011.583819.

Ratto, Matt, and Megan Boler, eds. 2014. *DIY Citizenship: Critical Making and Social Media*. Cambridge, MA: MIT Press.

Ratto, Matt, and Stephen Hockema. 2009. "Flwr Pwr: Tending the Walled Garden." In *Walled Garden*, edited by Annet Dekker and Annette Wolfsberger, 51–62. Amsterdam, Netherlands: Virtueel Platform. http://aaaan.net/walled-garden/.

Resch, Gabby, Dan Southwick, Isaac Record, and Matt Ratto. 2017. "Thinking as Handwork: Critical Making with Humanistic Concerns." In *Making Things and Drawing Boundaries: Experiments in the Digital Humanities*, edited by Jentery Sayers, 149–61. Minneapolis: University of Minnesota Press.

Rickert, Thomas. 2013. *Ambient Rhetoric: The Attunements of Rhetorical Being*. Pittsburgh, PA: University of Pittsburgh Press.

Ridolfo, Jim, and William Hart-Davidson, eds. 2015. *Rhetoric and the Digital Humanities*. Chicago, IL: University of Chicago Press.

Rieder, David M. 2017. *Suasive Iterations: Rhetoric, Writing, and Physical Computing*. Anderson, SC: Parlor Press.

Riley-Mukavetz, Andrea, Jacqueline Rhodes, Malea Powell, and M. Remi Yergeau. 2016. "Three Queer/Feminist/Indigenist Rants and a Critique of Heteropatriarchal Colonialism in Object-Oriented Theory." Presentation at the Cultural Rhetorics Conference, September 30–October 2, 2016, East Lansing, MI.

Rober, Mark. 2020. "Building the Perfect Squirrel Proof Bird Feeder." *YouTube*, May 24, 2020. https://youtu.be/hFZFjoX2cGg.

Sayers, Jentery. 2015. "Prototyping the Past." *Visible Language* 49 (3): 157–77.

Sayers, Jentery. 2017. "Introduction: 'I Don't Know All the Circuitry.'" In *Making Things and Drawing Boundaries: Experiments in the Digital Humanities*, edited by Jentery Sayers, 1–17. Minneapolis: University of Minnesota Press.

Sheridan, David M. 2010. "Fabricating Consent: Three-Dimensional Objects as Rhetorical Compositions." *Computers and Composition* 27 (4): 249–65. doi:10.1016/j.compcom.2010.09.005.

Sheridan, David. 2016. "A Maker Mentality toward Writing." *Digital Rhetoric Collaborative*, March 28, 2016. http://www.digitalrhetoriccollaborative.org/2016/03/28/a-maker-mentality-toward-writing/.

Sheridan, David M., Jim Ridolfo, and Anthony J. Michel. 2013. *The Available Means of Persuasion: Mapping Theory and Pedagogy of Multimodal Public Writing*. Anderson, SC: Parlor Press.

Shipka, Jody. 2011. *Toward a Composition Made Whole*. Pittsburgh, PA: University of Pittsburgh Press.

Somerson, Roseanne, and Mara L. Hermano, eds. 2013. *The Art of Critical Making*. Hoboken, NJ: Wiley.

Staley, David. 2017. "On the 'Maker Turn' in the Humanities." In *Making Things and Drawing Boundaries: Experiments in the Digital Humanities*, edited by Jentery Sayers, 32–41. Minneapolis: University of Minnesota Press.

Ulmer, Gregory. 1994. *Heuretics: The Logic of Invention*. Baltimore, MD: Johns Hopkins University Press.

Vee, Annette. 2017. *Coding Literacy: How Computer Programming Is Changing the World*. Cambridge, MA: The MIT Press.

Viglione, Giuliana. 2020. "Are Women Publishing Less During the Pandemic? Here's What the Data Say." *Nature* 581: 365–66. https://doi.org/10.1038/d41586-020-01294-9.

Wargo, Jon M., and Melita Morales. 2021. "Making Futures, Composing Worlds: Examining Young Children's Making as Speculative Design." In *Making Literacies and Making Identities in the Digital Age: Learning and Playing through Modes and Media*, edited by Cheryl A. McLean and Jennifer Rowsell, 133–148. London: Routledge.

Wysocki, Rick, Jon Udelson, Caitlin E. Ray, Jessica S. B. Newman, Laura Sceniak Matrevers, Ashanka Kumari, Layne M. P. Gordon, et al. 2019. "On Multimodality: A Manifesto." In *Bridging the Multimodal Gap: From Theory to Practice*, edited by Santosh Khadka and J. C. Lee, 17–29. Logan: Utah State University Press.

SECTION 1

Framing Critical Making

1
NOISE COMPOSITION
A Story of Co-Design and Relationality

Steven Hammer
Saint Joseph's University

THIS IS A STORY

This is a story about other peoples' bodies and about my own attempts to engage with bodies and persons with care, empathy, and creativity. Sometimes I tell this story as a researcher and teacher who wants to make my students more care-ful thinkers and (co)designers and citizens; sometimes I tell this story as a parent desperate to exchange expressions of love and affection with a child who does not speak traditionally, trying to untangle myself from the notion that the highest, most complex forms of expression are in the skillful arrangement and delivery of formal language. The salient characteristic of these voices, however, is that they are my own, and therefore extraordinarily limited both in understanding the experiences of others and regulating my own objectivity, if such a thing can even exist in our work. Therefore, I will attempt here to neither speak to others' experiences nor avoid emotional attachment to bodies and projects. Instead, I hope that I am able to tell stories that help convey what I have learned by listening to the bodies I find myself in relationship with.

ORIGIN STORIES

I was scheduled to teach a special topics course a few years ago, and I had long wanted to design a course in sound/musical instrument design; most of my research revolves around the ways that composition instruments shape our processes and finished works, and the ways that glitch art, dirty new media, and new materialisms can help us rethink composition, our bodies, and our perceptions of emerging technologies. I had also been able to connect with an arts organization in Philadelphia dedicated to creating an "inclusive and integrated environment" in which artists with disabilities can collaborate, learn, and network. I proposed a collaboration between members of the collective interested in musical instrument design and students in my special topics course, titled Physical Computing and Accessibility, and it was approved. The course description read:

> Physical computing grants us the ability to rethink the ways we interact with both digital systems and the physical world. In this course, students will first learn the basics of Arduino micro-controller hardware and programming in the context of music and sound-based art. Then, collaborating with persons with disabilities in Philadelphia, students will co-create Arduino-based instruments, apply and expand upon their knowledge of disability and accessibility, and assist in organizing a community performance event. No previous experience with physical computing or music is necessary to take this course, only interest in and commitment to working in a diverse and collaborative environment.

As I prepared the course—readings, scaffolding knowledge and skills, etc.—I wanted to first ensure that we carefully considered and reconsidered the notions of disability, noise, and bodies. This is, I hope, how I might contribute something meaningful to this collection, suggesting that before we build technologies meant to interface with bodies, we might first think very carefully and *relationally* about curatorial responsibility and ethics. The ways those bodies are framed, especially in relation to ideas central to technological concerns such as function(ality), brokenness, disability, adaptability, etc. Like so much careful work done before this project, including that of Michael McAllister, Elaine Yuen, and Stuart Bush (2012) working with many of the same participants in Philadelphia, I wanted to ensure that this course was not seen merely as an opportunity to serve (often read: save) Others, but instead as a co-design partnership.[1] That is what this chapter is really about: If we are to work with bodies—especially those bodies our culture has classified as "deficient"—how do we work with care? How do we work and co-design with people, not their diagnoses?

I mean to suggest that while teaching critical making certainly involves learning new technical skills (hardware, software, techniques, etc.), it is also an ethical practice. The ethical considerations vary contextually by course, focus, participant identity, researcher identity, and countless other factors, but here I will tell the stories of my own approach to ethical instrument co-design practices, calling out a few problematic-yet-prolific symptoms of our culture and suggesting alternatives. But this story—the story of the teacher trying to design a great course, or even the story of the researcher pointing out some problematic aspect of culture—is not where this story begins or why it is important enough to write about or pursue.

My son, Rowan, was born during a blizzard in Fargo, North Dakota as the year 2010 was nearing its end. His birth and early infancy were unremarkable from a medical-scientific perspective; he did most of the things that typical young humans do. He ate and slept, he brought joy and sleeplessness to his family, he discovered his body and began establishing relationships with other bodies and materials. And as is typical for those who find themselves in the culture of reproduction, his mother and I cautiously-but-eagerly discussed and researched questions like "What is happening at *x* months?" or "When do children start walking?" We understood that there is room for individuality, but we wondered and we talked with other parents and we valued things like early development. We inherit and (re)build norms and (re)tell stories of bodies, those that are *advanced* or *on track* or *behind*. What is *normal* and what is not, when to worry, when to seek help. After several months of life, Rowan began to *deviate*, a little at first, missing some of these checkpoints. Things like holding his head up and crawling. We compared his progress—and increasingly his *lack* of progress—to that of his older sister, wondering what was *wrong*. Doctors confirmed that his progress was indeed *deficient*, but not to worry.

When Rowan was about nine months old, we became increasingly concerned about his development and made an appointment for the following morning. But before the morning came, Rowan's seizures began. Hundreds of them per day at the beginning, so many that it became difficult to tell what was a seizure and what was not. *What was signal and what was noise.* After a few days in the hospital, a nervous doctor entered the room and said that after reviewing all the data, the team had confirmed Rowan's diagnosis as lissencephaly, or "smooth brain." He showed us scans of Rowan's brain with comparisons so we could see how it deviated from a typical brain. He handed us a warm, freshly printed handout explaining his diagnosis and prognosis and left the room.

> Diagnosis: The identification of the nature of an illness or other problem by examination of the symptoms. Origin: Late 17th century: modern Latin, from Greek, from diagignōskein "distinguish, discern," from dia "apart" + gignōskein "recognize, know." (Oxford University Press 2020a)

> Prognosis: The likely course of a disease or ailment. Origin: Mid 17th century: via late Latin from Greek prognōsis, from pro- "before" + gignōskein "know." (Oxford University Press 2020c)

This story is long and still-unfolding, but in short, we learned how Rowan's body, due to a rather mysterious and rare genetic characteristic, failed to typically position his brain tissue resulting in an observable "apartness" (diagnosis), and how that apartness would result in more concrete or functional/social apartness: He would not walk or talk, and his death was something we were told to "start thinking about" (prognosis). The piece of paper and the words from the doctor seemed to simultaneously ease and raise concern, comfort, and frighten. Or as Eli Clare (2017) so aptly articulates, "Diagnosis wields immense power. It can provide us access to vital medical technology or shame us, reveal a path toward less pain or get us locked up. It opens doors and slams them shut" (41).

This probably reads as a pretty sad story. It felt really sad at the time. Even as I write this chapter as a means to reframe concepts like disability, the deficiency model, adaptive technologies, and so on, I experience some sadness. An honest assessment, however, in the months and years that followed his diagnosis reveals to me that that much of my sadness results from my own implicit investment in what I will call here *the myth of noiselessness*: the persistent myth that systems, networks, and the nodes that comprise them somehow hold not only the potential for purity and functionality, but that this is their typical and preferred state of being. In other words, some[2] of my sadness was less rooted in empathy for Rowan's present state of wellness and more tied to the mourning of his typical development and abilities. That is rather difficult to admit, but it is honest and bears influence to my ideas here.

In this chapter, I will talk about some ideas I had to rethink to be a better parent, teacher, and instrument (co)designer. First, I will unpack the notion of the *myth of noiselessness*, problematizing what some call the deficiency model of understanding bodies and technologies. Second, I will offer a critique of "adaptive" technologies and suggest that instead of co-designing with the intention of helping deviant bodies perform traditional tasks on passive instruments, we co-design relationships between human and nonhuman bodies. Finally, I will venture deeper into instrument co-design philosophy-practices I have undertaken both with Rowan and in the Physical Computing and Accessibility course and, I hope, offer useful approaches to those interested in doing work in these contexts.

THE MYTH OF NOISELESSNESS

When I write that some of my sadness surrounding Rowan's diagnosis was a result of what I am calling *the myth of noiselessness*, I mean to say that the act of diagnosis—medically and legally changing his status of apartness—reveals a set of cultural values that both privileges and assumes that systems (from human bodies to laptops to social institutions) and their components possess the potential and preference for

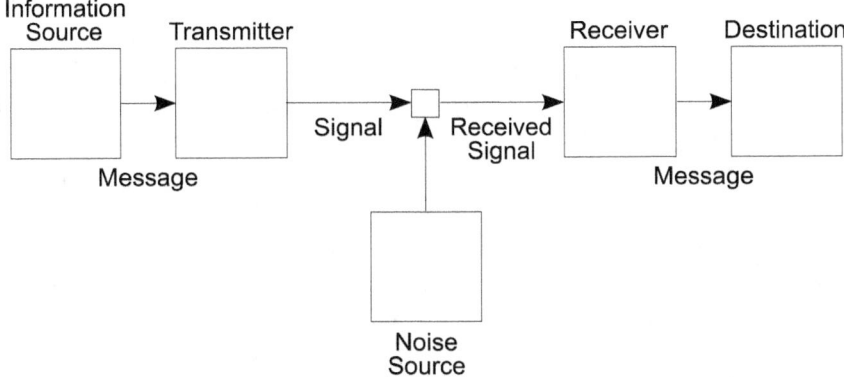

Figure 1.1. Claude Shannon's general communication model (courtesy of Wikipedia, public domain).

purity and functionality. I somehow believed that his body, maybe all bodies, had the potential for pure form and function. That these deviations were somehow unjust. A lot of people reacted this way when we told them about Rowan, that it was somehow unfair or surprising that this happened, that his body was this way. The myth of noiselessness has given folks a lot of strange ideas about what is fair and what is not.

We can see this myth performed and enforced in a variety of contexts, but perhaps we should begin with a discussion of just what noise is—or at least what it has come to mean in relation to technocultural desire and expectations. Etymologically, noise comes from the Latin *nausea*, to denote upset, malaise, seasickness, Anglo-Norman for din, disturbance, uproar, brawl, and disquiet, and Old French for quarrel or disturbance (Oxford University Press 2020b). A common working definition of noise is some variation of noise as a phenomenon that creates displeasure, or noise simply as an "unwanted or undesired sound" (Kerse 1975; Taylor 1970). These definitions, while not always useful,[3] seem fair enough for purposes of this discussion: Noise is that ever-present part of life that conflicts with our desires, that upsets some peace we have (temporarily) achieved. The problem arises, though, when we begin conceiving of noise not as "part of life" but as (a) a phenomenon separate from signal and (b) a phenomenon we can (and should) excise from that which we desire ("signal") with the right technologies.

Many discussions of noise begin with, or at least refer to, an influential paper that shaped not only conceptions of noise, but much of contemporary technoculture: Claude Shannon's (1948) article titled "A Mathematical Theory of Communication," in which he founds information theory, introduces the bit as a unit of information, discusses entropy and redundancy, and more. But of note here is that he illustrates and describes a "general communication system" in which messages are sent as signals from source to destination. The notable character, at least to the current discussion, is the "NOISE SOURCE."

The trouble with Shannon's model as applied to noise *broadly* is the still-prevalent idea of noise as an outside entity that disrupts an otherwise noiseless network of agents and processes. Perhaps you've heard the phrase "signal-to-noise ratio." There still exists a strong separation between the notions of signal and noise, in the same way that we distinguish mind–body, public–private, and so on. The separation of noise and signal, though, at least as we move from a mathematical to rhetorical model of communication, is both arbitrary and political. In other words, *noises are signals, signals are noises*. There is no inherent characteristic of sonic (or otherwise) phenomena that makes it noise, and there is no system that is free from interruptions, disruptions, corruptions, and the corresponding work it requires to resolve them. Yet many still talk of noise as though it is a distinct and separate event agent that spoils the communication party. This is the myth of noiselessness. It is pervasive, available in nearly every advertisement for contemporary digital devices. It is the promise of functionality, the increasingly clean and polished interfaces, and the decreasing ability of users to understand or modify digital tools.

Rosa Menkman (2010) writes of this cultural obsession with noiselessness in her formulation of glitch studies:

> The dominant, continuing search for a noiseless channel has been, and will always be no more than a regrettable, ill-fated dogma. Even though the constant search for complete transparency brings newer, "better" media, every one of these new and improved technologies will always have their own fingerprints of imperfection. While most people experience these fingerprints as negative (and sometimes even as accidents) I emphasize the positive consequences of these imperfections by showing the new opportunities they facilitate. . . . The user has to realize that improving is nothing more than a proprietary protocol, a deluded consumer myth about progression towards a holy grail of perfection.

Noiselessness, for Menkman, is not only impossible in systems, but it is a rhetorically constructed myth to ensure financial and ideological investment by users. She suggests that fingerprints of imperfection, what we call glitches, are ever present and offer opportunities to both understand and create within contemporary culture, so long as we acknowledge and embrace their existence and reject the myth of noiselessness.

In the biological-medical sense, the myth of noiselessness manifests as what some call the "deficit model," in which we "know" the body insofar as it deviates from pure functionality or form. Many root this in Aristotle's *Generation of Animals*, in which he argues that both animals and humans who depart from the able-bodied male are said to be a "'monstrosity' that, by its very essence, is less than human" (Wilson and Lewiecki-Wilson 2001, 13). This notion has been dominant in US society, and justified homophobia and misogyny in often fatal ways. Many, of course, have critiqued this model, including Rosemarie Garland-Thomson (2002), who might call this the "ability/disability system . . . [that] excludes the kinds of bodily forms, functions, impairments, changes, or ambiguities that call into question our cultural fantasy of the body as a neutral, compliant instrument of some transcendent will." (5). Garland-Thomson proposes a feminist disability theory that

> denaturalizes disability by unseating the dominant assumption that disability is something that is wrong with someone. By this I mean, of course, that it mobilizes feminism's highly developed and complex critique of gender, class, race, ethnicity, and sexuality as exclusionary and oppressive systems rather than as the natural and appropriate order of things. (6)

Robert McRuer's work on crip theory draws from a similar line of critique, investigating the ways that queer bodies have been historically—and more importantly to this discussion, *medically*—perceived as diseased, monstrous, or otherwise defective. McRuer responds with *Crip Theory*, asking that we understand compulsory able-bodiedness as linked closely to compulsory heterosexuality, and develop a critical approach that, "in contrast to an able-bodied culture that holds out the promise of a substantive (but paradoxically always elusive) ideal, crip theory would resist delimiting the kinds of bodies and abilities that are acceptable or that will bring about change" (2006, 31).

We can see this "medical" or "deficit" model, for instance, in published guidelines in which those eligible for special education benefits must provide "proof of intrinsic deficit" (Harry and Klingner 2007). Just as medical diagnoses are rooted in knowing the body via its apartness from functionality, deficit-focused conceptions legally and socially bind persons to some unattainable—or at least unsustainable—state of "health." While many of us may enjoy receiving a "clean bill of health" at least once in life, it never lasts long. All bodies fail eventually.

I mean to suggest here that, as I have written elsewhere, we must move past a model of noise that suggests it can, or should, be erased. Not only because it's not theoretically sound, but because it translates into a culture of curation in which we value bodies insofar as they are able to conceal their imperfection. We need to move past noise as the fly in the Modernist's soup. Because in the same way that we discard and discontinue unacceptably noisy technologies, we dismiss (or fetishize) similarly deviant bodies, though we typically encode such dismissals as *toleration, accommodation,* and *facilitation.* We ask how technologies can bridge the gap between the deviant body and noiseless body, instead of taking seriously the desires of the bodies themselves. We need to understand and work with our bodies in a way that foregrounds and celebrates *noise* as an essential component of being and composing, rejecting models of communicative fantasy in which noise can—and should—be erased in the interest of purity.

FROM ADAPTIVE TO RELATIONAL

Rowan receives a lot of therapies. Physical and occupational. Speech and mobility. Without question, however, music therapy was the most significant in terms of responsiveness and enjoyment. *Joy.* Rowan is also a singer. Sometimes in loud "ahh" sounds and sometimes percussive pops of his lips that I like to think of as kisses. And sometimes a "da da da" that I like to think of as calling my name.[4] I brought him to his older sister's gymnastics class one day. A large and echoing space. His singing there was met with a lot of attention, much to the embarrassment of his sister. It's hard to explain the myth of noiselessness to a then seven-year-old who mostly just wants to fit in. Because we are *so* deeply entrenched in that myth that noncompliant bodies become dangerous; maybe contagious, but certainly something best performed in private or designated spaces.

But his favorite instrument is an acoustic guitar.

The guitar is an instrument, a technology, a composition tool. A political and highly disciplinary tool. And the myth of noiselessness is apparent not only in expectations of bodies, but also of our technologies, or as Marshall McLuhan (1966) understood them, the extensions of our bodies. In other words, like bodies, technologies are politically and rhetorically framed as ideally noiseless. Deviations are outliers, problems to be fixed or upgraded or replaced or relegated to the trash heaps necessary to maintain Progress.

As you might intuit, Apple is a fairly easy target for this kind of critique, from campaign slogans such as "It just works" to increasingly rapid cycles of planned obsolescence that now seem essential to the brand identity. In many ways, critics have responded appropriately to this kind of technological rhetoric. Nearly three decades ago, Gail E. Hawisher and Cynthia L. Selfe (1991) published "The Rhetoric of Technology and the Electronic Writing Class," in which we are reminded not only that technological change influences how we write *and* teach writing, but also, and more importantly to this discussion, that our culture (both in popular consumer culture *and* in academia) most often veers sharply toward a perception of emerging technologies as empowering, democratic, hopeful, and visionary. This is often easy to spot in Apple advertisements, but perhaps it is more difficult to reflect on our own de facto endorsement of "new," "exciting," "cutting edge" tools in our classrooms.

Cynthia L. Selfe and Richard J. Selfe (1994) followed this work in "The Politics of the Interface," in which they argue that the interface of the contemporary computer—and its various components such as the ubiquitous Microsoft Word—are always political and never "just tools." They write: "Within the virtual space represented by these interfaces, and elsewhere within computer systems, the values of our culture—ideological, political, economic, educational—are mapped both implicitly and explicitly, constituting a complex set of material relations among culture, technology, and technology users" (485).

For Rowan, the acoustic guitar isn't a stringed and fretted instrument tuned to EADGBE. It's better described as a large, resonant drum with variable sustain, with optional string-holds to grasp, pull, and release. Further, his relationship with that material is guided not by a colonial-modernist notion of mastery, in which an instrument and player are distinct and locked into a human-object power dynamic. Instead, it is an immediate and honest engagement with the instrument-potentiality of any given object, and thus approaching each object as a relationship. A *relation*.

Here's where a materialist approach can be really helpful in thinking through sound, especially instrument design and performance. Various strains of object-oriented ontology, new materialism, speculative realism, affect theory, and more have influenced my thought and practice in many ways. People like Massumi, Barad, Haraway, Latour, Bryant, and others. But I want to venture outside typical intellectual channels here: So many artists and peoples have been dealing with things like entanglement and nonhuman agency for much longer than "new" materialists have been writing about them, as correctly noted in critiques of OOO by Indigenous and cultural rhetorics scholars (Powell et al. 2014).

What for many years was unjustly belittled as mere animism, several Indigenous understandings of the world rely on the same kind of relationality, at-handness, and nonhuman agency as those proposed by scholars noted above. Consider Ojibwe knowledges, for example. Long before nonmodern[5] philosophies began considering—though often only metaphorically—the interdependence of humans and nonhumans, Ojibwe peoples articulated and lived such relationships. Eddie Benton-Banai (1996) writes extensively about the relationship between humans and wolves, beginning with the Anishinaabe creation story:

> In his travels, Original Man began to notice that all the animals came in pairs and they reproduced. And yet, he was alone. He spoke to his Grandfather the Creator and asked, "Why am I alone?" "Why are there no other ones like me?" Gitchie Manito answered, "I will send someone to walk, talk and play with you." He sent Ma-en'-gun (the wolf).
> With Ma-en'-gun by his side, Original Man again spoke to Gitchie Manito, "I have finished what you asked me to do. I have visited and named all the plants, animals, and places of this Earth. What would you now have me to do?" Gitchie Manito answered Original Man and Ma-en'-gun, "Each of you are to be a brother to the other. Now, both of you are to walk the Earth and visit all its places." So, Original Man and Ma-en'-gun walked the Earth and came to know all of her. In this journey they became very close to each other. They became like brothers. . . . When they had completed the task that Gitchie Manito asked them to do, they talked with the Creator once again. The Creator said, "From this day on, you are to separate your paths. You must go your different ways. What shall happen to one of you will also happen to the other. Each of you will be feared, respected and misunderstood by the people that will later join you on this Earth." And so Ma-en'-gun and Original Man set off on their different journeys. (7–8)

Yet human–nonhuman relationships move far beyond creation stories; as Mary Hermes (2005) points out, Ojibwe knowledge, learning, and naming conventions are all rooted in "relational events" rather than stable, independent identities: "Henry explained: 'For example, if I put *asema* [tobacco] out, in English I would be putting a thing on the ground. But in Ojibwe, it is an event, a relational event.' The reference to 'a relational event' is marked by the relationships the actors have with one another and the process of establishing and maintaining those relationships. It is an event specific to a time, place, and people connected to past and future events through the language and the meanings people make of the event" (51). This kind of careful relationality embedded within the very language of the Ojibwe focuses on processes and events. *Relationships. Relations.*

Ojibwe knowledge involves an understanding of "persons" as a broad category that includes, but is not limited to, human beings. Stone-person, bird-person, wind-person, animal-person, human-person (Bird-David 1999, 71). These persons interact and rely upon one another in very real ways. Further, and perhaps most salient to my concern of understanding and designing and performing *with* instruments, A. Irving Hallowell (1964) recounted a conversation with an elder on the "aliveness" of nonhumans:

> Since stones are grammatically animate, I once asked an old man: Are *all* the stones we see about us here alive? He reflected a long while and then replied, "No! But *some* are." This qualified answer made a lasting impression on me. And it is thoroughly consistent with other data that indicate that the Ojibwa are not animists in the sense that they dogmatically attribute living souls to inanimate objects such as stones. The hypothesis which suggests itself to me is that the allocation of stones to an animate grammatical category is part of a culturally constituted cognitive "set." It does not involve a consciously formulated theory about the nature of stones. It leaves a door open that our orientation on dogmatic grounds keeps shut tight. Whereas we should never expect a stone to manifest animate properties of any kind under any circumstances, the Ojibwa recognize, *a priori*, potentialities for animation in certain classes of objects under certain circumstances. The Ojibwa do not perceive stones, in general, as animate, any more than we do. The crucial test is experience. (24)

In other words, all persons (human and nonhuman alike) have potentiality for movement, for action, for agency via interaction and relation. And whether or not a particular stone becomes a person, linguistically or cosmologically, is based on shared, interactive *experiences.*

Further, and more specific to musical instrument design, if we look to practices like the folk art known as circuit-bending, we understand that, as Bruno Latour (1993) later noted and others echoed, we have never been Modern. In other words, we have never actually been in a dominant position expressing a kind of mastery over nonhuman instruments. *We have always been co-authors.* Q. Reed Ghazala (1996) articulated the folk art known as circuit-bending as "the process of creative short-circuiting by which standard audio electronics are radically modified to produce unique experimental instruments" (11). Here's the gist: Go to the thrift shop. Buy an old Speak & Spell or Casio keyboard or Furby, anything that runs on batteries and makes sounds from digital circuits. Open it up. See its guts. While playing sounds, lick your finger and touch it. Make connections between points on the board that aren't supposed to connect. The sounds will change. Maybe the pitch, maybe the speed, maybe something strange that you don't understand. Find the connections that you like, then rewire the instrument to make these potentialities, these failures, its defining functions. Many use "body contacts" to trigger these malfunctions, using the conductive nature of the body to complete the circuit. Ghazala later reflected on this, the body literally becoming part of the instrument's circuit, noting that the distinction between the instrument and player was lost, that a new, momentary, hybrid being (what he calls a BEAsape, or BioElectronicAudiosapian) was in that moment at play:

> I felt that a new, albeit temporary, creature was created when a musician played a body-contact instrument—in this moment when the electricity of both bodies intertwines, the same essential electricity that if interrupted would cause each body to die. I was changed and the circuit was changed, and I had trouble deciding where each of us began and ended. I simply concluded that we were something new, and we were one. (2004, 101)

We exist and communicate and compose in momentary relationships with a range of persons. Persons with biases and politics and intentions and glitches and preferences and resistances. We enter into relationships with those around us, whether we call them persons or instruments or things or actants. We become relations. And if we act on our relations with care, stones are never just stones, diagnoses lose their definitional grip, and co-design obstacles become contemplative opportunities.

TOWARD A POST-NOISE PEDAGOGY

Thus far I have attempted to develop a few ideas as a means to set the stage for instrument design undertakings. First, neither our human bodies (regardless of diagnoses) nor the nonhuman bodies around us, nor the interactions between all of those bodies, are free of noise. But that noise is opportunity: to reflect and relate and express and create. Second, that the way we think about instruments—as passive, inert objects to use and master—is highly specific to a Modernist-colonialist framework that should be abandoned immediately if we are to enter into care-ful creative relations with bodies our culture opts to merely tolerate and accommodate.

I will close this chapter by applying some of these ideas pedagogically, synthesizing my experience into a kind of teaching/design manifesto that helps direct projects surrounding instrument and interface (co)design, whether they be formal partnerships or weekend projects with Rowan.

One: noiselessness in all persons, human and nonhuman alike, is a myth. Yet it is pervasive and works to maintain existing systems dominance and exclusion. As such, we must work very hard to understand and explore all of the beautiful deviations from those myths if we want to understand anything about ourselves or the world of persons we find ourselves entangled with. We must understand where dirt is being classified and excavated. When I say dirt, I'm calling on Mary Douglas's

(1996) work in *Purity and Danger*, in which she articulates dirt as "matter out of place. . . . It implies two conditions: a set of ordered relations and a contravention of that order. Dirt then, is never a unique, isolated event. Where there is dirt there is system. Dirt is the by-product of a systematic ordering and classification of matter, in so far as ordering involves rejecting inappropriate elements" (36). We must find, embrace, and track dirt on the clean carpet of traditional instrument design and performance. This will require openness, patience, and acceptance. Look and listen for and embrace the noise in your own body as you become tired, anxious, bored, thirsty, hungry, warm, cold, older, etc. Lean into that noise, make something with that noise.

Two: instruments are relations, not vehicles toward noiselessness. Your job is not to make tools to chase the ghosts of perfection and noiselessness via so-called adaptive technologies but to explore the relationship between bodies and facilitate a meaningful experience. This chase is both fruitless (noiselessness is a destination never reached) and work to reinforce the exclusion of deviant bodies. As M. Remi Yergeau notes, "To accommodate is to retrofit; it is to assume normative bodies as default and to build spaces and infrastructures around those normative default bodies; it is to deal with deviant bodily and spatial conditions as they bubble out at the seams" (Yergeau et al. 2013). Adaptive technologies not only work to reinscribe apartness, though; they impose extraordinary limits on what kinds of instruments and sounds and expressions are possible. Instrument co-design is not a practice of fixing, saving, or otherwise easing access to typical sonic experiences; it is exploring and sitting with the potential sensory relations at hand.

Three: negotiating desire must guide our co-design processes. Bodies—human or nonhuman—all have potentiality for function/personhood and failure/glitch. They all have desires (tendencies, actions met with little or no resistance) and resistances. Co-design is a process of exploring, negotiating, and implementing the shared desires of bodies-in-relationship.

This may be a good time to illustrate with stories from the course I taught. My students and their co-designers built and performed some really beautiful instruments, but not without frustrations. When presented with the project goal—to work together to create a musical instrument using an Arduino Uno and any combination of sensors—most in the room were overwhelmed by the open-endedness of both the direction and the undefined nature of the technology. We began by talking about music, dancing and moving our bodies to music, and making music with our bodies. Trying to abandon traditional ideas of music and dance. Thinking and talking about how our bodies *desired* to move to and make music. Co-designers shared and negotiated. Next, my students learned the basics of combining hardware and scripts to convert physical phenomena into digital sound. Most had no experience with either—and found themselves primarily learning about the possibilities (desires) and limitations (resistances) of the technology at hand. They found their own skills and abilities (programming, soldering, etc.) improving or stagnating. Co-designers met weekly to work on their instruments. Revising, rewiring, reprogramming. Negotiating desires and resistances.

PARTING STORIES

One co-design team used bowls of water to trigger bell-like tones. One team sewed flex sensors into a pair of hot-pink gloves. Another team used arcade buttons to play drum samples. Desires varied, and no team could fully realize those desires that existed in and between their imaginations. They did the best they could, given their relations and relationships, and the resistances they did not yet know how to negotiate. Our end-of-semester performance event certainly didn't produce any hit songs—in fact, there were some moments of uncomfortable bodily shifting in the audience because the music was never constructed in a familiar way, much like the kinds of art that David Sheridan, in chapter 2, illustrates and praises for prompting ontological questioning (*Is this music? Is that a musical instrument?*) and denial (*This is not music. That is not a musical instrument.*). These instruments, and their inextricable ties to their co-designers, asked everyone to reconsider what an instrument, a composition, a performance can be. And that evening performers demonstrated and talked about what it meant to feel and move and hear with their instrument, and most reported that they are probably the only person qualified to really play it and like the way it sounds. The performance of their instrument, maybe, is not an act upon an object/instrument, but an event of relationality. I am not sure I could ask for better feedback.

And like most toys given to most children, some of my instrument co-designs with Rowan have been more successful than others. Our favorite instrument is still putting our chests and necks together and sharing the vibrations. We usually last only a minute or two before we laugh at the whole performance. He also likes red buttons and low-frequency bass drums. He still loves an acoustic guitar. Pots and pans and tabletops and space blankets. But he finds instruments—the potential for sensory relationships and noise—most everywhere he goes, and I am grateful to be learning that way of being in the world.

NOTES

1. Co-design, though not directly addressed and expanded upon in this chapter, is an extraordinarily important part of these projects, especially as articulated by Elizabeth B.-N. Sanders and Pieter Jan Stappers (2008).
2. In no way do I mean to reduce the emotional complexities of illness, pain, or disability to a violation of our expectations of noiselessness. For instance, in my own experience, I feel sadness (and certainly a wide variety of emotions, including very positive feelings) for a variety of reasons including fear of loss, empathy during times of physical discomfort and pain, and so on.
3. I have written much more about the shortcomings of these conceptions of noise in the chapter "Writing Dirt, Teaching Noise" (Hammer 2018) in Courtney S. Danforth, Kyle D. Stedman, and Michael J. Faris's *Soundwriting Pedagogies*.
4. That I like to think of these performances as kisses and utterances of my name may be indicative of the practice I am critiquing here, specifically the tendency of normative bodies to map their preferences and desires onto deviant bodies.
5. By using "nonmodern," I refer to Latour's (1993) *We Have Never Been Modern* and subsequent works that dispute Modernist distinctions between culture and nature, human and nonhuman, etc.

REFERENCES

Benton-Banai, Edward. 1996. *The Mishomis Book: The Voice of the Ojibway*. Minneapolis: University of Minnesota Press.

Bird-David, Nurit. 1999. "'Animism' Revisited: Personhood, Environment, and Relational Epistemology." *Current Anthropology* 40 (1): 67–91. doi:10.1086/200061.

Clare, Eli. 2017. *Brilliant Imperfection: Grappling with Cure*. Durham, NC: Duke University Press.

Douglas, Mary. 1996. *Purity and Danger: An Analysis of Concepts of Pollution and Taboo*. New York: Praeger.

Garland-Thomson, Rosemarie. 2002. "Integrating Disability, Transforming Feminist Theory." *NWSA Journal* 14 (3): 1–32.

Ghazala, Qubais Reed. 1996. "The Casio SK-1: Escapist Sample Shuttle." *Experimental Musical Instruments* 12 (2): 11–16.

Ghazala, Qubais Reed. 2004. "The Folk Music of Chance Electronics: Circuit-Bending the Modern Coconut." *Leonardo Music Journal* 14: 97–104. doi:10.1162/0961121043067271.

Hallowell, A. Irving. 1964. *Ojibwa Ontology, Behavior, and World View*. New York: Columbia University Press.

Hammer, Steven. 2018. "Writing Dirt, Teaching Noise." In *Soundwriting Pedagogies*, edited by Courtney S. Danforth, Kyle D. Stedman, and Michael J. Faris. Logan: Computers and Composition Digital Press/Utah State University Press. http://ccdigitalpress.org/book/soundwriting/.

Harry, Beth, and Janette Klingner. 2007. "Discarding the Deficit Model." *Educational Leadership* 64 (5): 16–21.

Hawisher, Gail E., and Cynthia L. Selfe. 1991. "The Rhetoric of Technology and the Electronic Writing Class." *College Composition and Communication* 42 (1): 55–65.

Hermes, Mary. 2005. "'Ma'iingan Is Just a Misspelling of the Word Wolf': A Case for Teaching Culture through Language." *Anthropology & Education Quarterly* 36 (1): 43–56. doi:10.1525/aeq.2005.36.1.043.

Kerse, C. S. 1975. *The Law Relating to Noise*. London: Oyez Publishing.

Latour, Bruno. 1993. *We Have Never Been Modern*. Cambridge, MA: Harvard University Press.

McAllister, Michael, Elaine Yuen, and Stuart Bush. 2012. "Cultivating Design Partnership: A Participatory Design Exploration Engaging People with Disabilities." Presentation at Industrial Designers Society of America Education Symposium, August 15, 2012, Boston, MA. https://www.idsa.org/sites/default/files/McAllister.pdf.

McLuhan, Marshall. 1966. *Understanding Media: The Extensions of Man*. New York: Signet Books.

McRuer, Robert. 2006. *Crip Theory: Cultural Signs of Queerness and Disability*. New York: New York University Press.

Menkman, Rosa. 2010. "Glitch Studies Manifesto." *Amodern*. http://amodern.net/wp-content/uploads/2016/05/2010_Original_Rosa-Menkman-Glitch-Studies-Manifesto.pdf.

Oxford University Press. 2020a. "Diagnosis." *Lexico*. https://www.lexico.com/definition/diagnosis.

Oxford University Press. 2020b. "Noise." *Lexico*. https://www.lexico.com/definition/noise.

Oxford University Press. 2020c. "Prognosis." *Lexico*. https://www.lexico.com/definition/prognosis.

Powell, Malea, Daisy Levy, Andrea Riley-Mukavetz, Marilee Brooks-Gillies, Maria Novotny, and Jennifer Fisch-Ferguson. 2014. "Our Story Begins Here: Constellating Cultural Rhetorics." *Enculturation: A Journal of Rhetoric, Writing, and Culture* 25. http://enculturation.net/our-story-begins-here.

Sanders, Elizabeth B.-N., and Pieter Jan Stappers. 2008. "Co-Creation and the New Landscapes of Design." *CoDesign: International Journal of CoCreation in Design and the Arts* 4 (1): 5–18. doi:10.1080/15710880701875068.

Selfe, Cynthia L., and Richard J. Selfe Jr. 1994. "The Politics of the Interface: Power and Its Exercise in Electronic Contact Zones." *College Composition and Communication* 45 (4): 480–504.

Shannon, Claude E. 1948. "A Mathematical Theory of Communication." *Bell System Technical Journal* 27 (3): 379–423. doi:10.1002/j.1538-7305.1948.tb01338.x.

Taylor, Rupert. 1970. *Noise*. London: Penguin Books.

Wilson, James C., and Cynthia Lewiecki-Wilson, eds. 2001. *Embodied Rhetorics: Disability in Language and Culture*. Carbondale: Southern Illinois University Press.

Yergeau, M. Remi, Elizabeth Brewer, Stephanie Kershbaum, Sushil K. Oswal, Margaret Price, Cynthia L. Selfe, Michael J. Salvo, and Franny Howes. 2013. "Multimodality in Motion: Disability and Kairotic Spaces." *Kairos: A Journal of Rhetoric, Technology, and Pedagogy* 18 (1). http://kairos.technorhetoric.net/18.1/coverweb/yergeau-et-al/.

2

THE CIRCULATION OF TOUCH
Very Simple Machines for Creating Tactile Textual Experiences

David M. Sheridan
Michigan State University

MOVING BEYOND THE SURFACE

At a recent Computers and Writing conference, I gave a presentation on the way makerspaces can open up possibilities for writers. My basic claim was that the tools, skills, and practices found in makerspaces can help composers produce texts that operationalize possibilities elided by both traditional essays (black words on white paper) and by screen-based forms (like web pages). Composers working in makerspaces are producing "texts" that are three-dimensional, that have moving parts, and that elicit tactile engagement. Many artists, designers, and makers report that they think with their hands (see, for instance, Candy and Edmonds 1996; Ghent 2017; Kaplan et al. 2000; Lin 2000); the makerspace compositions I had in mind invite this kind of manual cognition when placed in the hands of "readers."

In my program at Michigan State University, for instance, we are experimenting with a flatbed digital cutter that uses a blade similar to a small X-Acto. In some ways, this cutter is analogous to a traditional printer, like a dot matrix printer or an inkjet printer. A traditional printer translates a digital design to an arrangement of ink on the surface of a piece of paper. Printers give designs made *inside* the machine a presence *outside* the machine. Similarly, digital cutters use a variety of instruments (blades, lasers, waterjets, routers) to translate digital designs into physical shapes. You draw a heart on your screen. The cutter cuts that heart out of some medium—such as paper, metal, wood, or plastic.

My argument, however, is that digital cutters are fundamentally different from traditional printers. Cutters do not merely inscribe the surface of a medium. They shape the medium itself. They cut, etch, and score. As such, they introduce a new array of possibilities—possibilities related to movement, dimensionality, tactility. A quick Google search will turn

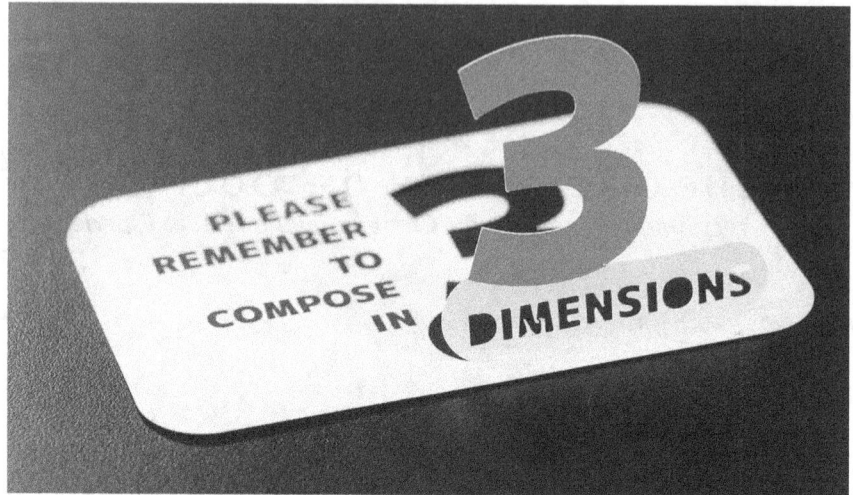

Figure 2.1. Prototype of card design meant to demonstrate the use of three-dimensional elements.

up designs for fully functional wooden-gear clocks made exclusively from laser-cut plywood parts that are held together without nails or glue.

I wanted to give those who attended my C&W presentation gifts that embodied some of the composing choices enabled by cutters—simple 3" x 5" cards that reveal the power of shape. My first card (figure 2.1) contains a basic pop-up element that performs the practice it advocates (Long 2013). I liked the design, though it felt somewhat literal-minded.

Experiments like this revealed the affinity between cut space and negative space—a design technique in which the absence of visual material introduces an additional element. An often-cited example is the arrow formed between the E and the x in the FedEx logo. Like shadows, elements coaxed into existence via negative space are interesting because they are present and absent at the same time. Eugen Gomringer (1967) uses negative space ingeniously in his concrete poem "Silencio" (figure 2.2).

Riffing on this paradoxical notion of presence and absence led to another card (figure 2.3). I liked the minimalist quality of this design: a single sentence cut out of black paper. The digital cutter produced letters that were crisp and precise, and there is something satisfying about holding this piece—cut from thick cardstock—in your hand. (I could talk all day about the beauty of rounded corners.) Most important of all, the fact that the text is cut rather than printed contributes to the meaning of the composition.

**silencio silencio silencio
silencio silencio silencio
silencio   silencio
silencio silencio silencio
silencio silencio silencio**

Figure 2.2. Eugen Gomringer, "Silencio" (1967).

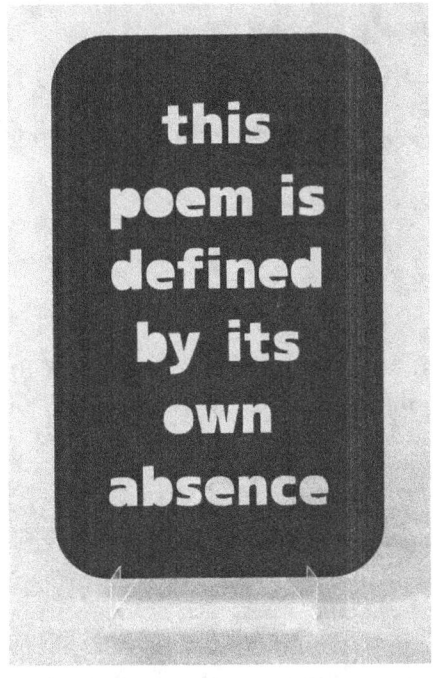

Figure 2.3. Prototype for a C&W gift card: THIS POEM IS DEFINED BY ITS OWN ABSENCE.

I was reminded of Henry Sayre's (1988) exploration of avant-garde literature, which I stumbled upon as an undergrad, pulling random books off library shelves. Sayre prompts us to think about the nature of textuality by examining works like Shusaku Arakawa's *Untitled 1969*—a large canvas inscribed with the declaration, "I have decided to leave this canvas completely blank" (Sayre 1988, 1195). As Charles Bernstein and Susan B. Laufer note, Arakawa's work reveals that "Language . . .

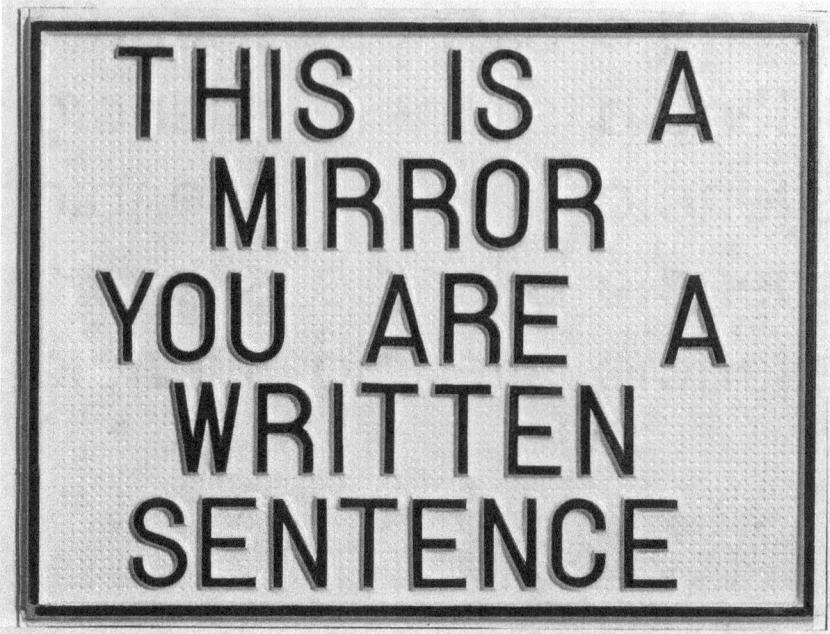

Figure 2.4. Luis Camnitzer. This Is a Mirror, You Are a Written Sentence, 1966–1968. Vacuum formed polystyrene. 19 1/8 × 24 4/8 × 0 4/8 in (48.42 × 62.49 × 1.48 cm). Courtesy Alexander Gray Associates, New York; Galería Parra & Romero, Madrid. © 2021 Luis Camnitzer/Artists Rights Society (ARS), New York. Photo source: P. Schälchli, Zürich.

cannot be thought of only as a verbal, word-bound system but is equally involved with the construction and mediation of visual seeing and of space" (qtd. Sayre 1988, 1195–6). Sayre also discusses Luis Camnitzer's self-reflexive mirror-sentence (figure 2.4). I began to wonder what Gomringer, Arakawa, or Camnitzer might do with a digital cutter. I was convinced they would use it to teach us new ways of thinking about art, language, and semiosis.

In this chapter, I offer a metaphorical reinterpretation of reprogrammable circuits. I propose a series of paper writing devices (PWDs)—very simple devices intended to encourage a community of writers to share and activate playful, experimental texts. I see these PWDs as analogous to hardware, such as the reprogrammable circuit. The texts designed to be loaded into these PWDs are analogous to software.

My mode in this chapter could be considered a kind of "think-aloud protocol." Like the writers Linda Flower and John Hayes (1981) studied, I describe a set of goals that inform my composing process. I frame these goals in terms of current conversations within the field—especially around rhetorical materiality and circulation. I am specifically concerned with the circulatory processes of compositions whose ultimate material form exists outside of digital environments—"physical" forms that work through shape, dimensionality, and tactile engagement. I see the digital cutter—a relatively new tool—as offering new possibilities for the circulation of touch. Ultimately, I see my chapter as an invitation to participate in a networked game of critical making: a multiplayer game in which we explore together tactile rhetoric via radically minimalist compositions. Will you join me?

MATERIALITY AS IMMERSIVE, EMERGENT, AND RELATIONAL

In his contribution to this collection, Steven Hammer critiques the "myth of noiselessness." One of the paths that leads to his awareness of how this myth operates is the experience of raising his son, Rowan, who, at an early age, received a medical diagnosis of lissencephaly—a brain condition that means Rowan will not walk or talk. Hammer explains that "the act of diagnosis . . . reveals a set of cultural values that both privileges and assumes that systems (from human bodies to laptops to families) and their components possess the potential and preference for purity and functionality." Hammer invites us to move past this myth of noiselessness to see other forms of possibility. For instance, in Rowan's approach, "the acoustic guitar isn't a stringed and fretted instrument tuned to EADGBE. It's better described as a large, resonant drum with variable sustain, with optional string-holds to grasp, pull, and release." In challenging society's focus on "purity and functionality," Rowan is able to find a different set of possibilities in the guitar. The instrument becomes a drum and the strings become handles. (This makes me wonder if Rowan would discern a kinship with percussive guitarists like Kaki King.) By rejecting the idea of "noise," we can see a range of possibilities that were not at first apparent in the guitar. Indeed, its very classification as a "stringed instrument" masks a range of percussive possibilities.

I want to bring Rowan's approach to a basic material that writers are familiar with—paper. As writers, we know that the potentials of paper are activated by inscription, just like we know that a guitar is an instrument activated by strumming. Paper is a surface. It is there to mark.

But what if it isn't? What if paper is not just there to mark but also to shape, cut, fold, and assemble?

This question leads us to the larger issue of rhetoric's materiality—something the field has become increasingly interested in over the past two decades or so. In his contribution to *Rhetorical Bodies*, Lester Faigley

(1999) writes that "literacy has *always* been a material, multimedia construct, even though we only now are becoming aware of this multidimensionality and materiality because computer technologies have made it possible for many people to produce and publish multimedia presentations" (175–76). In that same collection, Carole Blair (1999) explores the function of materiality using various memorials, including the AIDS Memorial Quilt and Montgomery's Civil Rights Memorial. These compositions are made of fabric, granite, and bronze, raising considerations that are often overlooked. As a counterbalance to the field's focus on rhetoric's "symbolicity," Blair invites us to think about the contributions that embodied experience, physical structure, and material properties (strength, hardness, weight, resilience, texture) play in shaping rhetorical meaning and experience (18). N. Katherine Hayles (2002) stresses that materiality is an "emergent property" that "depends on how the work mobilizes its resources as a physical artifact as well as on the user's interactions with the work and the interpretive strategies she develops" (33). Anne Frances Wysocki (2004), building on Hayles and others, argues that we should reserve the label of "new media" for compositions

> that have been made by composers who are aware of the range of materialities of texts and who then highlight the materiality: such composers design texts that help readers/consumers/viewers stay alert to how any text—like its composers and readers—doesn't function independently of how it is made and in what contexts. (15)

Hayles and Wysocki avoid simple binaries such as material/immaterial, digital/nondigital. Instead, they ask us to consider the staggering range of material possibilities available to composers.

Finally, I want to touch on the role of touch, which, as Matthew Davis and Kathleen Blake Yancey (2014) observe, is often overlooked, even in discussions of multimodal rhetoric (14). It is likely that touch is elided, in part, because of Eurocentric biases that order the senses hierarchically (see, for instance, Classen and Howes 2006). But the importance and power of touch as a tool for storing information and communicating cultural knowledge can be seen in numerous examples from various cultural groups throughout history. Rocío Quispe-Agnoli (2010) examines the tactile or "tangible" literacies activated in the pre-Hispanic Andean tradition of quipus—a system of storing and communicating information using knotted string. Angela Haas (2007) "traces a counterstory to Western claims to the origins of hypertext and multimedia by remembering how American Indian communities have employed wampum belts as hypertextual technologies" (77). She uses the label "digital

Figure 2.5. "Keep Austin Weird" bumper sticker. Courtesy BookPeople and Waterloo Records & Video.

rhetoric" to describe the way wampum activates the digits (fingers) (84). Richard Rath (2014) points to the haptic nature of "holding and passing" the wampum belt (315). Sonia Arellano (2017) uses an embodied methodology to explore the "tactile rhetoric" of the Migrant Quilt Project (13). Using the framework of disability studies, Shannon Walters (2014) explores rhetorical touch as "a potential for identification among bodies of diverse abilities that takes place in physical, proximal, and/or emotional contact" (3). These writers and others help us see the many possibilities associated with an overlooked rhetorical resource. I return to the role of touch in my discussion of artists' books below. For now, I want to extract from these discussions of materiality a set of design goals for my paper writing devices.

Design Goal #1: PWDs should seek compelling and strategic material arrangements.

Design Goal #2: PWDs should encourage awareness of those material arrangements.

Design Goal #3: PWDs should include material arrangements that invite relationality—the emergence of meaning through embodied interactions between composers, users/readers, and compositions.

SHORT TEXTS, LIMIT CASES, AND RHETORICAL CIRCULATION

The specific material form of a rhetorical composition is one factor that shapes the way the composition circulates. We can see this when we look at very short texts. I've been interested in the power of radically concise texts ever since reading Jenny Rice's article on rhetorical ecologies (Edbauer 2005). Rice relates the story of how, in response to the prospect of a Borders store opening across the street, an independent bookstore and an independent record store in Austin jointly ordered five thousand copies of a bumper sticker printed with the text KEEP AUSTIN WEIRD (figure 2.5). The stickers launched a trend that ultimately involved thousands of stickers, as well as T-shirts, coffee mugs, and other items. KEEP AUSTIN WEIRD is a three-word sentence that encapsulates conceptions of the city's identity, critiques the homogenizing effects of big box chains, asserts understandings of civic activism, and more. We might say this ingenious micro-essay functions as a limit case that reveals how concise a public-rhetorical composition can be.

But to understand the full power of KEEP AUSTIN WEIRD, we need to look past the text itself. Because this text is brief, it can be printed on a bumper sticker. While not a privileged genre within first-year composition, the bumper sticker has an advantage—namely the ability to stick to the bumper of an automobile. Cars are everywhere. And they move. By hitching a ride with a car, a text can find many eyes. The material properties of the document—its diminutive size, its horizontality, and its sticky backside—are as important as the text itself. Bumper stickers reveal the relationship between material form and circulation.

Conceptual artists leverage the material-circulatory strategies of the bumper sticker and related cultural forms. Camnitzer's sentence mirror (figure 2.4), for instance, emerged out of the New York Graphic Workshop, which was devoted to an ethic of FANDSO—free assemblable nonfunctional disposable serial object. The "mass produced art" of FANDSO aimed "to eliminate the high cost and pompous ritual that separate art from the public" (qtd. in Maroja 2015, 44). Camnitzer explains:

> We were Latin Americans, we were printmakers. We were aware of being segregated from the mainstream. One way of breaking that was making our own venue. And the cheapest, easiest venue was the mail. So we created our envelope gallery and mailed it to our audience. We were making a market and at the same time disputing ownership, disputing preciousness. Some pieces were mailed to names from telephone books, some pieces were to friends, whom we hoped would pass them on, some pieces were stuck in bathrooms and elevators. (qtd. in Rexer 2008, 55)

Camnitzer's sentence mirror was originally printed on a self-adhesive label for one of these mail pieces.

Similarly, Jenny Holzer draws inspiration from brief, commercialized texts. Her work often takes the form of T-shirts, tattoos, park benches, and posters. An exhibit at the Tate Modern features items such as condom wrappers printed with "MEN DON'T PROTECT YOU ANYMORE" (figure 2.6). Holzer is particularly known for her use of signs, such as billboards and scrolling LED screens. She explains,

> I came to the signs as another way to present work to a large public. . . . Because signs are so flashy, when you put them in a public situation you might have thousands of people watching. So I was interested in the

Figure 2.6. Jenny Holzer, Exhibition view: Artist Rooms: Jenny Holzer, Tate Modern, London, 2018. © 2021 Jenny Holzer, member Artists Rights Society (ARS), New York. Photo: Jack Hems.

efficiency of signs as well as the kind of shock value the signs have when programmed with my peculiar material. These signs are used for advertising and they are used in banks. (qtd. in Sayre 1989, 202)

To fully account for the meaning of Holzer's works, we have to confront their material form. As one art historian has put it,

> In Holzer's work, positions are assumed, identities delimited, meaning shaped and ideologies spoken *only* at the point that language takes physical form—poster, hat, billboard, type-face, colour, size, support, Japanese, English, German, engraved, offset-printed, hand-written, projected, cast in bronze, carved in stone—and *only* in relation to the contexts in which such forms are situated—the Las Vegas strip as opposed to Picadilly [*sic*] Circus in London, or St Peter's Church in Cologne. (Hughes 2006, 423–24)

Borrowing a phrase from William Carlos Williams, Sayre (1989) notes that Holzer's "rhythms are part and parcel of the contemporary 'American idiom.' Holzer inserts her language . . . into the visual rhythms of what Guy Debord has labeled the society of the spectacle" (199).

The work of Adrian Piper provides a particularly compelling example of the nexus of composition, material form, circulation, and meaning. Art historian Cherise Smith (2007) explains that, although Piper's early work received attention, the art establishment excluded her after discovering that she was both Black and a woman. Smith shows that Piper's subsequent work is best understood in that context. As part of a larger project called *The Mythic Being*, Piper circulated small advertisements in the *Village Voice*. These "ad-works" feature her own image, costumed in "an Afro wig, mustache, sunglasses, and 'working class' clothes" (Smith 2007, 47)—an image that echoes "the black American male nationalist" (50). These images are accompanied by thought clouds containing hand-lettered text. For example: "TODAY WAS THE FIRST DAY OF SCHOOL. THE ONLY DECENT BOYS IN MY CLASS ARE ROBBIE & CLYDE. I THINK I LIKE CLYDE. 9/21/61" (qtd. in Smith 2007, 47). According to Smith, Piper's mythic being is "an androgynous, liminal, or third-sex being. The figure looks like a man, but has thoughts that often seem those of a girl or woman. Piper's placing of a gender-, class-, and racially-indeterminate figure in the newspaper piques the audience's curiosity and challenges its assumptions" (51). Even as it interrogates societal norms, Piper's work critiques the narrow, exclusive practices of the art establishment by rejecting the gallery in favor of everyday mass-market channels (47–48). Piper herself has explained: "I don't make concrete, spatiotemporally unique, discrete objects that cannot be multiply reproduced. . . . I don't rely on discrete, spatially unique art contexts for presentation of the work. I utilize art contexts only in their information-disseminating capacities" (qtd. in Smith 49).

Like the authors of KEEP AUSTIN WEIRD, Camnitzer, Holzer, and Piper are eager to see their texts circulate. They favor cheap, minimalist forms that are easy to reproduce and distribute. The circulatory paths these works roam are, themselves, part of the message. Strategically selected lines of circulation allow the art object to dialogue with other compositions in the same cultural position (other signs, ads, T-shirts). They critique the practices of the art establishment by avoiding the establishment's preferred venues (museums, galleries) and by rejecting the one-of-a-kind art object. They convey an egalitarian ethos by favoring highly reproducible forms and everyday channels. They point to their own publicness, asserting the desire to be taken as a form of civic rhetoric. How these works *circulate* shapes how these works *mean*.

Moreover, the minimalist forms these works adopt force us to interrogate the nature of art, textuality, and the consumption practices we bring to them. (In what sense is "this is a mirror you are a written sentence"—printed on an adhesive label and circulated through the US mail—a work of art?) Their radical simplicity means that they function as limit cases. These compositions are akin to what Hayles (2002) calls a "technotext": "A term that connects the technology that produces texts to the texts' verbal constructions" (25). Such texts "play a special role in transforming literary criticism into a material practice" (25–26). An LED sign by Holzer asks us to reflect on the nature of LED-ness as a discursive resource.

Craig Saper (2001) explores the way artists deploy everyday channels of circulation to create art that properly can be seen as a network of multiple artifacts and participants rather than a single object. J. S. G. Boggs, for instance, paints replicas of the US dollar bill that are so convincing they have drawn the attention of agencies that police counterfeiting. According to Saper, the completed work of art is not the bill itself but the various transactions the bill facilitates. The final, framed assemblage comprises all of the artifacts that document these transactions: the "Boggs bill," cash register receipts generated when using the bill to buy goods, coins returned as change, etc. In one case, Boggs orchestrated an event that asked "all the participants to keep the Boggs bills in circulation for at least five exchanges (and to mark the bills with their thumbprints)" (xi). Saper claims that "Boggs's goal was to use the trappings of a bureaucratic system to invent a gift-exchange community in which the members would be involved in a more intimate sense with transactions usually considered impersonal" (xi). As they circulate, Boggs bills carve out a rhetorical-aesthetic-social space.

Saper cites Bob Brown as an example of an artist who devises a highly specialized apparatus in service of his network art. Brown (2014) proclaimed that "the written word hasn't kept up with the age. The movies have outmaneuvered it. We have the talkies, but as yet no Readies" (1). Brown (1931) envisioned

> a device whereby words pass before your eyes in a continuous uninterrupted stream. Books to be read on the reading machine are printed on rolls in microscopic type. The rolls unwind automatically at any speed desired, and the magnifying apparatus makes it possible for one to select the size of type most comfortable to read. (i)

Brown's reading machine is a humorous apparatus for highlighting the nature of reading as an embodied practice. Moreover, it functions as an invitation for others to participate in a network of textual production. Indeed, a wide range of writers, including Gertrude Stein and William Carlos Williams, submitted work for *An Anthology of Readies for Bob Brown's Machine.*

In the case of Boggs, Brown, and other examples Saper (2001) discusses in *Networked Art*, I see compelling precedents for the kinds of sharing envisioned by the editors of *Reprogrammable Rhetoric*, who propose sharing the code associated with the projects discussed in this collection's chapters. Readers can download the code, use it, modify it, upload mods, etc. Saper would ask us to reflect on the way in which all of the discrete instances of making and sharing might add up to something larger; all of us who participate in this process are linked in a network, and that network itself might be viewed as a rhetorical-aesthetic work. This is what excites me about the designs for PWDs I propose below: not the particularities of a single device or a single text that is fed into it, but the totality of the devices, texts, readings, sharings, and modifications engendered by the project. The invitation I offer here, then, is to participate in a networked process of "critical making" in the sense of the term popularized by Matt Ratto and others. A riff on critical thinking, critical making asks us to confront the possibilities associated with doing intellectual work through material processes of building, fabricating, and prototyping rather than through more traditional academic forms like articles and books (Ratto 2011). In Ratto's conception of critical making and in the minimalist art discussed here, the process—and what we learn from it—is emphasized over the final product. Making is a form of "conceptual exploration" (Ratto 2011, 253). Luis Camnitzer (1986) alludes to this when he writes, "If explanations exhausted my work, it would die and stop being art. The explanation would suffice and there wouldn't be any need for the art work. The art work would be no more than a redundant illustration of a theory. . . . If there is any part of it that survives beyond the reading of this text, it does so because of its inexplicability. Only this inexplicability is capable of an expansion of knowledge." The materiality of the work exceeds any theorizing of the work.

> **Design Goal #4:** PWDs should take a minimalist approach: small, simple, and cheap.
> **Design Goal #5:** PWDs should aspire to circulate wide and far.
> **Design Goal #6:** PWDs should use material form and circulatory channel as meaning-making resources.
> **Design Goal #7:** PWDs should invite the participation of others.
> **Design Goal #8:** PWDs should be shareable via networks using common file types.
> **Design Goal #9:** PWDs should rely on output devices that are highly accessible.

VISUAL-TACTILE READING EXPERIENCES: LEARNING FROM ARTISTS' BOOKS

As someone originally trained to write short stories and poems, I have long held the tacit assumption that the text was something independent of any specific material instantiation of it. In my MFA workshops, it would have been ridiculous to tell a writer: "I loved this story, but Times New Roman is a boring font choice," or "Great poem, but your selection of paper is uninspired." Now I teach in an interdisciplinary arts and humanities program that incorporates "book arts." Books produced by our students achieve their aesthetic and rhetorical effects by careful manipulation of all the things that I devalued as a short story writer: printing and binding techniques, choice of paper, page size and layout, and, yes, font choice. Since these books are often printed on letterpresses, font choice frequently involves consulting sets of movable type stored in cabinets. Artists' books as compositions are *engineered*, and their designer-makers are knowledgeable about the specific properties of relevant materials (paper, fabric, thread, adhesives), structural tactics (folding, cutting, binding), and printmaking techniques (letterpress, linocut, Risograph). Book artists use all of these resources and techniques to create visual-tactile experiences.

Books are inherently tactile, and book designers have integrated specialized tactile and interactive elements since the beginning. Volvelles—wheels meant to be turned by readers—are visual-mechanical tools for establishing relationships (think about the rotating depictions of the night sky that you can buy in planetarium gift shops). Krystina Madej (2016) surveys historical uses of volvelles and other "manipulable devices" (17) integrated into written documents to produce "tangible narratives" (22). She notes that Robert Sayer, an eighteenth-century printer, arranged flaps that "could be opened and closed to reveal images and text in various combinations and lead readers through a sequence of scenes that told the story" (23; see also Hayles 2002).

The work of Clarissa Sligh offers an excellent opportunity to see the way artists' books employ shape, dimensionality, and tactility. In her forward to *Transforming Hate: An Artist's Book*, Sligh (2016) writes, "I am a black woman. I am an artist. For many years I have been creating work to bring issues of social justice into the public discourse." She explains that this book grew out of a project in which she folded the pages of white supremacist texts into origami cranes, turning words of hate and violence into "a beautiful art object." I believe it is accurate to say that in this case dimensionality functions as a strategy of rhetorical transformation. The act of folding—which turns a two-dimensional page into a three-dimensional art object—physically disrupts the hate speech, rendering it partially unreadable. The final product, an origami crane, conveys creativity, beauty, and peace.

Sligh's (1988) *What's Happening with Momma?* adopts the shape of a house (figure 2.7). Its accordion-fold structure creates an unbroken

Figure 2.7. View of What's Happening with Momma? *Van Dyke Photographic Print by Clarissa Sligh at Lower Eastside Print, NYC, 1987.*

chain of panels, each of which contains a drawing of the house's façade overlaid with a different snapshot. A paper staircase unfolds from each panel, echoing the steps depicted in the photographs, and the horizontal surfaces of the steps are printed with text. This work uses three dimensions, requires a multifaceted reading process, and engages the hands by requiring multiple kinds of unfolding. The bookmaker's struggle to discover appropriate material configurations can be seen in the artist's statement that accompanies the book. Early on, Sligh (2021) writes, "I found myself groping for the physical form it would take." She notes that her "biggest challenge was to create a structure that would also provide a way for the viewer to interact with the book in order to 'read' it." She sought a form analogous to "the way that the Baptist preacher . . . and rhythm and blues singers and musicians used repetitive, rhythmic fragments to elicit the audience response necessary for the satisfactory completion of their work." Finally, "after agonizing over it for months," Sligh lands on a process of "unfolding" based on a performance by another book artist. The accordion fold structure she ultimately adopts allows the book to stand upright, revealing the visual repetitions and rhythms.

Bethany Collins (2015), a Black artist from Montgomery, Alabama, has created a series of artist's books that challenge societal constructions of race. Her *Colorblind Dictionary* consists of a copy of a Webster's dictionary in which all terms related to color have been erased. *America: A Hymnal* (figure 2.8), collects one hundred versions of "My Country 'tis of Thee." Collins explains that, spanning more than two centuries of history, the lyrics of this song have been adapted by many different

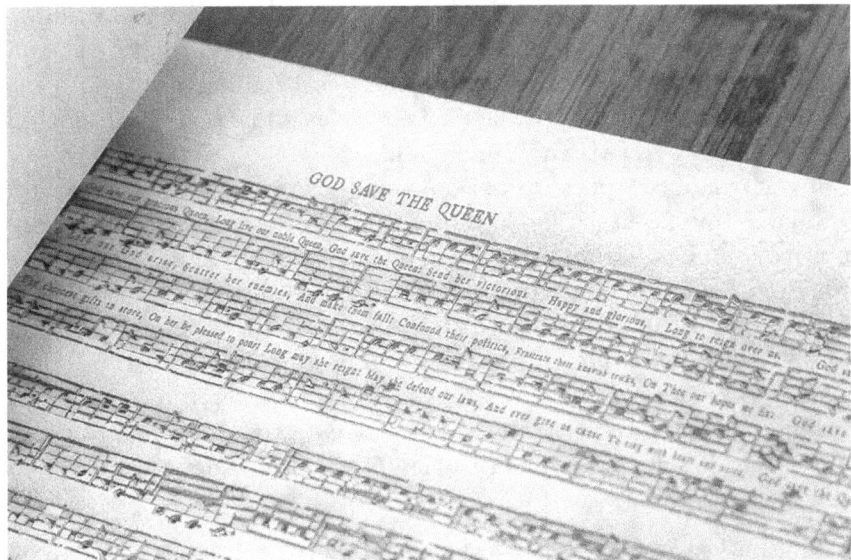

Figure 2.8. Bethany Collins, interior view of America: A Hymnal *(2017) 100 laser cut leaves, 6" × 9" × 1". Courtesy the artist, PATRON, Chicago and CANDOR Arts. Photography by Candor Arts.*

groups. An abolitionist version of the song, for instance, replaces "sweet land of liberty" with "stronghold of slavery" to expose the hypocrisy of a racist nation. Collins (n.d.) uses a laser cutter to remove the notes and staff of the song: "While the differing lyrics remain legible, the hymnal's unifying tune has been burned and etched away." The crisp outlines of the musical notation remain, creating holes that reveal subsequent pages. This tunneling creates a multilayered, highly dimensional, and tactile experience.

The works of Sligh and Collins are compelling examples of how shape, texture, movement, and dimensionality can be used to convey meaning and generate aesthetic-rhetorical experiences. These works require tactile engagement and cultivate new arrangements of embodied interaction. In the process, they cause us to revise our very definition of what a book is, just as Rowan causes us to revise our definition of what a guitar is.

Compositions that rely on touch, however, complicate processes of circulation. It's easy to share text documents via email, but if I want to convey information aimed at your fingers, even printing out those documents doesn't work very well. (Try to read an inkjet-printed image or sentence with your fingertips.) Traditional screens and inkjet printers are not well equipped to translate digital designs into forms that meaningfully activate touch. Yancey, for instance, relates the differences in encountering a scrapbook as a digitized object on the web and interacting with the same artifact in a physical setting. Sitting with the scrapbook in the library, the tactile became essential: "Without touch, I could not read" (Davis and Yancey 2014, 17).

In fact, in one respect artists' books are the opposite of the minimalist textual art discussed earlier. Because they are often intricate and laborious to produce, they are commonly released in limited editions of one hundred copies or less. Camnitzer, Holzer, and Piper are interested in highly reproducible works that circulate far and wide. The material form of the artist's book often limits both its reproducibility and its possibilities for circulation. In fact, libraries and museums often have difficulty making artists' books available for patrons to touch, requiring protective gloves and designated reading rooms (Clark 2013). Digital cutters shift this dynamic in important ways.

> **Design Goal #10:** PWDs should create experiences that are both visual and tactile; they should give the fingers something to do.
>
> **Design Goal #11:** PWDs should model the ways touch can be used to generate meaning.

THE CIRCULATION OF TOUCH

In skeletal form, my argument thus far looks something like this: (1) Compositions can take many different material forms and these forms inflect processes of semiosis. (2) The specific material form of a composition will influence the way that composition circulates. (3) Processes of circulation also inflect processes of semiosis. (4) Strategic circulation of compositions can be used to foster larger networks of meaning and participation. (5) Shape, dimensionality, and tactility are powerful but often overlooked components of material form. These points are not necessarily in harmony with each other. The deployment of shape, dimensionality, and tactility is, to some degree, at odds with the goals of circulation and network formation.

Digital cutters shift this dynamic, introducing new possibilities for the circulation of touch. If traditional printers inscribe designs on the surface of a medium, digital cutters allow composers to shape the medium itself. I draw a star, a digital cutter cuts the star out of wood, paper, plastic, or metal. Cutters can also create texture by etching and can enable precise folds by scoring. Moreover, because they start with a digitized design, these cutters enable familiar practices of electronic distribution. It is commonplace to email files and post them on the web.

End users can print out these files on laser printers and inkjet printers. Digital cutters enable a similar process. We share designs via email and the web; end users translate them into material form using increasingly common digital cutters.

I am not claiming that digital cutters make circulation frictionless or "easy" by some absolute standard. Cutters still require time, skill, money, and other resources. All technologies introduce complex considerations of access (see, for instance, Monroe 2004). A simple claim, such as "it's easy to circulate a text document," masks a wide range of considerations related to the computers, networks, software, and knowledge that allow such circulation. We need to ask, Easy for whom? Easy compared to what?

Moreover, some designs are easier to circulate than others. All technologies and materials have limitations. The cutter I routinely use, which cuts with a small blade, will rip the paper if the design is too intricate. The more complex the design, the more I need to fine-tune variables such as the pressure and speed of the blade, the mechanism used to mount the paper, the type of paper used, etc. Accordingly, the PWDs I'm proposing here are relatively simple, can be cut out of basic cardstock, and are suitable for relatively inexpensive and commonly available cutters that are increasingly found in schools, libraries, and makerspaces. These cutters are also increasingly found in the homes of individuals, having been integrated into the practices of scrapbookers and other makers. Many craft stores sell small cutters for about the same price as an inkjet printer.

The workflow that I am describing here also enables practices of collaboration—partly because the designs can be widely circulated online, but also because end users can modify designs as they see fit. If we're collaborating on an article, you don't need to retype the draft I send you; you can revise what's already there. The same workflow applies with compositions meant to be cut. Anyone can modify the shared files for my PWDs, enabling a network of users to transform, refine, hack, disrupt, parody, or resist as they see fit.

PAPER WRITING DEVICES

What I am proposing here is a set of minimalist paper writing devices—small, simple, cheap devices made from cut paper. In proposing these PWDs, I'm inviting you to join me in a game of critical making intended to explore the way movement, dimensionality, embodiment, and touch function in rhetorical compositions. I'm hoping that the minimalist approach of my designs will create low barriers for participation. At the same time, I'm hoping their radical minimalism results in

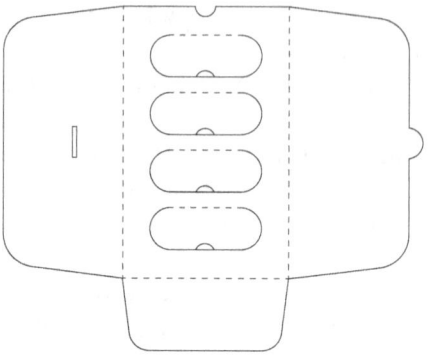
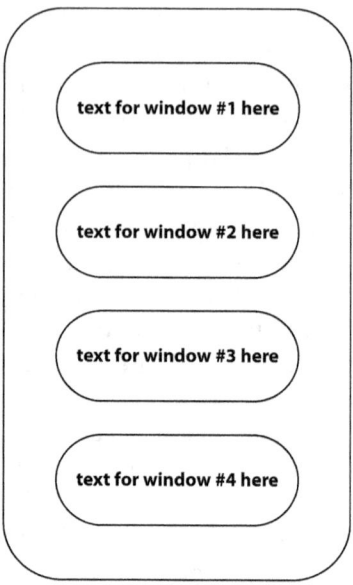

Figure 2.9. Design for four-window envelope.

limit cases that prompt reflection about the nature of art, textuality, and meaning-making. Together we might explore questions such as: How much can we get out of a single word or phrase? What is the simplest artist's book that we can imagine? What is the most elementary strategy for creating compositions that model semiosis as embodied and relational? How much can we do with how little?

Device #1: Four-Window Envelope

Book artists use flaps to introduce a two-phase dynamic of hide/reveal. To distill this dualistic operation to its most basic form, I propose a simple black envelope with four windows (figures 2.9 and 2.10). The envelope accepts 3" x 5" cards printed with text, allowing content to be swapped out easily. Once a card is inserted, a tactile response is required to access the contents of each window.

I see the flap-window strategy as an extension of the line break in poetry. In *Graphic Poetics*, Richard Bradford (2012) draws our attention to the line break as a *visual* (not just aural) device. "The placing of words within a visual structure," Bradford claims, "can demand the same attention to form and effect as the balancing of metre and rhyme against a syntactic sequence" (36). Line breaks can also be used to heighten our awareness of individual words and to defamiliarize language, as demonstrated by William Carlos Williams's (1938) "The Red Wheelbarrow."

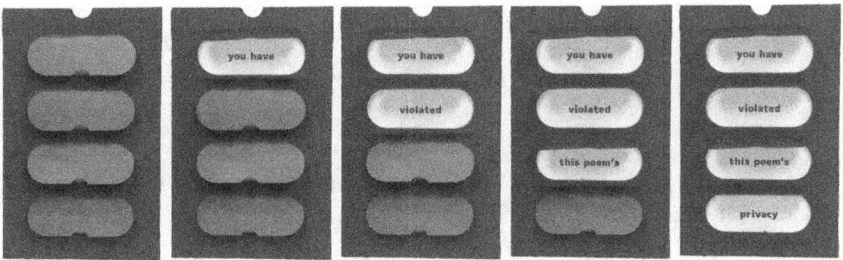

Figure 2.10. Four-window envelope. Animated version available at the collection's companion website: https://upcolorado.com/component/k2/item/6219-reprogrammable-rhetoric-supplemental-design-materials-and-programming-scripts.

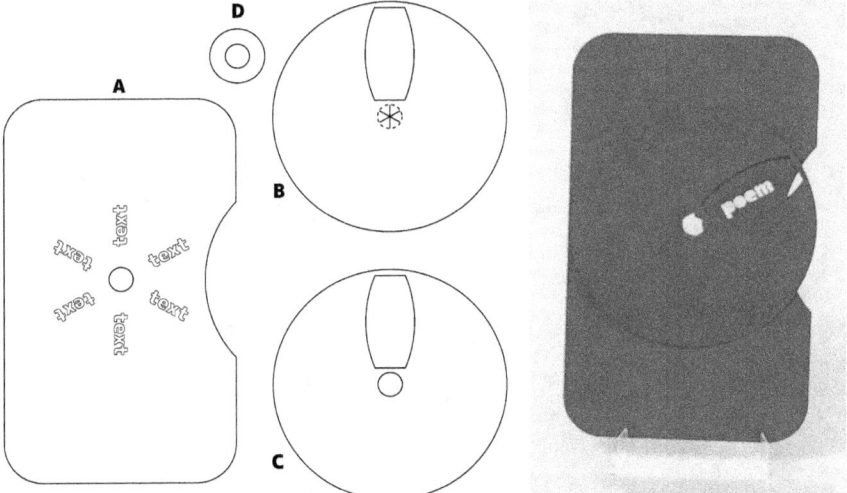

Figure 2.11. Design for a basic volvelle, showing (A) card with text cutout, (B) front wheel, (C) back wheel, (D) back reinforcement ring. The flaps on B are folded through the holes on A and C and then glued on the back of C.

Figure 2.12. Basic volvelle. Animated version available at the collection's companion website: [to come].

Dana Gioia (1987) notes that "the way Williams arranges the poem into brief lines and stanzas slows the language until every word acquires an unusual weight" (399). The flaps on my PWD intensify the disruption of line breaks, creating conditions suitable for exploring slowness, unfolding, revelation, secrets, looking, and privacy.

Device #2: Basic Volvelle

In its simplest form, a volvelle includes a wheel that reveals new information as it turns. The design I offer here requires four parts cut from

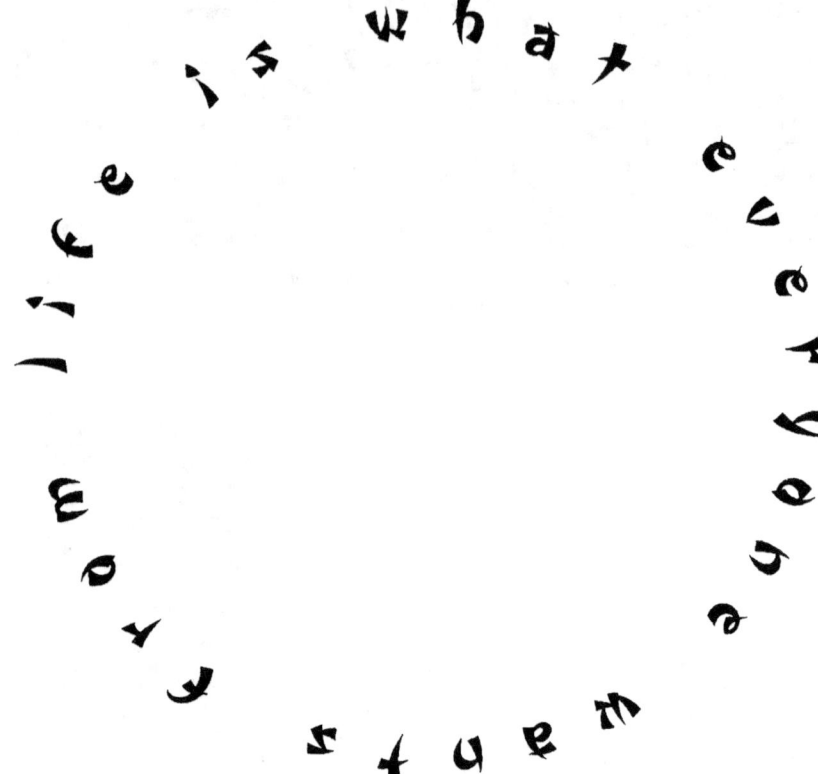

Figure 2.13. Richard Kostelanetz, "Infinities." Design by M. Zuray. © 1998, 2022 by Richard Kostelanetz.

paper, plus a little dab of glue (figure 2.11 and figure 2.12). This design invites themes related to cycles, loops, repetitions, and circularity. For instance, Richard Kostelanetz (1998) has published a series of circular poems that are designed to loop (see figure 2.13). They don't take the form of a volvelle, but they're suggestive of possible approaches.

Device #3. Mirror Text #1

In this design, text is cut out of flaps that are folded up to face each other (figures 2.14 and 2.15). It's as if the text is looking at itself in the mirror. Because the text is cut, readers can look through the first iteration to the second, and beyond, creating a multilayered or tunneled effect. I see this design as suitable for exploring themes of repetition, doubling, reflection, identity, and self-reference.

Figure 2.14. Design for mirror text #1.

Figure 2.15. Photo showing one iteration of mirror text #1: THIS POEM CONFRONTS ITSELF.

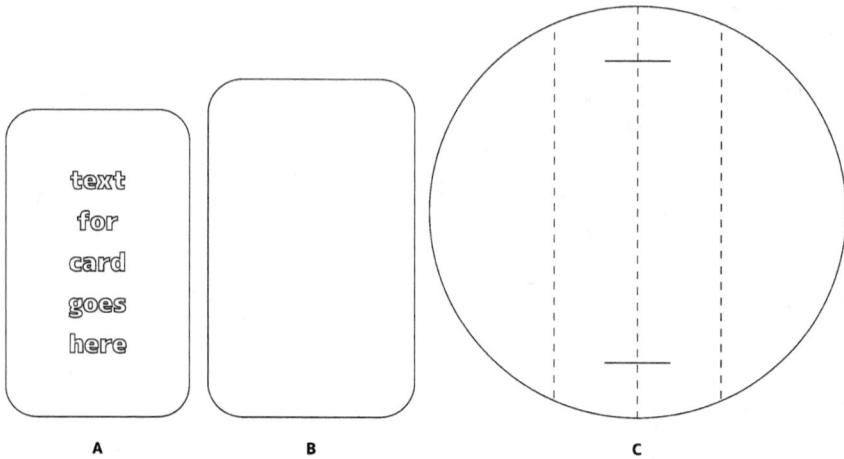

Figure 2.16. Design for mirror text #2, showing (A) cutout card, (B) card cut from cheap mirrored paper, and (C) stand.

Bonus Device: Mirror Text #2

We're all familiar with the way a mirror reverses the letters of printed text. Interestingly, when you hold cut text to a mirror, both the original text and its reflection appear in their proper orientation. This results in a strange doubling that is hard to describe. (When I showed this to a colleague, she asked if I was using a "trick mirror.") This last device (figures 2.16 and 2.17) incorporates a card cut from mirrored paper, cheaply available online. My attraction to this device is the way it moves a half-step toward something that feels more like a machine or science experiment. Like the last device, this design introduces ideas of doubling and self-referentiality, but also ideas of reversal. I can imagine playing with words like *mistake, echo, doppelganger*, iterations on "some things cannot be reversed."

THE THREE-DIMENSIONAL CLASSROOM

By now, we have many examples of creative teacher-scholars who ask their students to explore materiality and who have shared their experiences with the field (see, for instance, George 2002; Shipka 2005; Sirc 2004). One of my primary goals here has been to focus hard on the relationship between circulation and touch. It's much easier to produce a composition that uses touch in compelling ways but that isn't meant to go anywhere—an edition of one, produced through many hours of work. Making copies of the composition and sending them out into the

Figure 2.17. Photo showing mirror text #2.

world introduces challenges. I think students need to confront those challenges. Elsewhere, my co-authors and I describe asking students to devise "plans" that account for the strategic circulation of their compositions in order to achieve rhetorical goals (Sheridan, Ridolfo, and Michel 2012). These plans ask students to think about audience and purpose, but also budgets, venues, and channels: *Nice poster. How much will it cost to print 10,000 copies? Where will they be displayed? How will you distribute them?* I think that framework can serve as one starting point for exploring the circulation of touch.

Many students welcome the addition of the digital cutter to their toolkit because it allows them to make stickers, stencils, rubber stamps, and cards—cultural forms that they value and have adopted into their daily routines. I've been working with the students who staff the small lab I direct to establish a gift economy within our program—a tradition of designing and giving away stickers, artist trading cards, and mini booklets. At first it was difficult to gain momentum with this project, but now it has taken on a life of its own, and students often stop by to see what new offerings are available on our "take one" table.

Building on these experiences and the ideas discussed here, I am eager to explore with my students the uses of shape, dimensionality, and tactile engagement. I believe that we have only scratched the surface of what these resources offer rhetors. One student, upon seeing our experiments with volvelles, used this form to expose the destructive use of gerrymandering in Michigan. His device uses transparent overlays to show the ridiculous districting strategies deployed by our state. Given the digital workflow used by cutters, this design could be shared broadly and reproduced by citizens throughout the state—a viral machine for raising awareness. I'm looking for ways to help students go beyond this, to learn from Piper's, Holzer's, and Boggs's strategies for using tactile compositions to create networks of meaning and transformation. Smith (2007) relates the way Adrian Piper, having failed to gain admittance to New York's elitist art establishment, appropriated the practices of marketing. Piper mimeographed a booklet and mailed copies to established artists and critics, appropriating a list of addresses from the gallery where she worked. Smith observes that "it is difficult not to see the mail-art exhibition as a guerrilla act *on* the audience: after all, Piper's art literally infiltrated recipients' houses" (52). I am interested in exploring with my students ways that the creative, activist, and ethical deployments of shape, dimensionality, and touch can find fortuitous circuits of travel, penetrating through the barriers that prevent change.

AN INVITATION

My goal in this chapter has been to introduce a game focused on creating networks of shared compositions that integrate shape, dimensionality, and touch. As a critical making project, one goal of this game is to theorize—through the material process of making—what happens within a particular and evolving configuration of artifacts, circulatory channels, technologies, and collaborators. During the process of drafting this chapter, a colleague handed me a card that he made for the four-window envelope—the very first "app" written for a PWD by someone besides me. Will you write one?

NOTE

Designs for paper writing devices discussed in this chapter can be accessed on the collection's companion website: https://upcolorado.com/component/k2/item/6219-reprogrammable-rhetoric-supplemental-design-materials-and-programming-scripts.

REFERENCES

Arellano, Sonia Christine. 2017. "Quilting the Migrant Trail: Rhetorical Text(iles) and Rehumanizing Narratives." PhD diss., The University of Arizona.

Blair, Carole. 1999. "Contemporary U.S. Memorial Sites as Exemplars of Rhetoric's Materiality." In *Rhetorical Bodies*, edited by Jack Selzer and Sharon Crowley, 16–57. Madison: University of Wisconsin Press.

Bradford, Richard. 2012. *Graphic Poetics: Poetry as Visual Art*. London: Bloomsbury.

Brown, Bob. 1931. *An Anthology of Readies for Bob Brown's Machine*. Cagnes-sur-Mer, France: Roving Eye Press. https://brbl-dl.library.yale.edu/vufind/Record/3551716.

Brown, Bob. 2014. *The Readies*, edited by Craig Saper. New York: Roving Eye Press.

Camnitzer, Luis. 1986. "Chronology." *Luis Camnitzer: Retrospective Exhibition 1966–1990*. Uruguay: Museo de Artes Plasticas de Montevideo. http://www.lehman.cuny.edu/vpadvance/artgallery/gallery/luis_camnitzer/chronology.htm.

Candy, Linda, and Ernest Edmonds. 1996. "Creative Design of the Lotus Bicycle: Implications for Knowledge Support Systems Research." *Design Studies* 17 (1): 71–90. doi:10.1016/0142-694X(95)00026-N.

Clark, Amanda Catherine Roth. 2013. "The Handmade Artists' Book: Space, Materiality, and the Dynamics of Communication in Book Arts." PhD Diss., The University of Alabama.

Classen, Constance, and David Howes. 2006. "The Museum as Sensescape: Western Sensibilities and Indigenous Artifacts." In *Sensible Objects: Colonialism, Museums and Material Culture*, edited by Elizabeth Edwards, Chris Gosden, and Ruth B. Phillips, 199–222. Oxford: Berg.

Collins, Bethany. n.d. "America: A Hymnal." Bethany Collins (artist's website). Accessed July 31, 2020. https://bethanyjoycollins.com/artwork/4271823_America_A_Hymnal.html.

Collins, Bethany. 2015. "*Colorblind Dictionary*." Bethany Collins (artist's website). https://bethanyjoycollins.com/artwork/3867263-Colorblind-Dictionary.html.

Davis, Matthew, and Kathleen Blake Yancey. 2014. "Notes Toward the Role of Materiality in Composing, Reviewing, and Assessing Multimodal Texts." *Computers and Composition* 31: 13–28. doi:10.1016/j.compcom.2014.01.001.

Edbauer, Jenny. 2005. "Unframing Models of Public Distribution: From Rhetorical Situation to Rhetorical Ecologies." *Rhetoric Society Quarterly* 35 (4): 5–24. doi: 10.1080/02773940509391320.

Faigley, Lester. 1999. "Material Literacy and Visual Design." In *Rhetorical Bodies*, edited by Jack Selzer and Sharon Crowley, 171–202. Madison: University of Wisconsin Press.

Flower, Linda, and John R. Hayes. 1981. "A Cognitive Process Theory of Writing." *College Composition and Communication* 32 (4): 365–87. doi:10.2307/356600.

George, Diana. 2002. "From Analysis to Design: Visual Communication in the Teaching of Writing." *College Composition and Communication* 54 (1): 11–38. doi:10.2307/1512100.

Ghent, Sharlee. 2017. "Fashioning Identity and the Art of Bricolage: Studio-Based Research Methods and Reflection-in-Action in Fashion Design." *Scope Contemporary Research Topics: Art & Design* 15: 50–59. https://www.thescopes.org/journal/art-and-design/art-and-design-17.

Gioia, Dana. 1987. "Notes on the New Formalism." *The Hudson Review* 40 (3): 395–408. doi:10.2307/3851450.

Gomringer, Eugen. 1967. "Silencio." In *An Anthology of Concrete Poetry*, edited by Emmett Williams. New York: Something Else Press.

Haas, Angela M. 2007. "Wampum as Hypertext: An American Indian Intellectual Tradition of Multimedia Theory and Practice." *Studies in American Indian Literatures* 19 (4): 77–100. doi:10.1353/ail.2008.0005.

Hayles, N. Katherine. 2002. *Writing Machines*. Cambridge, MA: MIT Press.

Hughes, Gordon. 2006. "Power's Script: Or, Jenny Holzer's Art after 'Art after Philosophy.'" *Oxford Art Journal* 29 (3): 419–40. doi:10.1093/oxartj/kcl017.

Kaplan, Janet A., Bracken Hendricks, Geoffrey Hendricks, Hannah Higgins, and Alison Knowles. 2000. "Flux Generations." *Art Journal* 59 (2): 6–17. doi:10.2307/778097.

Kostelanetz, Richard. 1998. "Infinities." *Red Cedar Review*, 34 (1): 33. https://n2t.net/ark:/85335/m5qv3c54x.

Lin, Maya. 2000. *Boundaries*. New York: Simon & Schuster.

Long, Christopher P. 2013. "Performative Publication." *Digital Humanities, The Long Road*, July 19, 2013. http://cplong.org/2013/07/performative-publication/.

Madej, Krystina. 2016. "Print Narrative, Interactivity, and Collaboration." In *Interactivity, Collaboration, and Authoring in Social Media*, 15–29. Cham, Switzerland: Springer.

Maroja, Camila. 2015. "Framing Latin American Art: Artists, Critics, Institutions and the Configuration of a Regional Identity." PhD diss., Duke University.

Monroe, Barbara Jean. 2004. *Crossing the Digital Divide: Race, Writing, and Technology in the Classroom*. New York: Teachers College Press.

Quispe-Agnoli, Rocío. 2010. "Spanish Scripts Colonize the Image: Inca Visual Rhetorics." In *Rhetorics of the Americas: 3114 BCE–2012 CE*, edited by Damián Baca and Victor Villanueva, 41–67. New York: Palgrave Macmillan.

Rath, Richard Cullen. 2014. "Hearing Wampum: The Senses, Mediation, and the Limits of Analogy." In *Colonial Mediascapes: Sensory Worlds of the Early Americas*, edited by Matt Cohen and Jeffrey Glover, 159–75. Lincoln: University of Nebraska Press.

Ratto, Matt. 2011. "Critical Making: Conceptual and Material Studies in Technology and Social Life." *The Information Society* 27 (4): 252–60. doi:10.1080/01972243.2011.583819.

Rexer, Lyle. 2008. "Roundtable: New York Graphic Workshop." *Art on Paper* (September/October): 50–59.

Saper, Craig J. 2001. *Networked Art*. Minneapolis: University of Minnesota Press.

Sayre, Henry M. 1988. "The Avant-Garde and Experimental Writing." In *Columbia Literary History of the United States*, edited by Emory Elliott, Martha Banta, and Houston A. Baker, 1178–99. New York: Columbia University Press.

Sayre, Henry M. 1989. *The Object of Performance: The American Avant-Garde Since 1970*. Chicago: University of Chicago Press.

Sheridan, David M., Jim Ridolfo, and Anthony J. Michel. 2012. *The Available Means of Persuasion: Mapping a Theory and Pedagogy of Multimodal Public Rhetoric*. Anderson, SC: Parlor Press.

Shipka, Jody. 2005. "A Multimodal Task-Based Framework for Composing." *College Composition and Communication* 57 (2): 277–306.

Sirc, Geoffrey. 2004. "Box-Logic." In *Writing New Media: Theory and Applications for Expanding the Teaching of Composition*, by Anne Frances Wysocki, Johndan Johnson-Eilola, Cynthia L. Selfe, and Geoffrey Sirc, 111–46. Logan: Utah State University Press.

Sligh, Clarissa T. 1988. *What's Happening with Momma?* Rosendale, NY: Women's Studio Workshop. http://wsworkshop.org/artist_books/pdf/f2.pdf.

Sligh, Clarissa. 2016. *Transforming Hate: An Artist's Book.* Asheville, NC: Clarissa Sligh.

Sligh, Clarissa. 2021. "Making Artist's Books." Under *What's Happening with Momma?* Vamp & Tramp, Booksellers, LLC. http://www.vampandtramp.com/finepress/s/clarissa-sligh.html.

Smith, Cherise. 2007. "Re-Member the Audience: Adrian Piper's Mythic Being Advertisements." *Art Journal* 66 (1): 46–58. doi:10.1080/00043249.2007.10791239.

Walters, Shannon. 2014. *Rhetorical Touch: Disability, Identification, Haptics.* Columbia: University of South Carolina.

Williams, William Carlos. 1938. "The Red Wheelbarrow." In *The Collected Poems of William Carlos Williams, Volume I, 1909–1939*, edited by Christopher MacGowan. New York: New Directions Publishing Corporation.

Wysocki, Anne Frances. 2004. "Opening New Media to Writing: Openings and Justifications." In *Writing New Media: Theory and Applications for Expanding the Teaching of Composition*, by Anne Frances Wysocki, Johndan Johnson-Eilola, Cynthia L. Selfe, and Geoffrey Sirc, 1–41. Logan: Utah State University Press.

SECTION 2

Text Mining as Critical Making

3

THE WOMAN WHO TRICKED THE MACHINE
Challenging the Neutrality of Defaults and Building Coalitions for Marginalized Scholars

Cana Uluak Itchuaqiyaq
Virginia Tech

There once was a woman who lived at the margins and who felt lost and alone. She decided to make a map to find others who also needed support, that way she could be part of a group of friends and would no longer be an outsider. However, she did not know how to make a map on her own; she needed help. She found a machine and asked if it could make a map that showed where she and her friends were.

"All you need to do is feed me the right kind of information and follow my instructions, and I will make you a good map that shows everyone and their friends," the machine said.

Elated, the woman gathered the right kind of information and fed it to the machine, careful to follow the machine's instructions diligently. When the machine was done eating, she waited for the promised map. She was excited to finally have a map that showed her place among others, and to know where she could find support. The machine, satisfied after digesting all the information, finally gave her the map. The woman studied it, looking closely to find the place where she was on the map. She could not find herself and she could not find her friends. Instead, all she saw were the names of people at the center, large and bold on the map.

Annii, she thought, *surely this isn't right*. This machine had promised to make a map that would show everyone but had made a map that made her feel more lost and alone than ever.

She decided to try again, double-checking that she fed the machine all the information carefully and according to its instructions. The map stayed the same. The machine tried to convince her that she was taking this map too personally. The woman showed the map to a passerby, and they found themselves and their friends on the map right away.

https://doi.org/10.7330/9781646422586.c003

"Maybe you're just too far out at the margins to be on this map," they said.

"But I *am* here, just the same as you," she replied, exasperated.

Finally, the woman realized that the machine was built to only highlight those people located at the center, and that she must trick it into showing herself and her friends located at the margins. She realized that she must gather a list of her friends and somehow force the machine to make their names the prominent feature of the map. She hid the list inside the packet of information she had been feeding the machine, and carefully fed the machine once again. The machine gobbled this information up and gave her a new map. She studied it, and her eyes grew wide.

"There I am!" she exclaimed, easily finding herself on the map.

"There are my friends!" she said with relief.

Aarigaa, she thought, and no longer felt like she was lost, alone, or an outsider.

"See, I told you I make good maps," the machine gloated, and the woman smiled.

BUT I *AM* HERE, JUST THE SAME AS YOU: TECHNICAL COMMUNICATION AND PARAMETERED REALITIES

As the above story illustrates, programs are not neutral. Default settings pretend at neutrality for the "average user"—but that assumption is based on cultural expectations. John Cheney-Lippold (2017) describes how HP faced backlash for their computer's inability to easily recognize black skin tones with their facial recognition software. This issue, stemming from the computer's built-in settings, demonstrates a cultural phenomenon: the presupposition of whiteness attached to users of technology and, by extension, makers of technology via "seemingly neutral technological and infrastructural projects with (often white supremacist) racial logics at their most ground level" (18). If we view technology (and its accompanying technical communication) as a method of telling stories about the users, relationships, purposes, and priorities that exist within our society (refer to Legg and Sullivan [2018] for discussion of technical communication's relationship with story), what story does the HP computer tell? What story does the map-making machine tell? What stories need to be told? This chapter examines how critical making can act as a bridge between critical inquiry and technological innovations for social justice activism in technical communication. Specifically, this chapter demonstrates how I, an outsider in academia who felt lost and alone, became a "maktivist" (Mann 2018) and figured out how to trick a mapping program into showing me where the coalitions supporting multiply marginalized and underrepresented scholars—scholars like me—were in my field. While this study focuses primarily on technical communication scholarship, the coalitional map-making process that is described can (and should) be replicated in other fields, such as rhetoric and composition, in order to determine what coalitions of support exist for marginalized scholars.

Technical communicators are called to act in collaboration with makers in designing communication that transmits and shapes reality (Miller 1979). Through a recognition of the rhetoric implicit within the default settings of the programs technical communicators interact with in their work, we can better exert our authorial agency (Colton and Holmes 2018; Slack, Miller, and Doak 1993)—and even act as makers—via customization of these programs in order to promote more equitable outcomes. Users are often tricked into believing programs are interactive because of the various settings choices offered—which they are—but only within the program's own defined parameters. In other words, we are persuaded by the rhetoric of the interface via a procedural enthymeme where "we convince *ourselves* that we are actively making decisions about how to participate in a given system when, in reality, we accept options made apparently available to us from a set of constrained possibilities" (Brock and Shepherd 2016, 21). This type of persuasion, a tactic of assimilation into the reality offered by the program, makes us at once dependent on the program (to perform a task) and at the mercy of the program. Aaron Beveridge and Nicholas Van Horn state in their chapter within this collection, "Data-driven research often requires that researchers employ an inventive *maker* approach," which highlights the fact that technologies are constructed, parametered realities that can be directed, or manipulated, for executing unintended tasks. Ian Bogost (2007) states, "Interactivity guarantees neither meaningful expression nor meaningful persuasion, . . . choices do not necessarily entail all possible choices in a given situation; rather, choices are selectively included and excluded in a procedural representation to produce a desired expressive end" (44). However, an aspect of critical making is the recognition of what is and what isn't possible within the defined parameters of a program. This critical making happens when one is able to recognize a program's parameters as well as imagine potential workarounds, or hacks (in the DIY/reverse engineering sense as Mann [2018] describes), to redefine the possibilities within that system. This chapter argues that small changes and simple hacks can shift the paradigm inscribed into the program by its manufacturer and change the meaning of the results produced by the program. This chapter describes the customization process the author followed in order to change how an existing mapping program performed, yielding a product that highlights coalitions of support for multiply marginalized and underrepresented (MMU) scholars within the field of technical and professional communication (TPC) rather than a map that simply highlights scholars with power.

Critical race theorist Richard Delgado's multiphase citation analysis project in civil rights literature discovered that an inner circle of twenty-six white men were cited most frequently by authors. Once he called this practice out (1984), he found that the subsequent changes to citation practices were problematic (1992). Instead of meaningfully incorporating MMU scholars into work, Delgado found that many were cited in what he identified as marginalizing ways (1350). Delgado's study may raise a few questions for the reader, such as: What are the citation practices like in our field? Do we have our own version of an inner circle? Technical communication is linked deeply to issues of social justice, especially with regards to how our work affects MMU populations. Much like Delgado, technical communication scholars Rebecca Walton, Kristen R. Moore, and Natasha N. Jones (2019) recognized a theme in the justifications used for not incorporating more minority and marginalized voices in TPC scholarship: "There are just not enough Black, Indigenous, minority, transgender, or scholars with disabilities, etc. in our field" (169). In response to this pushback, the authors provided a list of eighty-six MMU scholars alongside a strong call to amplify these voices in our classrooms and in our scholarship.

There have been multiple studies explicitly concerning citation in TPC scholarship (Smith 2000a, 2000b, 2003; Smith and Thompson 2002). However, these studies focus broadly on citation patterns related to topics and neglect to discuss issues of diversity in citation (in terms of the identity of the cited authors). Though little research exists regarding the impact of citation practices in TPC scholarship upon issues related to diversity and inclusion, there does exist established, if not formally written, parameters of what is expected in citational work in TPC. For example, *Communication Design Quarterly* reviewer guidelines ask, "Are there any sources the reader should be citing (but is not) in examining the ideas covered in the manuscript?" What's important to note is that this question places the onus on reviewers to decide which *relevant* materials *should* be cited. Emily January Petersen and Rebecca Walton (2018) state, "In social justice scholarship, gaps and silences are important sites of study to investigate who is left out of discourse and decision making, how silence functions as a power strategy, and which groups might need recognition for work already accomplished" (423). One method of revealing the gaps and silences in our field is to thoroughly investigate inclusion in citation practices. Delgado (1992) asserts that his research is not aimed to condemn his colleagues, rather he hopes that his research "will provide the occasion for mainstream writers who recognize themselves in its pages to reevaluate their scholarly practices with respect to insurgent scholars. At the same time, I hope it will assist insurgent scholars in articulating their criticisms of unregenerate attempts to keep them on the margin" (1351). As our field looks for more ways to enact social justice in our scholarly practices—ways that move beyond discussions of diversity towards actually making our field more inclusive (Williams and Pimentel 2012)—meaningfully incorporating scholarship from current and future members of Walton, Moore, and Jones's (2019) MMU "list" through our citations is an action attainable by all.[1]

While a systematic investigation of inclusion in our field is necessary—that is, an investigation about how it *feels* to be a diverse body in our field—there does not exist an easy method to meaningfully perform this investigation. One potential method might ask MMU scholars how often their presence in our field has felt erased, objectified, or even risky. However, just answering those types of questions in an honest manner involves emotional labor and real risk of further marginalization. For example, I have had well-meaning people ask about my experience as an Inuit scholar in a predominately white field, but then turn on me when my stories are negative. (Now, all I say is, "It's so *interesting*"—but I mean "interesting" in the same way that my mom taught me to discuss food I didn't like.) One method that can assist with such an investigation is to map the networks of support that MMU scholars create with other MMU scholars and with allies/accomplices.

Building coalitions—essentially building a network, however large or small, of support—is integral to social justice work, especially activism that calls for shifts in paradigms upheld by institutions invested in tradition. If one is to make social change, finding and attracting like-minded individuals and organizations also dedicated to making that change is paramount (Chávez 2011). In the story about the map machine, the woman needed a map to find others who were also outsiders in order to create a coalition of support. Karma R. Chávez (2013) asserts that "women of color feminists have long advocated for the necessity of coalitional politics to address oppression and power at its roots and to utilize difference as a resource rather than a hurdle to be overcome" (7). As an Inuit scholar, I come from a highly underrepresented population in academia and tend to seek other BIPOC (Black, Indigenous, and people of color) scholars to form a community of support. It happens almost instinctively when we enter a classroom or academic space: We scan the room, and *if* there are any other BIPOC individuals, we tend to gravitate towards one another. For example, I attended Dr. Natasha Jones and Dr. Miriam Williams's Narrative Inquiry workshop at the 2019 Association of Teachers of Technical Writing conference and performed this auto-scan when I entered the room. When the twenty or so workshop participants were instructed to break into small groups, the majority of BIPOC participants (including myself) formed a large group silently and automatically. These groups are born out of a still-present need for security[2] and are formed via glances and low-key nods and smiles. Often occurring behind the scenes, coalition-building involves an understanding of the rhetorical situation and using it in building bridges with allies (Chávez 2011). Not only are BIPOC and other marginalized scholars seeking one another out for building coalitions of support, but they are also looking to allies for support. Citational maps that highlight MMU scholars, and those that cite MMU scholarship, are a useful tool in MMU coalition-building because, like the woman in the story, sometimes you just need to know where your friends are.

Using mapping techniques to show the relational aspect of MMU citation—who is citing MMU scholars and how frequently—is one method to demonstrate how inclusion efforts are enacted within common scholarly practices. Mapping the physical locations of coalitions of support for MMU scholars and using critical cartography methods is not a new idea to technical communication or rhetoric and composition (Faris and Selber 2013; Ridolfo 2020; Sullivan and Graban 2010; Unger and Sánchez 2015). However, most of the existing mapping research focuses on "marginalized" topics, like queer rhetorics, rather than centering on marginalized scholars themselves. Regardless, this vein of mapping research indicates that locating potential support is key to coalition-building. Like the story illustrates, while existing mapping shows author publication rates and scholarly coalitions bound together through citations and keywords, they are designed to only highlight information that fits neatly within the program's parameters. In other words, because the programs are designed to report back only information contained in the data used to make the map, such as bibliometric data, "inner circle" citation issues are not easily visible. This limitation occurs because demographic data is not meant to be inputted into the program. An interface's "educational" effect is the enactment of colonial structures that train users to carry out certain prescribed procedures in order to carry out desired tasks (Selfe and Selfe 1994). Cynthia L. Selfe and Richard J. Selfe discuss computer interfaces as "maps that enact—among other things—the gestures and deeds of colonialism, continuously and with a great deal of success" (482).

In this chapter, I discuss the use of the citation mapping program VOSviewer (2020) to make coalitional maps. Though entering data into the program according to the available data settings within its interface, it is implied that the user should not consider demographics (figure 3.1) when making their maps. One is taught to trust the interface as presenting solutions to their needs, and not critically examine how the "user-friendliness" of an interface masks the limited choices presented within that space. Through this sort of training, the interface becomes a site of forced assimilation into a particular reality. Because a program's code is rhetorical and culturally based, issues like bigotry can affect how the program operates and perpetuate oppressive outcomes like erasure (Bogost 2007; Brock and Shepherd 2016; Noble 2018). Safiya Noble (2018) states, "While we often think of terms such as 'big data' and 'algorithms' as being benign, neutral, or objective, they are anything but. The people who make these decisions hold all types of values, many of which openly promote racism, sexism, and false notions of meritocracy" (1–2). Critical making provides a way to resist such structures and force programs to perform outside of their defined parameters. Matt Ratto describes critical making as "embedded in practices of socio-technical reflection and critique, [providing] the possibility for truly innovative thinking and making, the result of which is not just more of the same but includes novel and more comprehensive understandings as to the relationships between social life and technical work" (2011, para. 6). My hack uses the mapping program's interface to force the "neutral" program to reveal the demographic context that is left out of traditional citation mapping. This hack is simple and requires no coding skills. To do this type of critical making, one must simply trick the program as the woman in the story does. The first step in this simple critical making process is to locate the barriers written into a program. These barriers, such as the inability to input demographic data, are often not obvious and require the user to critically consider the interface to determine what's missing or what's unavailable. The next step is to locate gates, or avenues of input. These gates may be obvious, such as the thesaurus file input in figure 3.1, or can be located within the program's file settings or preferences. It is important to understand the mechanisms of gates (such as how information must be organized in the thesaurus file) in order to determine how to repurpose them. By locating barriers and

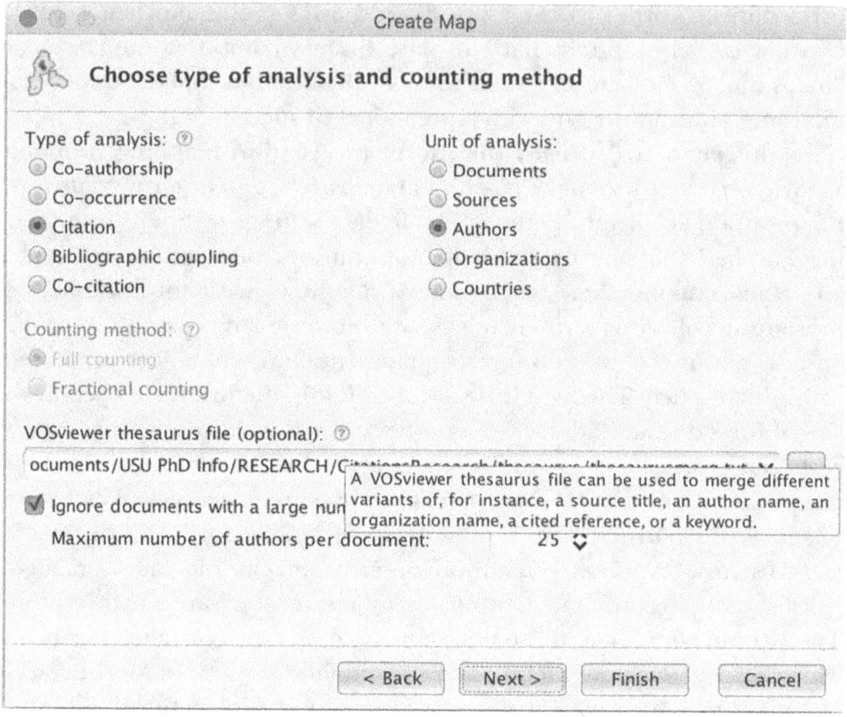

82   CANA ULUAK ITCHUAQIYAQ

Figure 3.1. Screenshot of VOSviewer interface showing customization options. The dialogue box displaying the description of optional thesaurus files is visible.

gates, one locates both the problem and the solution to a critical making technique. It is much easier to identify how to get past a fence when you know what kind of fence it is and where its gates are located.

HOW THE WOMAN TRICKED THE MACHINE

As you might have guessed, the woman in the story is based upon my own experiences attempting to use citation data to find coalitions of support during my graduate program. Being a BIPOC scholar often comes with multiple vectors of precarity and frustration (Steele 2010). Like the woman in the story, I desperately needed a way to find other BIPOC scholars so that I didn't feel isolated. I also needed to find ally-scholars who demonstrated *through their scholarly practice* (instead of merely voicing) that they valued the ideas and arguments generated by scholars like myself. If I needed a guide to potential coalitions of support, such as a citation map, I knew that others would need it, too.

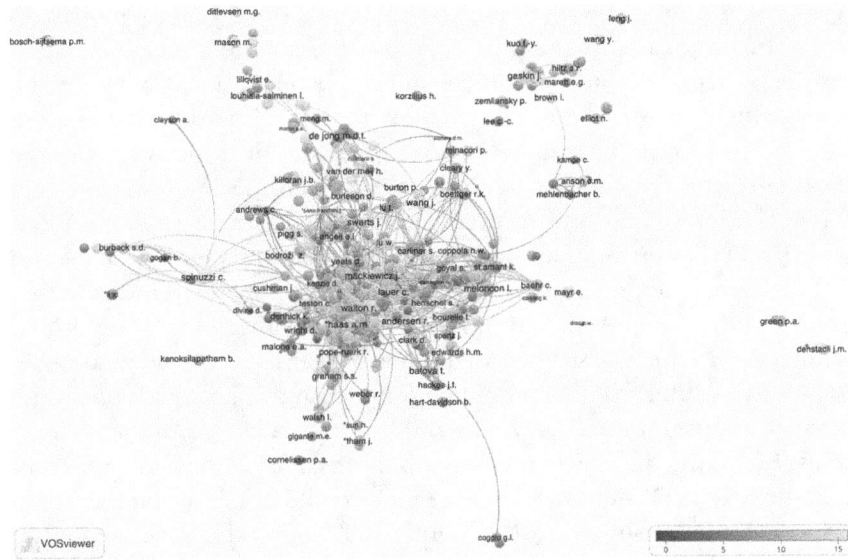

Figure 3.2. Sample citation map from VOSviewer featuring TPC authors linked through citation in five TPC journals since 2000. This map contains only authors who fit the user-defined thresholds: minimum two publications and minimum zero citations. The different size nodes indicate higher frequencies of citation. The node colors indicate the average number of citations attributed to each author. The interactive feature of the live version of this map would allow you to zoom in and/or click on different authors to see connections more clearly (as shown in figure 3.3).

Citation mapping is the process of visually representing relationships between authors, publications, and citation practices. Citation maps are made using downloaded bibliometric data collected from abstract and citation databases like Scopus. This data is then uploaded into mapping programs, like VOSviewer, in order to make interactive maps. VOSviewer is a mapping program available for free online that visually identify networks within bibliometric data. These networks, or clusters, act as "solutions" to problems like, "Which authors have a relationship through direct citation in TPC journal scholarship since 2000?" (figure 3.2[3]). VOSviewer maps are interactive, so users can select a specific node, like an author's name, to highlight its associated relationships.

Although generating maps in VOSviewer is relatively simple, it requires a lot of finessing and preparation in order to make these maps meaningful. The first step is to choose what the nodes represent. One method to help with this step is to ask, "What is the subject of the map?" For example, for the question above (Which authors have a relationship through direct citation in TPC journal scholarship since 2000?), *authors* would be the subject of the map and therefore should be the nodes. As the number of an author's publications increase, the larger their node becomes. The next step is to determine the relationship between the nodes, which are represented by lines that vary in width depending on the links strength. In my map examples, the purpose is to visualize coalitions through citation, so the lines between the nodes represent a direct citation relationship between authors. It is important to point out that in VOSviewer the map does not distinguish between "parent" and "child" publications. In other words, the lines on my map indicate that either/both Author A or Author B has cited the other in their published scholarship. One has a limited number of customizations within the VOSviewer's interface but can make some customizations such as using color to indicate information like publication dates or number of citations. In conducting my citation mapping research, it became apparent that the maps VOSviewer was built to make could not easily highlight the citational relationships of MMU scholars. Through the development of critical making strategies, the possible customizations in VOSviewer can be increased through tactics such as repurposing thesaurus files. Much like the woman in the story, I tricked the program into indicating which nodes represented MMU scholars. What I did was simple: I wrote a basic thesaurus .txt file that changed the appearance of MMU scholars' names by adding an asterisk (figures 3.2 and 3.3). This addition of an asterisk to their names visibly differentiates them from other authors. If one were to rely on just the standard options of this mapping program, the only way to differentiate authors would be via their individual publication statistics, which makes their names and nodes bigger, as well as changes to the meaning of their node's color to either indicate numbers of citations (as in figure 3.3) or the average year of publication.

My citation mapping study aims to map MMU scholar networks and therefore relies on representing authors accurately. There are two identifying markers that my study is concerned with: name and MMU status. However, authors sometimes publish using different versions of their name. For example, technical communication scholar Natasha Jones has published as Natasha Jones and Natasha N. Jones. Because bibliometric data condenses names using initials for first and middle names, Natasha Jones is categorized under "N. Jones" and "N. N. Jones." Fortunately, programs like VOSviewer are built to allow users to edit the names of nodes within the map. It is recommended that users "clean" their data through creating a thesaurus file that acts as a find-and-replace function within the mapping program. A thesaurus file is necessary to tell the mapping program to combine the bibliometric data for "N. Jones" with

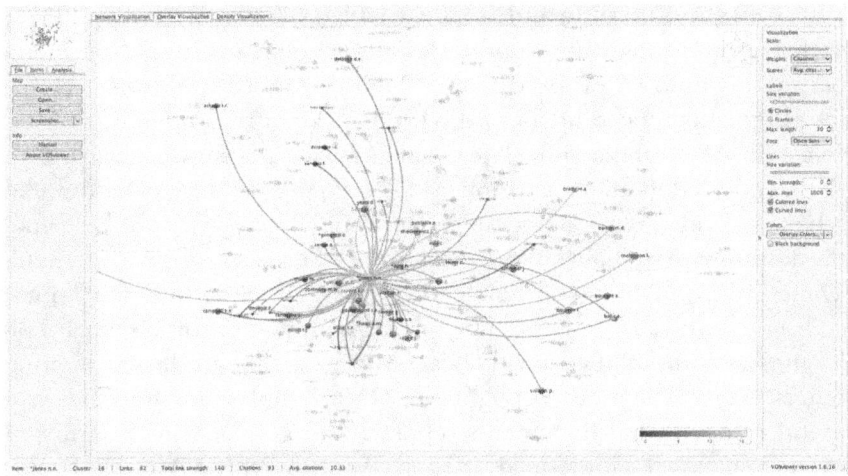

Figure 3.3. Screenshot of the VOSviewer interface displaying the citation coalition of technical communication scholar Natasha N. Jones. Note the asterisk (*) next to the names of MMU scholars. This image shows a zoomed-in version of the map from figure 3.2. The node colors indicate the average number of citations attributed to each author. Note that some nodes connected to Jones are not visible at this level of zoom.

the data for "N. N. Jones" and produce a single node that represents all that data together: "N. N. Jones." In other words, all the publications, citation relationships, co-authorship relationships, keywords, and abstracts that were filed under Natasha Jones and Natasha N. Jones, will now be filed only under "N. N. Jones." It is through hacking this "cleaning" feature that MMU authors' names can be differentiated from other authors. In order to make MMU author names distinct on the map I changed how the names themselves were written. By default, authors' names are displayed in the following format: "lastname a.b." I used a thesaurus file to subsume "jones n." citations into "jones n.n." citations *and* to change "jones n.n." to read as "*jones n.n." to indicate her status as an MMU scholar.[4] This may seem like a minor change—and, really, it is—but it is a change that completely shifts the meaning of the maps created in VOSviewer from one that highlights those who have high publication statistics to one that also highlights MMU scholars.

Like was discussed, the second identity marker used in my citation mapping study was the author's status as MMU scholar. Because marginalizing factors like non-white racial identity, disability identity, and LGBTQIA2+ identity can be deeply personal, invisible, and stigmatized in society leading to (un)intentional oppressive outcomes (Bonilla-Silva

2018; Cox 2019; Delgado and Stefancic 2017; Del Hierro, Levy, and Price 2016; Samuels 2003), it is important to let scholars self-identify as members of any number of these groups. Walton, Moore, and Jones (2019), in recognizing a need to compile a listing of minority and marginalized voices in TPC scholarship, "developed through a snowball approach that requested both permission and recommendations of additional scholars. [They] developed the collection of scholars here to signal the multiply marginalized and underrepresented groups in our field. These groups include colleagues who are racial and ethnic minorities; colleagues who are lesbian, gay, queer, transgender, or bisexual; colleagues with disabilities and colleagues who are neurodiverse; among others" (171). Their list of eighty-six MMU scholars in the field[5] is a useful resource that documents scholars who gave specific permission to be publicly acknowledged as such. Applying this MMU list to the list of authors in my map through adding an asterisk to their name in a thesaurus file is an act of critical making and is an effective way to highlight authors as multiply marginalized and underrepresented in an ethical manner. It is important to note that using a thesaurus file to add a visual element to MMU names retains the functionality of the standard VOSviewer features. This move demonstrates that critical making is a useful ally to technical communication's interest in enacting social justice activism because it at once acknowledges the moral responsibility of nonhuman agents (Johnson and Johnson 2018) and answers the call from David Gaertner's 2019 keynote speech about the maker movement and its relationship with marginalized identities at the Digital Humanities Summer Institute Congress asking, "How can we reorient our politics of citation so that Indigenous peoples and people of colour figure more prominently in our critical genealogies? How can we use the tools of [Digital Humanities] itself to amplify those voices and disrupt the predominantly white hand of techne?" (para. 16).

THE WOMAN WHO FOUND A MAP

Once there was a woman who lived alone at the margins. One day she made a map that showed where she could find the other people who were located at the margins. Encouraged by this map, she decided to set forth to find them and ask if they would like to travel with her on the long and risky journey to the center. She always wanted to go to the center but was a little scared to go alone because there were so many obstacles along the way.

Following the map she made, she met another person from the margins who had a lot of important things to say but just needed someone to listen.

"I'll listen to you," she said, and she sat down and listened and learned many things she hadn't thought of before. It turns out that not all people living at the margins are the same! She thanked her new friend for sharing their knowledge with her and asked if they wanted to join her in the journey to the center.

"I'd love to come," they said, and together they looked at the map and decided where to go next.

The two became three, then four, then five, and so on. The friends traveled together, slowly inching their way toward the center of the map. They helped one another move around or even tear down the obstacles they encountered along the way. They worked together to build bridges over large chasms that were once impassable. On their journey, they met a lot of kind people who were from the center and whose locations were shown on the map. These allies listened to their stories about living at the margins and how the woman and her friends wanted to get to the center.

"Thanks!" the kind people said after hearing their stories. "What can I do to help get you to the center?" they asked. The allies wanted to become accomplices!

The woman and her friends told the kind people that listening to and believing their stories about the margins was a great way to help, and that sharing these stories with others (with proper attribution, of course!) also helped. The friends asked their accomplices to hold open the gates they encountered and help tear down the obstacles that were blocking the way to the center so that others may follow the friends' path more easily. The accomplices happily agreed to do this work.

On an on the friends traveled, following their map and clearing the way on their long journey toward the center.

THE END

Critical making offers technical communication a rich avenue for problem solving. To be honest, it really wasn't until I started making citation maps and wondered about how to highlight MMU scholars on these maps that Chávez's (2011, 2013) and Walton, Moore, and Jones's (2019) concepts about coalition building were brought to light. Through this process, I was better able to understand how seemingly benign things like what is and isn't offered as a setting in a program can be complicit in perpetuating oppressive norms. These concepts were not new to me, but their meaning—their *true* meaning—was veiled by my own inexperience and lack of critically tinkering with their on-the-ground expressions. By identifying what was missing within a mapping program, I was able to clearly recognize the fences installed into the program itself and discover successful workarounds through existing gates.

Much like Delgado (1984, 1992) found, sometimes segregatory citational practice is intentional. However, sometimes it acts like a default setting where authors tend to cite the same authors because it is just how it has always been done (Hemmings 2011), and authors have been trained not to question it. Likewise, the defaults of the mapping program I used to make my maps perpetuate inner circle citation issues instead of challenging them. By adding the asterisk—a simple but rhetorically powerful change—I changed the meaning of the map from one that identifies power to one that identifies coalitions of support for MMU scholars in technical communication. This type of critical making reframes the map to highlight how diversity and inclusion efforts play out in our field's scholarship practices.

I wish I had some neat way to tie the above story up with an "and they lived happily ever after." While I'm not even sure how much this saccharinely idealistic story is grounded in reality, I do know that stories like these are useful in illustrating possibility. Academe is a brutal business that tends to infect even the strongest personalities with terrible feelings of insecurity (Sano-Franchini 2016), and this lack of security is especially true for MMU scholars. By critically using the tools we have at hand, even if it means repurposing them to perform different tasks, we can create paths together toward a more socially just future.

NOTES

1. I am going to make this clear, especially considering the recent arguments made about multiply marginalized and underrepresented (MMU) scholarship's merit (see social media posts using #communicationsowhite for more information): I *do* mean to suggest that one should engage with MMU scholarship *simply because it is MMU scholarship*. This says nothing about the merit of our scholarship; it says everything about the conditions of academe. These perspectives need *to be read, engaged with, and validated* if we are going to make any real progress toward being inclusive and to make reparations for generations of systemic abuse of MMU communities (Ahmed 2017). (See social media posts using #BlackintheIvory for examples of the implicit bias and racism facing Black scholars.)
2. Refer to Claude M. Steele's (2010) *Whistling Vivaldi: And Other Clues to How Stereotypes Affect Us* for more information about how and why BIPOC students tend to seek one another out.

3. This map was made from a sample of all articles published between the years 2000 and 2018 from five leading TPC journals available on Scopus: *IEEE Transactions on Professional Communication, Journal of Technical Writing and Communication, Journal of Business and Technical Communication, Technical Communication,* and *Technical Communication Quarterly.*
4. Though Natasha Jones did not include herself on the MMU list presented in her book *Technical Communication After the Social Justice Turn: Building Coalitions for Action* (Walton, Moore, and Jones 2019), she gave me permission to include her name on the MMU list and to discuss her identity as a Black woman in this chapter.
5. This list yields a large body of authors and scholarship, but it is not complete. The author is currently examining the results from her IRB-approved survey to add names to Walton, Moore, and Jones's (2019) list. Currently, as of early 2021, the list of MMU scholars is at 120 scholars.

REFERENCES

Ahmed, Sara. 2017. *Living a Feminist Life.* Durham, NC: Duke University Press.

Bogost, Ian. 2007. *Persuasive Games: The Expressive Power of Videogames.* Cambridge, MA: MIT Press.

Bonilla-Silva, Eduardo. 2018. *Racism Without Racists: Color-Blind Racism and the Persistence of Racial Inequality in America* (5th ed.). Lanham, MD: Rowman & Littlefield.

Brock, Kevin, and Dawn Shepherd. 2016. "Understanding How Algorithms Work Persuasively Through the Procedural Enthymeme." *Computers and Composition* 42: 17–27. doi:10.1016/j.compcom.2016.08.007.

Chávez, Karma R. 2011. "Counter-Public Enclaves and Understanding the Function of Rhetoric in Social Movement Coalition-Building." *Communication Quarterly* 59 (1): 4–30. doi:10.1080/01463373.2010.541333.

Chávez, Karma R. 2013. *Queer Migration Politics: Activist Rhetoric and Coalitional Possibilities.* Urbana: University of Illinois Press.

Cheney-Lippold, John. 2017. *We Are Data: Algorithms and the Making of Our Digital Selves.* New York: New York University Press.

Colton, Jared S., and Steve Holmes. 2018. "A Social Justice Theory of Active Equality for Technical Communication." *Journal of Technical Writing and Communication* 48 (1): 4–30. doi:10.1177/0047281616647803.

Cox, Matthew B. 2019. "Working Closets: Mapping Queer Professional Discourses and Why Professional Communication Studies Need Queer Rhetorics." *Journal of Business and Technical Communication* 33 (1): 1–25. doi:10.1177/1050651918798691.

Delgado, Richard. 1984. "The Imperial Scholar: Reflections on a Review of Civil Rights Literature." *University of Pennsylvania Law Review* 132 (3): 561–78. doi:10.2307/3311882.

Delgado, Richard. 1992. "The Imperial Scholar Revisited: How to Marginalize Outsider Writing, Ten Years Later." *University of Pennsylvania Law Review* 140 (4): 1349–72. doi:10.2307/3312406.

Delgado, Richard, and Jean Stefancic. 2017. *Critical Race Theory: An Introduction* (3rd ed.). New York: New York University Press.

Del Hierro, Victor, Daisy Levy, and Margaret Price. 2016. "We Are Here: Negotiating Difference and Alliance in Spaces of Cultural Rhetorics." *Enculturation: A Journal of Rhetoric, Writing, and Culture*, no. 21. http://enculturation.net/we-are-here.

Faris, Michael J., and Stuart A. Selber. 2013. "iPads in the Technical Communication Classroom: An Empirical Study of Technology Integration and Use." *Journal of Business and Technical Communication* 27 (4): 359–408. doi:10.1177/1050651913490942.

Gaertner, David. 2019. "Decolonial DH?: The Maker Movement across Indigenous Studies and the Digital Humanities." Novel Alliances (website). June 7, 2019. https://novelalliances

.com/2019/06/05/decolonial-dh-the-maker-movement-across-indigenous-studies-and-the-digital-humanities/.

Hemmings, Clare. 2011. *Why Stories Matter: The Political Grammar of Feminist Theory*. Durham, NC: Duke University Press.

Johnson, Meredith A., and Nathan R. Johnson. 2018. "Can Objects Be Moral Agents? Posthuman Praxis in Public Transportation." In *Posthuman Praxis in Technical Communication*, edited by Kristen R. Moore and Daniel P. Richards, 121–40. New York: Routledge.

Legg, Emily, and Patricia Sullivan. 2018. "Storytelling as a Balancing Practice in the Study of Posthuman Praxis." In *Posthuman Praxis in Technical Communication*, edited by Kristen R. Moore and Daniel P. Richards, 23–45. New York: Routledge.

Mann, Steve. 2018. "Maktivism: Authentic Making for Technology in the Service of Humanity." In *DIY Citizenship: Critical Making and Social Media*, edited by Matt Ratto and Megan Boler, 29–51. Cambridge, MA: MIT Press.

Miller, Carolyn R. 1979. "A Humanistic Rationale for Technical Writing." *College English* 40 (6): 610–17. doi:10.2307/375964.

Noble, Safiya U. 2018. *Algorithms of Oppression: How Search Engines Reinforce Racism*. New York: New York University Press.

Petersen, Emily January, and Rebecca Walton. 2018. "Bridging Analysis and Action: How Feminist Scholarship Can Inform the Social Justice Turn." *Journal of Business and Technical Communication* 32 (4): 1–31. doi:10.1177/1050651918780192.

Ratto, Matt. 2011. "Critical Making." In *Open Design Now*, edited by Bas van Abel, Roel Klaassen, Lucas Evers, and Peter Troxler. Amsterdam, The Netherlands: BIS Publishers. http://opendesignnow.org/index.html%3Fp=434.html.

Ridolfo, Jim. 2020. "Rhet Map: Mapping Rhetoric and Composition." Rhet Map (website). Accessed June 21, 2020. http://rhetmap.org/.

Samuels, Ellen. 2003. "My Body, My Closet: Invisible Disability and the Limits of Coming-Out Discourse." *GLQ: A Journal of Lesbian and Gay Studies* 9 (1): 233–55.

Sano-Franchini, Jennifer. 2016. "'It's Like Writing Yourself into a Codependent Relationship with Someone Who Doesn't Even Want You!' Emotional Labor, Intimacy, and the Academic Job Market in Rhetoric and Composition." *College Composition and Communication* 68 (1): 98–125.

Selfe, Cynthia L., and Richard J. Selfe Jr. 1994. "The Politics of the Interface: Power and Its Exercise in Electronic Contact Zones." *College Composition and Communication* 45 (4): 480–504.

Slack, Jennifer D., David James Miller, and Jeffrey Doak. 1993. "The Technical Communicator as Author: Meaning, Power, Authority." *Journal of Business and Technical Communication* 7 (1): 12–36. doi:10.1177/1050651993007001002.

Smith, Elizabeth Overman. 2000a. "Points of Reference in Technical Communication Scholarship." *Technical Communication Quarterly* 9 (4): 427–53. doi:10.1080/10572250009364708.

Smith, Elizabeth Overman. 2000b. "Strength in the Technical Communication Journals and Diversity in the Serials Cited." *Journal of Business and Technical Communication* 14 (2): 131–84. doi:10.1177/105065190001400201.

Smith, Elizabeth Overman. 2003. "Points of Reference Contributing to the Professionalization of Technical Communication." In *Power and Legitimacy in Technical Communication, Volume II: Strategies for Professional Status*, edited by Teresa Kynell-Hunt and Gerald J. Savage, 51–72. Amityville, NY: Baywood Publishing Company, Inc.

Smith, Elizabeth Overman, and Isabelle Thompson. 2002. "Feminist Theory in Technical Communication: Making Knowledge Claims Visible." *Journal of Business and Technical Communication* 16 (4): 441–77. doi:10.1177/105065102236526.

Steele, Claude M. 2010. *Whistling Vivaldi: And Other Clues to How Stereotypes Affect Us*. New York: W. W. Norton & Company.

Sullivan, Patricia A., and Tarez S. Graban. 2010. "Digital and Dustfree: A Conversation on the Possibilities of Digital-Only Searching for Third-Wave Historical Recovery." *Peitho* 13 (2): 2–11. https://cfshrc.org/peitho-the-newsletter/

Unger, Don, and Fernando Sánchez. 2015. "Locating Queer Rhetorics: Mapping as an Inventional Method." *Computers and Composition* 38: 96–112. doi:10.1016/j.compcom.2015.09.011.

VOSviewer. 2020. VOSviewer version 1.6.15, April 1, 2020. Leiden University, Leiden, The Netherlands.

Walton, Rebecca, Kristen R. Moore, and Natasha N. Jones. 2019. *Technical Communication after the Social Justice Turn: Building Coalitions for Action*. New York: Routledge.

Williams, Miriam F., and Oscar Pimentel. 2012. "Introduction: Race, Ethnicity, and Technical Communication." *Journal of Business and Technical Communication* 26 (3): 271–76. doi:10.1177/1050651912439535.

4

CRITICAL TEXT MINING
Ethical Paradigms for Determining Emoji Frequency in #blacklivesmatter

Kellie M. Gray and Steve Holmes
Texas Tech University

#BlackLivesMatter ✊🏽✊🏾✊🏿

INTRODUCTION

Stephen Ramsay (2011) published *Toward an Algorithmic Criticism* in 2011. Around that period in time, "digital humanities" hires who could do any sort of computational data analysis were only beginning to emerge in English departments. He patiently defended the use of empirical quantitative methods alongside and not against the subjective hermeneutics of literary scholars. In the ensuing decade, the situation has changed. By now, many rhetoric and composition scholars have engaged the digital humanities (Ridolfo and Hart-Davidson 2014), text mining, corpus linguistics methods, natural language processing, and countless other related methods and tools. One common thread lies in the effort to detect patterns in largescale corpora of texts that human interpreters alone would be unlikely to discover ("distant reading" as Hayles [2012] called it). As Ramsay (2011) notes, one contribution lies in fact-checking previous hermeneutic assumptions or *topoi* (11). As a case in point, Zak Lancaster (2016) has notably employed electronic concordancing software and methods drawn from functional linguistics to determine that the templates used in Gerald Graff et al.'s popular (and debated) *They Say / I Say* textbook "for the most part" are not reflected in his large corpus of professional research writing samples (455).

However, as Ramsay (2011) importantly observed, computers can do more than just fact check non-computationally derived forms of interpretation (11). Indeed, they can generate new and unknown interpretative possibilities as well (10). To demonstrate the value of Ramsay's (2011) second claim, this chapter explores how the concept of *critical text mining* can help unpack some of the potential for using computational analysis to participate in critical making conversations. Text mining refers to "the process of analyzing a semantically rich document or set of documents to understand the content and meaning of the information they contain" (Shi and Kong 2009, 4167). Text mining is a form of analysis which, as the name implies, encompasses all manner of efforts to program computational devices to process and analyze language from the basic (frequency) to the complex (P values).

Since text mining is a method grounded in computer science, linguistics, data science, quantitative methods, and (often) empirical research, the former is too often subject to generalized algorithmic critique by humanities scholars or, conversely, critiqued or defended as an objective apolitical method. By contrast, critical text mining is the explicit articulation of an ethical framework (positive or critique-driven) for enacting some sort of activist text mining research method or practice. Echoing tenets of this edited collection and critical making as a whole, critical text mining can help situate theory and practice in reciprocally emerging relations that are guided by a positive but contingent and situational ethics.

Second, we offer what we think is a flexible ethical framework for making normative ethical claims in Jacques Rancière's (1999) politics of dissensus and equality. Briefly, Rancière's thinking is compatible with critical making since it is an individualistic enactment that enables concrete and varied forms of resistance to support political equality while never being fully programmable. Third, we briefly gesture toward how we use this Rancièrian framework to webscrape and analyze emoji frequency on the #blacklivesmatter hashtag on Twitter through the information science programming language R. The #blacklivesmatter hashtag initially appeared on Twitter in July 2013 in response to George Zimmerman's acquittal after he shot and killed Trayvon Martin. The Pew Research Center (Anderson et al. 2018) drew on the Crimson Hexagon software suite to study five years and thirty million uses of this hashtag through May 1, 2018. Any consideration of emoji was noticeably absent, and the role of emoji for this hashtag has not been investigated by rhetoric and composition scholars.

As we have argued elsewhere (Gray and Holmes 2020), emoji have a secondary status in text mining and many digital rhetoric conversations as a whole. Emoji data scientist Jessica Peterka-Bonetta (2017), whose emoji decoder code we modify as part of this chapter, declares, "Emojis can help easily identify positive content, but they're not so good at identifying negative or serious, business related content . . . I'd rather suggest performing traditional sentiment analysis and enrich it with emoji data." Yet, she does not consider the opposite case: Nearly 80 percent of all internet communication contains an emoji (Shaul 2015). Emoji are much more than a secondary or supplementary feature to text, and they are not only positive as the black circle emoji used by the #alllivesmatter hashtag demonstrates or our previous analysis of the toy-ification of the pistol emoji has demonstrated (Gray and Holmes 2020).

In response, we have reprogrammed an open-source version of Peterka-Bonetta's (2017) emoji decoder, which is available on this collection's companion website at https://upcolorado.com/component/k2/item/6219-reprogrammable-rhetoric-supplemental-design-materials-and-programming-scripts and included our instructions for how to use a basic Twitter API in R to scrape and perform frequency analysis on emoji on the #blacklives hashtag. Beyond analysis, our chapter and supporting materials on this collection's companion website also enact an ethic of creating an ongoing public Twitter database that we have been webscraping since July 2020 on the #blacklivesmatter official Twitter hashtag.

DEFINING CRITICAL TEXT MINING

Many readers in our field may think of critical making scholarship more in terms of physical computing and wearables. Glossed over in much of this focus on physical computing in critical making is often the use of the software itself to fashion not physical outputs but still primarily digital outputs like datasets or the results of analyzed datasets. Yet, there is certainly nothing in the original definition of the term that would be inhospitable to purely digital forms of critical making. "At a meta-level," write Matt Ratto and Stephen Hockema (2009), "Critical Making aims to focus attention of the ways in which materially-engaged activities provide cognitive resources for thinking through complex individual, social, and societal issues" (52). Numerous scholars have since established coding, hacking, software development, and so forth as important forms of critical making. Kate Milberry (2014) argues that the self-reflexive work of tech activists–including "hackers, coders, and self-described geeks"—"invokes the spirit of critical making as both an activity and a site for deepening a transformative sociotechnical praxis" (53).

To take a step back, it is fair to ask if text mining is actually a form of making let alone critical making. As an excellent example, Ryan Omizo's chapter on the Faciloscope 2.0 in this collection demonstrates that developing text mining programs and enabling others to reprogram them for new ends is clearly aligned with making cultures' reprogrammable *ethos*. Similarly, Aaron Beveridge and Nicholas Van Horn's chapter also explores how the inexpensive Raspberry Pi computer can be used to engage online webscraping and archiving as an activist practice. Furthermore, Ratto and Hockema (2009) argue for a combination of theory and practice as it combines "critical thinking,' often considered as abstract, explicit, linguistically based, internal and cognitively individualistic; and 'making,' typically understood as material, tacit, embodied, external and community-oriented" (52). Coding is material process seen in arguments like Wendy Hui Kyong Chun's (2011) exploration of how female ENIAC (Electronic Numerical Integrator and Computer) programmers were the original programmers.

Similarly, the specific output of a given making activity should not matter in the sense of being digital or nondigital. Creating new processes to think through can be as important as new products for end users. A dataset produced or analyzed for an audience to view, analyze, or repurpose is always already a material thing of this world that can extend networks of agency and grow new ones. As a timely illustration of this, consider a group of Harvard University economists' kairotic response to the 2020 COVID pandemic. They set up the website *Track the Recovery* ("Opportunity Insights" 2020) based upon publicly available and anonymized spending datasets from credit card companies in order to help determine that a lack of spending by wealthier zip codes was negating the initial taxpayer stimulus checks of $1,200 for qualifying individuals. Simply compiling and posting a new dataset or the results of studying a dataset in a public space can intersect with critical making depending on the ethical cause that motivates it.

Some digital humanities scholars *have* linked critical making and text mining. However, our field hasn't really engaged critical text mining per se. It is far more natural for many in our field to critique ideologies such as the broader state and/or corporate use of algorithms for oppressive purposes while cloaking this oppression in the terms of apolitical technology (Cheney-Lipphold 2017; Finn 2017). Safiya Noble (2018), for example, productively notes how Google's page rank algorithm, which many users assume is a neutral tool, is actually laden with racism, which Kevin Brock and Dawn Shepherd (2016) have also observed in the idea of "procedural enthymemes."

It is also possible that some of the generalized scientific and empirical use of methods related to text mining can seem more value-neutral and somehow divorced from the sorts of more directly activist concerns that motivate a great deal of critical making scholarship (Long 2018). In a comment we in no way mean as critical, the Faciloscope 2.0's ability to detect genre signals in, say, scientific journal articles probably will not contribute as much to current public debates *unless* the texts that researchers input and the research questions that they seek to answer are critical in the first place. In other words, part of defining *critical text mining* lies in identify how and in what ways ethical and political values inform the use of text mining as a methodological framework.

While there is a good deal of work that would resemble critical texting in all but name, we searched Google Scholar for two linked keywords in January 2021—"critical making" and "text mining"—to see what we could locate. We returned a whopping thirty-seven results, most of which referenced either term a handful of times but never specifically investigated or outlined a link between the two. The majority of references came from literature in the form of MLA conference proceedings and not published research articles. For example, in her 2019 MLA presentation, Whitney Sperrazza discusses her work on using algorithms to analyze an archive of sonnet sequences written by sixteenth- and seventeenth-century English poets. For Sperrazza, then, "critical making" is employed specifically as a sort of generalized experimental practice to help her reenvision her dataset (i.e., to tweak her algorithms) instead of just fact checking them (in Ramsay's [2011] sense). Matt Wilkens (2009) made a direct appeal for what he calls "critical text mining": Computational methods "offer a new kind of evidence that's especially well-suited to support the kinds of extracanonical cultural, historical, and sociological claims that have come to occupy a central place in the discipline over the last few decades" (Wilkens 2009; see also Düring 2017).

RANCIÈRIAN ETHICS AND CRITICAL TEXT MINING

While we should be wary of certain algorithmic forms, to paraphrase the great African American lesbian poet Audrey Lorde, sometimes we have to use the master's tools to dismantle the master's house. We would remind readers that there is nothing inherently progressive or left-leaning about *critical making*. As a case in point, the (at times) craftivist knitting community, Ravelry, who helped produce the pink pussy hat protest against the allegations of Trump's leaked audio recording describing what sounded to many like a possible enactment of rape culture, took on a progressive bent. However, when Ravelry banned Trump supporters, conservative knitters formed their own communities (Basu 2020). One frustrated user, Deplorable Knitter, founded a blogging community called "The Adventures of a Politically Incorrect Knitter." This community hosts knit-along sessions to produce cowls and hats that support President Trump that are undeniably a form of critical making from a politically conservative perspective (Basu 2020).

Simply put, progressive critical text mining requires a progressive ethic to motivate it. One method that is compatible with both the process of critique and the normative work of using positive ethics to inform critical text mining lies in having a shift from passive to active equality (May 2010). Using the philosopher Rancière, Todd May (2010) describes how a great deal of how we talk about activism lies in "passive equality models." For example, in a liberal distributive justice system, we signal the presence of inequality, and we lobby government officials or employers to redistribute equality in this sense. While such constitutes an important mechanism for progressive politics, May demonstrates that equality remains passive since by definition only the institution or state can distribute it. By comparison, *active* equality is a form of dissensus (Rancière's term) against inegalitarian "partitions of the sensible" or the "police order," which maintains symbolic and actualized forms of oppression (Colton and Holmes 2017). Dissensus makes the form of oppression visible either by someone who is part of the invisible "count of those who have no count" or, as his later work makes clear, by their allies. Jared Colton and Holmes (2017) give an example of how legally segregated African American activists in Greensboro, North Carolina walked into a whites-only café in a Woolworth's and asked to be served as equals regardless of the law. These activists made visible (disturbed) a partition of the sensible.

Rancière's work also works at the aesthetic level. Partitions of the sensible can be anything from laws to unequal or offensive cultural representations that do not allow individuals to verify their equality with others in a community. Here, we see a natural connection between Rancière and foundational critical making concepts such as John Hartley's "DIY citizenship," which Ratto and Megan Boler (2014) draw on. DIY citizens use the materials at hand (bricolage, tactics) to create a political identity: " 'Citizenship' is no longer simply a matter of the social contract between the state and subject . . . DIY citizenship is a choice people can make for themselves" (Hartley, quoted in Ratto and Boler 2014, 11). Ratto and Boler (2014) complain that Hartley works from an atomistic individual theory of classical liberalism. However, their critique of Hartley misses what is still a relevant contribution by Hartley that overlaps with Rancière: Anyone anywhere can resist the partitions that maintain us unequally. Hartley may have been thinking more about how materiality can participate in consensus, but Rancière can help push ideas like Hartley's toward material dissensus.

Related to the issue of DIY citizenship and critical text mining is Rancière's ethical structure of normativity. He unapologetically declares that political equality is a normative good. We might not be able to describe as a universal ethical axiom (like Kant's categorical imperative) what dissensus will be for each singular individual and their allies at all times, but we can readily identify when equality is not allowed to occur. For example, the Black Lives Matter enactments of dissensus following the death of George Floyd are clearly acts of dissensus against literal police orders that inflict oppressive violence disproportionately upon black bodies. Furthermore, the value of Rancière's framework exists only in actions that participate in disturbing a partition of the sensible. He says that politics in his idiosyncratic discussion *only* refers to acts of dissensus against partitions of the sensible (and occurs very rarely). We don't have to wait for permission or redistribution of justice to use all the tools and means around us to enact dissensus, including text mining. The point is that it is a situational and dispositional ethic: one cannot come up with universal Kantian moral imperatives to legislate what constitutes dissensus for each marginalized person or their allies in each circumstance. Partitions of the sensible will be multiple, overlapping, and divergent. #Blacklivesmatter is important, as is the combination of #blacklivesmatter and the #trans hashtag.

CONNECTING RANCIÈRE TO CRITICAL TEXT MINING

Here, we will turn to a brief illustration of how we use Rancière's thinking to guide text mining to study emoji research as critical text mining. Tacit Rancièrian approaches to critical text mining already exist in all but name. Since 2015, the DocNow (Documenting the Now) collective (2020) has been housing Twitter datasets scraped by everyday internet users as well as developing tools for them to use to analyze their datasets. DocNow is a form of DIY citizenship; as the "About" section of their website describes, "Documenting the Now has a strong commitment to prioritizing ethical practices when working with social media content, especially in terms of collection and long-term preservation" (Documenting the Now 2020). Of course, not all of their datasets will include motives by an ethic of political equality. However, insofar as their participants strive to make datasets visible that call attention to invisible forms of social impression, there is a loose Rancièrian ethic at play.

As a classroom assignment for critical text mining, one could assign students to scrape Twitter for a particular activist hashtag that is trying to create a dissensus-like change (like #blacklivesmatter or #occupy) and then contribute that dataset to DocNow or host it themselves on a public repository like Git. It is important to note that Rancière's ethic applies to all stages where we need to verify our own and each other's equality in relation to any ongoing partitions of the sensible involved in a text mining research project. One might be tempted to jump at this point straight to how to form a research question about or a technical question about how to webscrape emoji using R or Python, but Rancière's framework requires us to start with partitions of the sensible posed by emoji themselves. Emoji are controlled by the Unicode consortium. They work with large mobile phone providers to determine which emoji (typically 100–200) are added each year. In the past two years, relations of representation have improved somewhat dramatically with more gender diverse and transgender emoji. However, the 2019 emoji release contained new emoji that reflected how deeply Unicode's initial emoji offering reinscribed intersectional problems of race, class, and gender. Part of 2019's sixty-five new emoji were gender-inclusive. The police officer emoji, for example, was default male. The person-getting-a-haircut emoji default was female. "So it goes," as Kurt Vonnegut invariably reminds us.

What is interesting for critical text mining is to think about limitations: anyone is welcome to compose their own emoticons, for example. Yet only Unicode-adopted emoji are officially considered emoji due to ensuring cross-platform compatibility. Unicode has to impose a sort of universal logic. This logic is a huge problem and opportunity for text mining concerns. Let us return to non-emoji examples for a second. Researchers can and should pause and ask how their word banks and scripts are equipped (or not equipped) to handle different languages and nonconforming text. Issues of racism and word banks for sentiment analysis, for example, have been well documented (as have all of the facial recognition machine learning programs) (Earl 2019). A researcher might need then to determine who can and cannot verify equality within the input dataset itself before they even begin to initiate a research question and text mining application. For example, the activist Jennifer 8. Lee who founded the grassroots diverse emoji lobbying organization Emojination (Gray and Holmes 2020) actually joined Unicode's technical communities to help shape the institution in a "passive" sense (May 2010).

Thus, the answer to questions that researchers might want to know, such as "What are the more common emoji tri- and bi-grams in popular activist hashtags?" to understand how activists use emoji is to realize that emoji are already constrained in a way that most other forms of digital communication are not, at least in terms of algorithm. Thus, there are unique protocological constraints of inequality to studying emoji before we even get to analyzing the learned structural biases of users.

Let's take a very basic text-mining application, frequency analysis. (Readers do not even need sophisticated software tools for frequency analysis, by the way. Voyant's open-source suite text mining functions very well for non-coders.) In rhetoric and composition, frequency analysis is important. We have keywords in the field. In turn, Dylan Dryer (2019) has used large-scale analyses to determine whether our keywords in keywords collection (like Heilker and Vandenberg's [1996] *Keywords in Composition Studies*) are actually representative of the keywords that the field employs. For the purposes of this more introductory example, we'll start with a basic webscraping of Twitter using R, a programming language designed for statistics and data analysis. In comparison to learning how to build programs in Python or JavaScript, most of R involves learning how to install package scripts like the "TwitteR" package which, in turn, comes with specific commands that users execute without having to build their own programs.

As a research question, Black Lives Matter resurged again on Twitter and social media as well as with on-the-ground protests in response to the summer 2020 death of George Floyd at the hands of four police officers. There is probably no clearer enactment of political and racial inequality in the United States during the summer of 2020 than well-documented police brutality against black bodies (Herndon and Searcey 2020). Racial inequality is so visible at the moment that polls have shown that historically reluctant white voters are *finally* in agreement with the basic premise that the police as a structural institution have supported a regime of violence and denial of rights and bodily integrity to African Americans. The latter are clearly the "count of those who have no count" as Rancière (1999) would call them in this social situation.

Imagine that we are interested in supporting racial equality causes by collecting some sort of data that might shed light on how users of this hashtag employ emoji (or do not employ emoji) in ways to communicate their cause. Or, maybe more broadly, we might be interested in asking our students to do some webscraping as part of a digital rhetoric/social media rhetoric assignment on public rhetoric and activism. Frequency analysis can help to establish forms of communication being used by the majority of composers and audiences who retweet, like, etc., content. Critical text mining motivated by an ethic of being an ally toward political equality can help provide an analysis of the rhetorical tools than identify frequently used forms of communication in this rhetorical situation. For example, Hamden Azhar (2017) analyzed sixty thousand tweets in activist Twitter campaigns in 2016 and discovered that the raised fist emoji (and the red heart, tears of joy, and the American flag) were the most commonly used. In turn, it is worthwhile to contrast how these previous forms work within and against contemporary movements like #blacklivesmatter to determine how activist work occurs online.

HOW TO BEGIN IN R

We highly recommend consulting introductions to R resources online. For text mining, Julia Silge and David Robinson's (2021) *Welcome to Text Mining with R* manual is particularly useful for introductory text parsing in R and methodological explanations. The instructions below are provisional, and we offer fuller instructions, the code we used, and explanations on this book's companion website. Both authors have successfully taught this code to our students in a number of different contexts related to digital rhetoric, social media analysis, and public rhetoric. To webscrape emoji with R, you will need:

1. A Twitter developer account with API access
2. A computer (not a mobile phone) and a reliable internet connection
3. Rstudio (https://rstudio.com/)
4. Patience (it's a *hexis*/ethical disposition)
5. Our public tutorial with code and dataset repository, found on this collection's companion website

These instructions on the book's companion website (https://upcolorado.com/component/k2/item/6219-reprogrammable-rhetoric-supplemental-design-materials-and-programming-scripts) will lead you through the basics of package installation and some of the commands in R you will need in order to get started with basic webscraping related to #blacklivesmatter and emoji frequency analysis.

As a warning—and speaking of equality with respect to Garnet Hertz's (2016) comments—Twitter's API only allows the past seven days and a limited number of tweets (n=18,000) to be scraped. The complete archive and additional pull requests are hidden behind a paywall. This area is another workflow site at which to think about how the critical intersections of partitions of the sensible and power operate, but we digress.

EMOJI AND CODING

If you webscrape and attempt to decode emoji from Twitter in R using the standard TwitteR package, the output is a strange hexadecimal-looking code. As an example:

ANGRY FACE";"☺";"\xF0\x9F\x98\xA0";"<ed><a0><bd><ed><b8><a0>

In order to analyze emoji as text, it is necessary to import an external emoji bank (a "word" bank with emoji), which converts this hexadecimal code to readable emoji names to then be parsed by R. In our example, we have modified slightly with attribution an open-source emoji word bank from Peterka-Bonetta's (2017) emoji analysis script on R-Bloggers and GitHub. To date, her script is still one of the best open-source scripts available on GitHub and R-Blogger, until someone programs an emoji analytics package for R.

If it seems slightly disingenuous to use someone else's code, we will add that for those of us who are in the middle ground of coders (self-taught with no formal instruction), using R scripts on the internet—let alone employing basic functions—seldom work as advertised. R packages update constantly and even the most well-meaning academic and industry open-source code contributors have day jobs. R coders do not always update their older code or include working instruction sets. When R subpackages update, they can render entire scripts useless if a required subpackage disappears. In other words, to work with R even at basic levels is already to be engaged actively in reprogramming. Thus, we've included our troubleshooting tips on GitHub ways to work with Jessica's code—reprogramming-*lite*, if you will, that is aimed at complete beginners. As a case in point, Peterka-Bonetta's (2017) instructions leave off a few packages (Diplyr, readr) that were necessary for her code grammar to work.

A final word of advice before you try our tutorial or pursue your own webscraping efforts in R is to Google your error returns. It might be that a subpackage has changed. R packages often need other packages to help use certain commands (%>%) or grammars to make the user queries and inputs easier (albeit potentially more confusing). There is a wealth of information online in the R-Blogger forum and elsewhere about fixing R codes.

RESULTS- AND DISCUSSION-LITE

The code hosted on the book's companion website will help you scrape your own tweets using R on any hashtag and run emoji frequency analysis. We have also provided daily and aggregated webscraping of eighteen thousand tweets using the #blacklivesmatter hashtag between August 1, 2020, and August 31, 2020, that are available for download and analysis if the reader wants to use it. We have the individual daily files plus a cleaned aggregate file with duplicate tweets removed (using the Grep () function). There are other interesting functions like which words

Table 4.1. Frequency table for #blacklivesmatter emoji scraped on July 13, 2020

Rank	Emoji name	Number
	<chr>	<int>
1	raised fist	588
2	sparkles	118
3	double exclamation mark	76
4	red heart	70
5	balance scale	35
6	male sign	30
7	high voltage	29
8	white medium star	25
9	female sign	21
10	exclamation question mark	18

Table 4.2. Frequency table for #BLM emoji scraped on July 13, 2020

Rank	Emoji name	Number
	<chr>	<int>
1	raised fist	210
2	right arrow	133
3	red heart	79
4	white heavy check mark	57
5	warning	56
6	male sign	36
7	radioactive	31
8	female sign	19
9	cross mark	18
10	double exclamation mark	14

Table 4.3. Frequency table for #alllivesmatter emoji scraped on July 13, 2020

Rank	Emoji name	Number
	<chr>	<int>
1	heart suit	3,230
2	raised fist	852
3	red heart	66
4	male sign	30
5	ballot box with check	25
6	black circle	22
7	sparkles	17
8	female sign	16
9	white heavy check mark	15
10	latin cross	13

pair with which emoji. Sentiment and emoji can be run as well as frequency with Peterka-Bonetta's (2017) code. However, as with all things text mining, it is useful to start with datasets and someone taking the time to scrape them and compile them as we've started to do with #blacklivesmatter in alliance with those like DocNow and other activists and data scientists who have already enacted this sort of work.

Tables 4.1, 4.2, and 4.3 display some of the frequency results that we ran on the #blacklivesmatter hashtag on July 13, 2020.

We have included three different results from three different hashtags: #blacklivesmatter (table 4.1) or the official hashtag, #BLM (table 4.2) or the unofficial shorthand hashtag, and #alllivesmatter, or the adopted pro-police hashtag. We lack the space in this short chapter to adequately explain the sample size issues (too small to be representative), but even some basic results here on a single day are revealing. First, emoji similarities (raised fist emoji) and differences emerge depending on the hashtag used (#BLM, #blacklivesmatter). In terms of differences, it is likely that the white heavy check mark means that more policy declarations in #blacklivesmatter are likely because this symbol (as we found in another study currently under review) is used by US senators on Twitter in this sense. In other words, perhaps #blacklivesmatter is a more political or public policy-oriented hashtag versus the shorthand #BLM, which may have a more informal set of rhetorical purposes. Nevertheless, the two hashtags share so many emoji in common, that the differences may be inconsequential.

In terms of #alllivesmatter, heart emojis expressing support for the police were the overwhelming more popular emoji, followed by a negative emoji at #6: the black circle, which often means eclipse or blind. In other words, we can see love and support expressed for the police and critical emoji of being blind to the true issues (allegedly) in #alllivesmatters tweets. Again, these examples are hardly representative or defining, but they do hopefully communicate to the reader the importance of undertaking this sort of analytical work. If we hope to understand let alone intervene in public spheres on social media that form around hashtags, then we need to understand the full range of communicative practices within the aggregate community.

CONCLUSION

For more information about the value of frequency (and much more sophisticated methods of largescale computationally driven corpus studies), we point the reader to Lancaster's (2016) study of *They Say/I Say* as well as Dryer's (2019) study of keywords in rhetoric and composition. At the minimum, studying something like emoji frequency can help to communicate how individuals enact rhetorical practices for particular internet audiences on Twitter. Simply put, we hope that this chapter encourages readers to think about tool learning in text mining as first and foremost an ethical invitation to think about what motivates any individual's desire to do, well, *anything* maker-oriented. Thus, even if Rancière is not every reader's favorite political theorist, we hope that this example of how one can use even a bare-bones political philosophy to guide an ethical activity of critical making is the sort of structure that we'd encourage readers to consider in their own making practices.

This point is one that our colleague, Cana Uluak Itchuaqiyaq, has raised in her chapter in this section. Whereas many humanities academics by now are used to paying at least lip service to inclusivity of multiply marginalized peoples, Itchuaqiyaq notes that text mining can reveal the extent to which racism and colonialization actually do pervade the field in invisible ways like citation practices. Scholars like Malea Powell, whom Itchuaqiyaq cites in her chapter, already know this type of generalized claim to be true. However, Itchuaqiyaq's analysis—like text mining in general—is a productive way of showing the *degree of change*. For example, John Gallagher and Steve Holmes (n.d.) have found that the bigram "African American" is the most frequently employed word (along with "student" and "writing") across twenty years' worth of articles in seven major rhetoric and composition journals. Topically, these findings may make it seem as though decolonization in rhetoric and composition is occurring. However, as Itchuaqiyaq's work highlights, topical inclusion *may not* be proportionate to citational inclusion in a meaningful sense. Her chapter in this edited collection section is an outstanding enactment of critical text mining in all but name.

REFERENCES

Anderson, Monica, Skye Toor, Lee Rainie, and Aaron Smith. 2018. "An Analysis of #BlackLivesMatter and Other Twitter Hashtags Related to Political or Social Issues." *Pew Research Center*, July 11, 2018. https://www.pewresearch.org/internet/2018/07/11/an-analysis-of-blacklivesmatter-and-other-twitter-hashtags-related-to-political-or-social-issues/.

Azhar, Hamden. 2017. "The Resistance Will Be Emojified." *Emojipedia*, Feb 6, 2017. https://blog.emojipedia.org/the-resistance-will-be-emojified/.

Basu, Tanya. 2020. "How a Ban on Pro-Trump Patterns Unraveled the Online Knitting World." *MIT Technology Review*. Mar. 6, 2020. https://www.technologyreview.com/s/615325/ravelry-ban-on-pro-trump-patterns-unraveled-the-online-knitting-world-censorship-free/.

Brock, Kevin, and Dawn Shepherd. 2016. "Understanding How Algorithms Work Persuasively through the Procedural Enthymeme." *Computers and Composition* 42: 17–27. doi:10.1016/j.compcom.2016.08.007.

Cheney-Lippold, John. 2017. *We Are Data: Algorithms and the Making of Our Digital Selves*. New York: New York University Press.

Chun, Wendy Hui Kyong. 2011. *Programmed Visions: Software and Memory*. Cambridge, MA: MIT Press.

Colton, Jared S., and Steve Holmes. 2017. "A Social Justice Theory of Active Equality for Technical Communication." *Journal of Technical Writing and Communication* 48 (1): 4–30. doi:10.1177/0047281616647803.

Documenting the Now. n.d. Accessed June 1, 2020. https://www.docnow.io/.

Dryer, Dylan B. 2019. "Divided by Primes: Competing Meanings among Writing Studies' Keywords." *College English* 81 (3): 214–55.

Düring, Marten. 2017. "C²DH Receives Funding for a Project on Critical Text Mining in Historical Newspapers." *C²DH*, June 2, 2017. https://www.c2dh.uni.lu/de/news/c2dh-receives-funding-project-critical-text-mining-historical-newspapers.

Earl, Carl. 2019. "Gender and Racial Bias in Cloud NLP Sentiment APIs." *Data for Breakfast*, August 21, 2019. https://data.blog/2019/08/21/gender-and-racial-bias-in-cloud-nlp-sentiment-apis/.

Finn, Ed. 2017. *What Algorithms Want: Imagination in the Age of Computing.* Cambridge, MA: MIT Press.

Gallagher, John, and Steve Holmes. n.d. "Stability, Unity, and Fragmentation: A Quantitative Analysis of Seven Writing Studies Journals from 2000–2019." Unpublished manuscript.

Gray, Kellie M., and Steve Holmes. 2020. "Tracing Ecologies of Code Literacy and Constraint in Emojis as Multimodal Public Pedagogy." *Computers and Composition* 55 (March). doi:10.1016/j.compcom.2020.102552.

Hayles, N. Katherine. 2012. *How We Think: Digital Media and Contemporary Technogenesis.* Chicago, IL: University of Chicago Press.

Heilker, Paul, and Peter Vandenberg, eds. 1996. *Keywords in Composition Studies.* Portsmouth, NH: Boynton/Cook.

Herndon, Astead W., and Dionne Searcey. 2020. "How Trump and the Black Lives Matter Movement Changed White Voters' Minds." *New York Times,* June 27, 2020. https://www.nytimes.com/2020/06/27/us/politics/trump-biden-protests-polling.html.

Hertz, Garnet. 2016. "What Is Critical Making?" *Current,* no. 7. https://current.ecuad.ca/what-is-critical-making.

Lancaster, Zak. 2016. "Do Academics Really Write This Way? A Corpus Investigation of Moves and Templates in *'They Say/I Say.'*" *College Composition and Communication* 67 (3): 437–64.

Long, Seth. 2018. "Changing Words into Numbers: Rhetoric, the Digital Humanities, and Methodological Transparency." In *Rhetorics Change/Rhetoric's Change,* edited by Jenny Rice, Chelsea Graham, and Eric Detweiler. Enculturation Intermezzo Epub. South Carolina, SC: Parlor Press. http://intermezzo.enculturation.net/07-rsa-2016-proceedings.htm.

May, Todd (2010). *Contemporary Political Movements and the Thought of Jacques Rancière: Equality in Action.* Edinburgh, UK: Edinburgh University Press.

Milberry, Kate. 2014. "(Re)Making the Internet: Free Software and the Social Factory Hack." In *DIY Citizenship: Critical Making and Social Media,* edited by Matt Ratto and Megan Boler, 53–63. Cambridge, MA: MIT Press.

Noble, Safiya. 2018. *Algorithms of Oppression: How Search Engines Reinforce Racism.* New York: New York University Press.

"Opportunity Insights Economic Tracker." 2020. Track the Recovery (website). Accessed June 1, 2020. https://tracktherecovery.org/.

Peterka-Bonetta, Jessica. 2017. "Emojis Analysis in R." *R-bloggers,* March 23, 2017. https://www.r-bloggers.com/emojis-analysis-in-r/.

Ramsay, Stephen. 2011. *Reading Machines: Towards an Algorithmic Criticism.* Urbana: University of Illinois Press.

Rancière, Jacques. 1999. *Disagreement: Politics and Philosophy.* Translated by Julie Rose. Minneapolis: University of Minnesota Press.

Ratto, Matt, and Megan Boler, eds. 2014. *DIY Citizenship: Critical Making and Social Media.* Cambridge, MA: MIT press.

Ratto, Matt, and Stephen Hockema. 2009. "FLWR PWR—Tending the Walled Garden." In *Walled Garden,* edited by Annet Dekker and Annette Wolfsberger, 51–62. Amsterdam, Netherlands: Virtueel Platform. https://criticalmaking.com/wp-content/uploads/2009/10/2448_alledgarden_ch06_ratto_hockema.pdf

Ridolfo, Jim, and William Hart-Davidson, eds. 2015. *Rhetoric and the Digital Humanities.* Chicago, IL: University of Chicago Press.

Shaul, Brandy. 2015. "Report: 92% of Online Consumers Use Emoji (Infographic)." *Adweek,* September 30, 2015. https://www.adweek.com/digital/report-92-of-online-consumers-use-emoji-infographic/.

Shi, Guoliang, and Yanqing Kong. 2009. "Advances in Theories and Applications of Text Mining." *2009 First International Conference on Information Science and Engineering*, 4167–70. doi:10.1109/ICISE.2009.214.

Silge, Julia, and David Robinson. 2021. "*Welcome to Text Mining with R.*" Tidy Text Mining (website). https://www.tidytextmining.com/index.html.

Sperrazza, Whitney. 2019. "Patterns of Violence: Critical Making and the She/Her/Hers of Early Modern Poetry." Presentation at the Annual Meeting of the Modern Language Association, May 2019, Chicago, IL. http://dx.doi.org/10.17613/beec-4j32.

Wilkens, Matt. 2009. "Critical Text Mining, or Reading Differently." *Work Product*, December 8, 2009. https://mattwilkens.com/2009/12/08/my-mla-talk-critical-text-mining-or-reading-differently/.

5

REPROGRAMMING THE FACILOSCOPE
A Software Development Story

Ryan Omizo
Temple University

The Faciloscope (Omizo et al. 2016) represents an early effort to use computational tools to automate the application of rhetorical theory for analysis, specifically rhetorical theory descended from rhetorical genre theory (Miller 1984) and Swalesean move analysis (Swales 1990). In this sense, the Faciloscope represents a form of critical making in that it was an experiment in encoding well-established rhetorical methodologies to computational methods as well as translating the practical work of a team of facilitation researchers into a functioning web application that would, ideally, accelerate their work. As with any novel endeavor, there were successes, failures, and limitations. While the Faciloscope identifies a constrained palette of rhetorical moves with reliable accuracy, it leaves opaque many other significant aspects of genre such as the topicality of texts. As a critical tool of rhetorical genre theory, there is still much more to explore and many more points to map between computation and rhetoric to create a more robust computational rhetoric software program.

Coincident with the work of Kellie Gray and Steve Holmes (this collection), I describe recent efforts to enhance the range of genre analysis offered by the Faciloscope and as a path to new, theoretically grounded "interpretive possibilities" (Gray and Holmes). As such, I invite readers to the development bench to reflect on the experience of designing a computational tool to be responsible to both rhetorical traditions and natural language processing and analysis. My aim is to demonstrate to the makers and users of computational tools for textual analysis the work involved behind the scenes so that they might better manage their own computational rhetoric projects. Although the Faciloscope is narrowly tailored in its results, the process of building and building out the Faciloscope can still offer important object lessons for theoretically minded making—specifically, qualitative coding research.

https://doi.org/10.7330/9781646422586.c005

The software development story of this chapter will describe the origins and operations of the first version of the Faciloscope. I will then recount significant revisions to the app created specifically for this chapter as two "reprogramming steps." I demonstrate the impact of these reprogramming steps by applying the revised Faciloscope app to conversations circulating on an online discussion forum, focusing on the ethical disposition of cultural relics. I conclude this chapter by reflecting upon how such accounts of software development can themselves contribute to the process of critical making in the fields of computational rhetoric and the digital humanities.

WHAT IS THE FACILOSCOPE?

The Faciloscope (http://faciloscope.cal.msu.edu/facilitation/) is a supervised machine learning web application that uses natural language processing and support vector machine learning libraries to automatically classify rhetorical moves in texts. The Faciloscope app emerged as an extension of a previous, qualitative coding project[1] (Grabill and Pigg 2012; Pigg et al. 2016; Sackey, Ngyuyen, and Grabill 2015) aimed at helping science museum facilitators foster learning in an informal way through meaningful and moderated online conversations on a range of scientific and social topics.[2] In this precursor project, researchers coded moves that trained online facilitations deemed valuable for promoting engagement about science among informal learners.

As many have observed (see Baumer et al. 2017; Chen et al. 2018; Crowston et al. 2012; Geisler 2018; Guetterman et al. 2018; Lauer, Brumberger, and Beveridge 2018; Leetaru 2012; Muller et al. 2016; Yan, McCracken, and Crowston 2014; Yan et al. 2014; Yilmazel et al. 2007), qualitative coding can be labor- and time-intensive—a fact which the members of the original facilitation research team confirmed throughout their work. The Faciloscope was created to accelerate the qualitative/rhetorical coding these researchers were producing through the application of a truncated codebook (see Boettger and Palmer 2010) of three rhetorical moves that were considered the most impactful to informal learning: Staging, Inviting, and Evoking.

Staging moves refers to sentences that describe entities, events, or processes, operating to "set the stage" or establish the topicality of communication. Inviting moves are sentences that promote interactions among communicators by making requests or offers of participation. Evoking moves refers to sentences intended to trigger affectual response from interactants, the most notable example being insults or thank-yous.

The Faciloscope's theoretical underpinnings hew to the assumptions of North American rhetorical genre theory (Devitt 2004; Freedman and Medway 1994; Paré 2014; Miller 1984; Russell 1997; Schryer 1993; for a disciplinary review see Bawarshi and Reiff 2010). In the words of William Hart-Davidson (2015), rhetorical genre theory (henceforth, RGT) conceives genre as

> habitual responses to recurring socially bounded situations. Regularities of textual form most lay people experience as the structural characteristics of genres emerge from these repeated instances of action and are reinforced by institutional power structures. Genres are constructions of groups over time usually with the implicit or explicit sanction of organizational or institutional power. (39–40)

"Move analysis" is derived from John Swales's (1990, 140–66) work on genre analysis. Move analysis describes a method for modeling textual communication as a series of goal-directed messages or "signals" that advance argumentation through a step-wise procedure among discourse communities. Because discourse communities are united by shared forms of messaging, rules, conventions, and situations, moves that vehiculate these forms while addressing community expectations are often treated as constituents of legitimated genres. Moves, thus, represent the "habitual responses to recurring socially bounded situations."

In this case, the "recurring socially bounded situation" refers to the online informal learning environments while Staging, Inviting, and Evoking moves indicate discrete "responses" within these environments that contribute to their situated legibility among communicants. By aggregating these moves, the Faciloscope can visualize the larger patterns that individual moves create over time. Using the Faciloscope's various data views (Hart-Davidson 2005), people can also see how moves are distributed throughout the conversation ("bands" visualization) and read the source text with their attendant facilitation tags in tables. The process is intended to be hermeneutic, obliging the reader to vacillate between global or distant reading to more targeted or close reading (see Jockers 2013; Long and So 2016; Piper 2018; Ramsay 2011; Sinykin, So, and Young 2019).

REPROGRAMMING STEP 1: EXPANDING TRAINING DATA

The original supervised machine learning algorithm (Pedregosa et al. 2011) was trained on over thirteen thousand sentences categorized by two human coders of the Faciloscope team (Ian Clark and Minh-Tam Nguyen) from discussion forum data sourced from backyardchickens

.com (Beekissed 2014; coops gone crazy 2013), a platform that allows chicken enthusiasts to share stories, ideas, and advice about chickens. The coders achieved an interrater kappa score of .92. When tested on an unbalanced set of three thousand sentences held out of the original training set, the Faciloscope achieves kappa scores in the range of .70–.80 (see Omizo et al. 2016).

However, feedback from an early demonstration to stakeholders questioned the overreliance on backyardchickens.com for training and testing data. The concern, shared by the Faciloscope development team, was that the classifier was "overfitting" to the data (see Dietterich 1995; Domingos 2012), rendering it accurate for the identification of chicken-related Staging, Inviting, and Evoking moves and less equipped to handle new data unrelated to chickens. Practically speaking, the Faciloscope's vocabulary would overestimate the influence of chicken terms in new data and undervalue the influence of terms outside of the original training data.

To address the lack of generalizability, I have expanded the training data with the use of publicly available and researcher-curated datasets. Because the original Faciloscope training data contained a disproportionately high volume of Staging moves (essentially descriptive sentences), Inviting and Evoking moves suffered from a lack of representation and a lack of validation samples. To bolster Inviting moves, questions collected by Xin Li and Dan Roth (2002), Julian McAuley and Jure Leskovec (2013), and McAuley et al. (2015) have been added to the training and testing data. In this case, interrater scoring was not used because questions are definitively Inviting moves based on the Faciloscope codebook. To bolster Evoking moves, sentences from the Kaggle Insults Dataset ("Detecting Insults in Social Commentary" n.d.) were used because insults represent a dimension of the affectual valence that defines Evoking moves in the Faciloscope codebook. The reprogrammed version of the Faciloscope[3] is trained on 19,444 sentences and tested on 8,455 sentences, resulting in a precision score of .936, a recall score of .905, and F1 score of .920.

To give readers a sense of the Faciloscope's capacity for move analysis, I discuss the results that the app returns and how these results can be used to diagnose and enhance facilitator engagement with online conversations through a case study of the Science Buzz (Sam 2005) discussion forum, "Should Ancient Artifacts Return Home?" (henceforth, "Ancient Artifacts").

"Ancient Artifacts" contains 190 posts. The initial post appeared on July 19, 2005. Although the scale of "Ancient Artifacts" does not offer an opportunity for "big data" analysis, the relatively low volume of posts does offer a more tractable means to correlate close rhetorical readings with statistical results.

"Ancient Artifacts" begins with a long story about the efforts by Egypt to secure the return of ancient Egyptian artifacts through lawsuits (the discussion thread is apparently based on a currently defunct Yahoo News item). Sam, the Science Buzz poster, references other cases of disputed ownership rights such as Greece's request for the return of the Elgin Marbles and Italy's return of a stolen monument to the Ethiopian city of Axum (Sam links to a BBC news story, "Rome Obelisk Set for African Return," to demonstrate this point). The post ends with an overture to the Science Buzz community. Sam asks, "What do you think? Where do these objects belong?" (Sam 2005). The responses to this initial Inviting move cohere thematically around the question of "where do these objects belong," whose answers are positioned as binary oppositions through Sam's introduction of the issue. The debate about the property rights of ancient artifacts revolves around those countries in possession of artifacts and those claiming dispossession of artifacts (i.e., Great Britain, the possessor of many ancient Egyptian artifacts, versus Egypt, the country who has lost its artifacts). Respondents generally express agreement with the notion that artifacts should be returned to their regions, countries, and cultures of origin or express disagreement with the prospect of returning artifacts. Example posts advocating for the return of artifacts to their places of origins include the following:

> I belive artifacts should be put pack and never taken in the first place.
>
> Historical places will start losing value if they become empty. (Sam 2005)
>
> The artifacts MUST be returned home to their countries of origins. (Sam 2005)

"Ancient Artifacts" posts that argue for restoring artifacts to their places of origin frame the acquisition of artifacts by museums and/or exogamous entities (i.e., nations such as the United States and Great Britain removing artifacts from their homelands) are framed as theft. Returning artifacts now would serve as reparations.

Posts that disagree with the prospect of returning artifacts rely upon several core rationalizations. One key rationalization is that current challenges over the ownership of ancient artifacts are unjustified by virtue of being ipso facto claims. An anonymous poster writes:

> You can't demand things back retrospectively if they were were legitimately at the time. No society can survive when you never know what you do today will be made illegal tomorrow and you get carted off to prison. (Sam 2005)

Another anonymous poster writes:

> But you can't cry about what happened 100 years ago. Then the whole world is paying the whole world for damage and lost and robbery. (Sam 2005)

For this argument, the passage of time has eroded property claims on the part of the looted countries. The laws were different back then. The countries were different back then. Cultures were different back then.

Other arguments for not returning artifacts to their places of origin revolve around racist beliefs in Western exceptionalism. Poster D.K. writes:

> The items should not be returned. Look what happened to the Bagdad museum! As soon as the country destabilized, the townsfolk looted the museums and took everything. now all those wonderful artifacts from Mesopotamia are gone. (Sam 2005)

Writing in 2014, an anonymous poster states:

> Headlines today: ISIL is destroying artifacts in its control. Fortunately, many other artifacts are safe in the British Museum. (Sam 2005)

Such posts spur binaries that cast Western entities such as the British Museum as noble stewards of world culture and those countries and peoples who have historical and cultural claims to their artifacts are depicted as unable to be trusted with their own artifacts due to some cultural flaw such as political chaos or religious fundamentalism.

"Ancient Artifacts" Results after Reprogramming Step 1

The Faciloscope returns the following move distribution for "Ancient Artifacts" on 805 sentences.[4]

- Staging moves: 579
- Inviting moves: 119
- Evoking moves: 107

Because Staging moves most often designate descriptive sentences, they predominate in most cases as they do for "Ancient Artifacts." Because facilitation is invested in promoting engagement and interaction among communicants, Inviting and Evoking merit special attention because both moves elicit responses from others. For example, in the following passage, a Science Buzz poster challenges the remarks by a previous poster with an Inviting move:

> Posted on sun, 06/24/2007—11:43pm reply freddy says: what say european countries would not engage in war? in fact europe has seen war twice

Faciloscope Move per Sentence

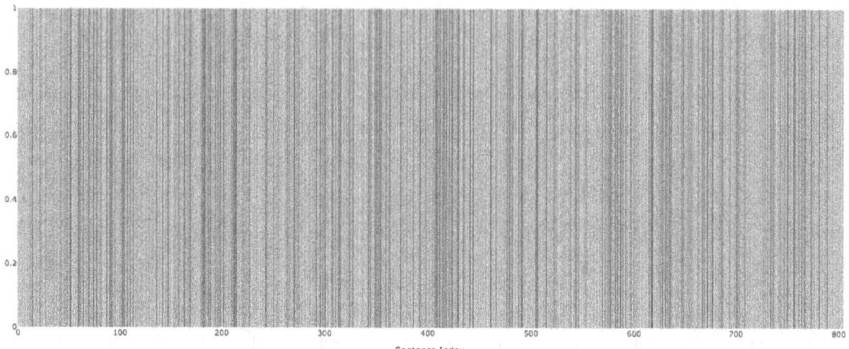

Figure 5.1. Facilitation move distribution in "Ancient Artifacts."

in the last century and paris was nearly destroyed by hitler. the fact that somebody can take better care of the artifact does not warrant stealing it. theft remains theft. it is like stealing rich people's money because you think you can better spend/keep/invest it. western countries contribute just as much if not more than middle east countries to the instability of the region. it is like me keep raiding your home and wouldn't return what i have stolen because i believe "your home is unsafe." (Sam 2005).

Evoking moves proliferate in the latter portions of "Ancient Artifacts." For example, we see an intense discussion of "raiding":

Seriously! think about it why should you leave the artifacts in the places were you found them??!! havent you heard of "raiding"? if an archaeologist didnt pick it mise as well just be sold off to the highst bidder!! because people dont really care about the artifacts they just want the $$ that comes with it! its safer at a museum than were it was found!! (Sam 2005)

As mentioned above, global patterns of move distributions can indicate how much participants in online discussions are engaging with their communities. Inviting moves occur with regularity in "Ancient Artifacts," with the highest collection of Inviting moves appearing at the ends of the thread (see figure 5.1). That participants continue to make inquiries and solicit responses from others across the thread might explain the duration of the thread. Facilitators or analysts can immediately discern how often different moves are deployed. For "Ancient Artifacts," Inviting feedback is a routine practice, which is indicative of the messaging tolerated by the forum.

As an experiment in reconciling human applications of RGT and Swalesean move/step modeling to computational methods, the Faciloscope's charting of rhetorical moves represents a "top-down" approach to analysis (see Biber, Connor, and Upton 2007; Upton and Cohen 2009). This "top-down" approach presumes a static generic framework for facilitation, whose purpose is to promote conversation among communicants to foster learning, and is constituted by three general rhetorical moves, Staging, Inviting, and Evoking. While adequate to the premises of RGT and move/step modeling, the above results are blunt and oblige users to sift through sentences to understand what high-value Inviting and Evoking moves are referencing to flesh out their interpretations of the global move patterns.

Delivering more detailed information about the topicality and language of individual moves can be accomplished in various ways, which include latent Dirichlet allocation topic modeling (Blei, Ng, and Jordan 2003) or vector space models, all of which make use of word frequency counts. Information about significant terms would provide a more ground-level view of the rhetorical development of "Ancient Artifacts" over the life of the thread. In terms of Swalesean move analysis, this term frequency information would illuminate the "signals" used to trigger steps within moves.

At the same time, such an addition to the Faciloscope's analytical pipeline still leaves the "meso-level" of genre (Spinuzzi 2003) uninspected. While the surfacing of significant terms can provide information about what an online discussion is about and the global patterning of rhetorical moves in the facilitation genre can suggest how people are participating in the conversation, other important dimensions of genre formation about the vicissitudes of genre stability and topical development elude computational processing. Following the top-down approach to move analysis discussed by Douglas Biber, Ulla Connor, and Thomas A. Upton (2007) and Upton and Mary Ann Cohen (2009), the issue of connecting global move patterns and significant words is one of unit size—what unit can represent a point of articulation between term distributions and global move distributions?

Germinal work in RGT by Carolyn R. Miller's (1984) "Genre as Social Action" and Thomas Frentz and Thomas Farrell's (1976) "Language-action: A Paradigm for Communication" provide useful concepts for modeling meso-levels of genre. Frentz and Farrell (1976) propose a hierarchical model of communication they call the "language action paradigm." The "language action paradigm" consists of three levels: context, episodes, and symbolic acts. Context includes wide-ranging "forms of life" that condition knowledge and social relationships (334–40). They are ways of thinking and doing; ways of making sense of reality by drawing and observing boundaries around concepts. After "forms of life," there are "encounters," which can be glossed as an interaction among people. Episodes refers to sequences of communicative exchanges that are rule-dependent, goal-directed, and communally organized. Key to the episode as a unit of discourse is its "explanatory significance of their developmental structure" (337), meaning that nature of the episode grants meaning to the utterances employed during the episode. In this sense, an episode is a pocket of conventions and engagement patterns created by communicators that allow meaning to inhere. On the most fundamental level of Frentz and Farrell's "language action paradigm" are "symbolic acts," which, significant for our purposes, they define as "verbal and/or non-verbal utterances which express intentionality."

Miller's (1984) hierarchical model of genre explicitly adopts some of Frentz and Farrell's theoretical assumptions. The primary difference is that Miller adds genre formation into the communicative dynamic (161). Miller's explanation of genre formation and performance is also more interested in the anabolic relationship between "substance" and "form." "Substances" refer to units of communication—utterances, speech acts, claims, questions, gestures, and, like Frentz and Farrell (1976), "intentions." "Form" indicates the frames of understanding that render "substances" coherent given the rhetorical situation (Miller 1984, 159). Form and substance interact at multiple levels of communicative experience. Drawing on Kathleen Hall Jamieson and Karlyn Kohrs Campbell (1982), Miller argues that when substances and forms "fuse" (i.e., are recognized to have relevant meaning in the issuance of substance and form to audiences), they produce "social actions" in the world.

The Faciloscope would seem to distill information at the lower levels of Miller's (1984) model. Facilitation moves might equate to the "propositions" or "speech act" substances in Miller's diagram of genre operations (160). The form that facilitation moves resemble include "illocutionary force." These combinatoric levels are located near the base of the diagram, far from the level of genre. Again, this is not to say that the Faciloscope removes users' ability to track these rhetorical functions themselves. The matching of the original text with its annotated version was a deliberate design choice meant to account for the extended interpretive work required of uses. This is to say, however, that changes can be made to the Faciloscope so that we can advance through Miller's hierarchy and return more insights into genre performance to users. To start this process, I plan to add "substance" or increase the granularity (Hart-Davidson 2005) of the Faciloscope's results so that it includes language information about facilitation moves in the form of named entity recognition. This addition would, in reference to Miller's (1984) diagram, deliver us to a higher order of complexity within the genre hierarchy, moving us to the "locution" zone of "substance" and "speech act" level of "form." This combination would target the hows and whats of utterances, whose association might elucidate more compounded "forms" of discourse.

REPROGRAMMING STEP 2: EPISODES

To better fit the Faciloscope with Frentz and Farrell's (1976) and Miller's (1984) models of communication and genre formation, we can compile "episodes," which can then be analyzed for their more complex rhetorical intentions. I approach these two theoretical dimensions via two practical computational techniques: text segmentation and term weighting. Text segmentation will be explored as a means to excising "episodes" from online discussion threads. Term weighting will then be used to summarize the content ("locution") of these segments after excision. I use TF-IDF term weighting to extract the most salient terms per section. Because TF-IDF is a well-established practice (see Lavin 2019; Manning, Raghavan, and Schütze 2008), I will devote most of this discussion to the segmentation of online discussion forum texts because of the novelty of the proposed method. I conclude my computational analysis by conducting named entity classification on derived episodes.

What constitutes an "episode?" How can we tell when one episode ends and another begins in a textual infrastructure that allows for multiple, asynchronous conversations and for people to respond to posts out of order? Here we see why Frentz and Farrell (1976) note the importance of rules for the emergence and sustainability of episodes. As contained sequences of communication, episodes require rules for starting and stopping and for constraining content. Text segmentation occurs based on social agreement about what serves as useful textual information online and is vehiculated by various semantic markers such as margins, which would separate one paragraph from another.

The pursuit outlined above has precedents in the work of Marti Hearst's (1994) TextTiling algorithm and TileBars text analysis program. Hearst's TextTiling procedures and TileBars visualization software seek to identify topics and subtopics. Topics refers to Gilian Brown and George Yule's (1983, 69–70; see also Hearst 1997, 39) definition of a topic as a "unifying principle which makes one stretch of discourse 'about' something and the next stretch 'about' something else." Following Brown and Yule's (1983) observation that topical segments might be best found by tracing the key textual features bordering topics, Hearst (1997) identifies variations in vocabulary uses across texts in order to demonstrate "topical shifts" (Brown and Yule 1983, 94–95). For example, overlapping vocabulary terms might suggest a subtopical extension of a major topic. Non-overlapping sections may indicate new topics. In this case, episodes are characterized by the rules that the differentiation of topics subtends from the differentiation of ideas, which is manifested through the differentiation of word usages. Hearst's work is supported by Jane Morris and Graeme Hirst (1991). Based on the theory of lexical cohesion (Halliday and Hasan 1976), Morris and Hirst (1991) use lexical chains to determine the organization of textual segments. Lexical chains are constituted by the repeated use of content terms, which creates conceptual relationships among sections of texts through identity or relatedness. Moreover, as Morris and Hirst (1991) explain, "Any structural theory of text must be concerned with identifying units of text that are about the same thing. When a unit of text is about the same thing there is a strong tendency for semantically related words to be used within that unit. By definition, lexical chains are chains of semantically related words. Therefore, it makes sense to use them as clues to the structure of the text" (35).

If we understand moves as discrete, purposeful acts that must be performed in a recognizable sequence to be meaningful (Swales 1990), then it stands to reason that a latent structure is extant. Jan Van Kuppevelt's (1995; 1996) "topic/comment" theory offers a productive means to explain episodic structure in terms of facilitation moves. Van Kuppevelt's theory is that communicators establish topics through indirect or direct questioning. Commentary follows this indirect or direct questioning until parties are satisfied that the indirect or direct question has been resolved.[5] New topics and subtopics can and do emerge through additional questioning and answering. Put in relation to Frentz and Farrell (1976), Van Kuppevelt's topic/comment model provides a means to explain the exigencies of episodes, the beginnings and endings of episodes, and the necessity of interactions—communicators broach topics through implicit or explicit questions due to a problem that can only be addressed through the responses of others. When commentary satisfies the opening question, commentary ends until new questions (new topics) are asked. Inviting moves, by definition, perform this rhetorical function when they are used to ask questions and/or solicit responses from others (a question awaits an answer to end questioning; a request awaits actions to fulfill needs). Consequently, we can use the position and substance of Inviting moves to segment more undifferentiated online discussions. Inviting moves represent openings of topics, which suggests that each Inviting move that occurs after the first also could indicate the closing of a previous topic. Those moves that occur between Inviting moves would then be assumed to function as commentary on the Inviting move.

The process for using the Faciloscope's assignment of Inviting moves includes the following:

1. The index values (position of sentence within the linear sequence) of all Inviting move sentences are recorded.[6]
2. The indices of the sentences that intervene between one Inviting move and the next Inviting move are recorded as a span, e.g., the indices between (0,3) are (0,1,2,3). Sentences that fall within the intervening spans of Inviting moves are captured as segments.
3. TF-IDF weighting is used to reveal the most defining terms in segments as a function of the total term frequencies in the collection of documents.[7]
4. spaCy named entity recognition is conducted on the Inviting move spans.

"Ancient Artifacts" Results after Reprogramming Step 2

The 118 inviting moves in "Ancient Artifacts" is converted into 117 segments based on the procedure outlined above. The first segment by default is the leading "story" post by Sam. The last segment is the text that follows the final Inviting move in the corpus. TF-IDF vectorization method[8] was applied to the 117 text segments and the top ten term weights are harvested, which should render the most characteristic words of each segment. To reduce noise that this scenario may produce, adjacent Inviting moves are merged into single Inviting moves. The results of this additional processing step reduce our corpus to eighty-eight Inviting move segments.

Space limits me to the discussion of a few sample Inviting move segments.[9] However, we can discern from the TF-IDF results possibly useful condensation of "Ancient Artifacts" content. The TF-IDF results for Segment 1 (S1) capture the initial story posted by Sam and surface key terms of the discussion, which include:

[(0.228, 'possible'), (0.223, 'object'), (0.212, 'stone'), (0.212, 'rosetta'), (0.205, 'like'), (0.159, 'museum'), (0.152, 'egypt'), (0.141, 'unlocking'), (0.141, 'quickly'), (0.141, 'promote')]

From this sampling of the top ten weighted terms, we can infer traces of general and specific topicality. "Museum" is one of the guiding terms of Sam's argumentative frame. Meanwhile, "rosetta," "stone," and "egypt" suggest a concretizing of the general concern with an example of the controversy (in this case, Egypt's attempts to legally petition for the return of ancient Egyptian artifacts). The top terms for S2 signal a subtopical shift, signaling a focus on NAGPRA—the Native American Graves Protection and Repatriation Act, which governs the restoration of human remains to first peoples of the United States when lineage is proven between the remains and claimants (National Park Service 2019, 2020):

[(0.252, 'object'), (0.233, 'artefact'), (0.211, 'protection'), (0.211, 'nagpra'), (0.193, 'condition'), (0.179, 'native'), (0.174, 'american'), (0.159, 'cultural'), (0.130, 'wooden'), (0.130, 'temptation')]

The TF-IDF for S7 also demonstrates a clear topical shift from NAGPRA to Egypt's history experiences with tomb raiding (especially by British archaeologists) with the terms "tomb," "egypt," and "england":

[(0.379, 'tomb'), (0.196, 'carter'), (0.193, 'egypt'), (0.168, 'send'), (0.15944571371780547, 'decision'), (0.152, 'artifact'), (0.140, 'greece'), (0.12356489143348325, 'example'), (0.123, 'england')]

As one might expect from a discussion topic motivated by the dispossession and repatriation of cultural treasures, numerous geographic and institutional entities are cited. Utilizing spaCy's named entity recognition classifier, the reprogrammed Faciloscope can aggregate individual entities into the following generalized entity tags, reproduced below from spaCy documentation ("Annotation Specifications SpaCy API Documentation" n.d.) for exactness:

PERSON People, including fictional
NORP Nationalities or religious or political groups
FAC Buildings, airports, highways, bridges, etc.
ORG Companies, agencies, institutions, etc.
GPE Countries, cities, states
LOC Non-GPE locations, mountain ranges, bodies of water
PRODUCT Objects, vehicles, foods, etc. (not services)
EVENT Named hurricanes, battles, wars, sports events, etc.
WORK_OF_ART Titles of books, songs, etc.
LAW Named documents made into laws
LANGUAGE Any named language
DATE Absolute or relative dates or periods
TIME Times of less than a day
PERCENT Percentage, including "%"
MONEY Monetary values, including unit
QUANTITY Measurements, as of weight or distance

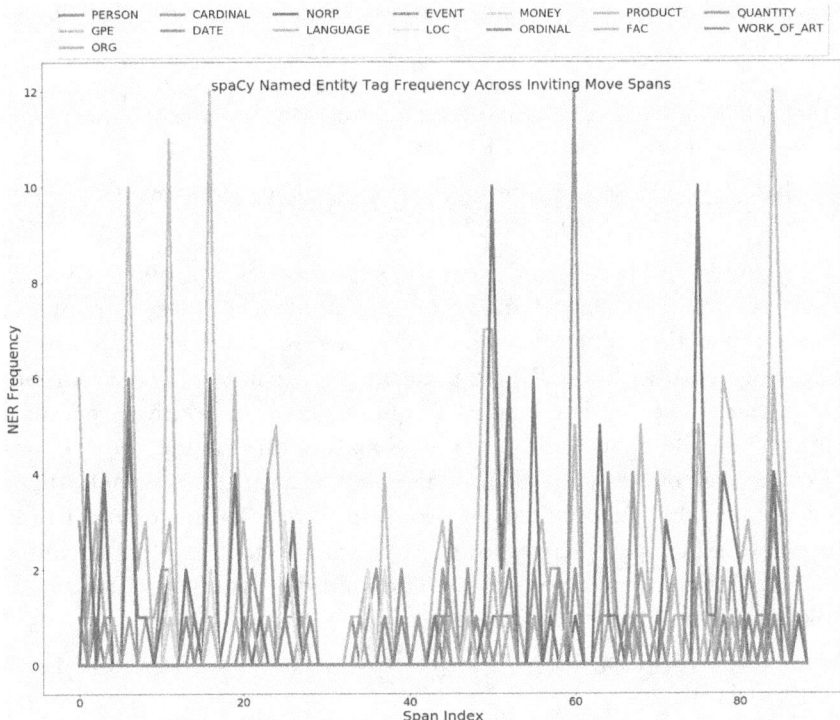

Figure 5.2. spaCy Named Entity Classification Tag Frequency on Ancient Artifacts Inviting Move Spans.

ORDINAL "First," "second," etc.
CARDINAL Numerals that do not fall under another type

By abstracting from the term level to a class level, we gain a more synoptic view of the Inviting move spans or the "comment" component of Van Kuppevelt's topic/comment schema. Figure 5.2 charts the frequency counts of each entity tag across the eighty-nine Inviting move spans.

The GPE tag, referring to countries, cities, and states is the most regular and most frequent across the spans, as we would expect given the subject of "Ancient Artifacts." Moreover, the NORP tag, referring to "nationalities or religious or political groups" positively parallels GPE. Span 3 contains an equal count of GPE and NORP (4). This is reflected in Span 3's reference to "the Haida of British Columbia, Canada" and "a group of indigenous people of New Zealand, the Maori." Span 16 demonstrates a similar pattern, featuring twelve GPE and six NORP tags; and like Span 3, Span 6 features collocations between places and their peoples:

> Japan stole enormous amount of Korean history while they controlled Korea.
>
> Now Japan is restricting Korean tourists to get even closer to Temples fearing some Korean tourists will still the artifacts which originally belong to Korea and made by Korean ancestors.
>
> When you Go to ROM in Toronto, The only genuine pieces made by real Canadian are First Nation section.

The status of Spans 3 and 16 as the extension of topicality into commentary and their shared frame of argumentation (place + peoples) demonstrate the value such charting can offer for tracking episodic shifts—especially when NER tag patterns diverge as they do in Span 60. Here, the DATE tag is the most frequent (12) and positively parallels GPE (5). While Span 60 is generally debating on whether the removal of cultural artifacts by forces endogamous to that culture is theft, there is a notable shift in the form of proof displayed. Rather than a tropic cupola of place/people, we can discern a frame that involves narrative and explicit markers of periodicity. The following is a list of sentences in Span 60 that evidence this chronological emphasis:

> I have an article that was purchased in 1200 AD by my ancestor, from the original maker.
>
> It was used in the family for 500 years and shelved around 1720 right here in the good old USA.
>
> Along about 1890, my Great Grandfather sold it to a collector where it stayed until 1920. It was sold back to my grandfather and brought back to the old farm house where it sits today.
>
> In 1978, some stupid lawyer tried to claim it had been stolen from a museum, and had it not been for an identifying mark that had luckily been placed on the item more than three centuries earlier we would not have it today.
>
> This was started back in 1983 so all artifacts taken before then should be left where they are but the ones that were taken after that should still be left in the us and the british museums.

Another topical shift occurs in Span 75, where NORP (10) exceeds GPE (5) (although this shift is consistent with other instances in which NORP positively parallels GPE). Again, the general discussion about the rightfulness of claims made on ancient artifacts by contemporary museums distanced from the inheritors of that cultural legacy continues; here, focusing on the case of the Elgin Marbles. Span 75 deliberates on the "Greek" response to the acquisition of the Elgin Marbles by the British Museum:

Greeks lost the marbles when the country was under the Ottoman empire. Elgin stole them and I have to visit the British museum to see them.

Here, the places/people fusion discernible in Spans 3 and 16 has been inverted to emphasize people/places.

CONCLUSION

This chapter has described an effort to reprogram the Faciloscope so that its natural language processing aligns with its theoretical roots in RGT and move analysis. I offer two methodological takeaways for this work.

First, the Faciloscope build featured in this chapter suggests a novel extension of RGT methodologies within natural language-processing frameworks as well as the potential for established rhetorical theories to innovate natural language-processing methods.

Second, as a story of software development, this chapter documents the messy choices and activities that computational rhetoric methodologies entail, but which are often occluded by a polished user interface or sanitized in the crafting of "methods" sections in academic articles as the work of Bruno Latour and Steve Woolgar (1986) in *Laboratory Life* demonstrate.[10] By describing the backstage work of reprogramming the Faciloscope, I aim to guide and gird researchers interested in computational rhetoric methodologies for the journeys ahead. From a more polemical perspective, I also offer this story of development to disrupt the pretenses of positivism that often lacquer statistical analyses in social science and the digital humanities (see Law 2004, 1–16) and focus attention on the very human lineaments that join theoretical and technological components and the propositional nature of computational rhetoric research despite its use of what Anna Lowenhaupt Tsing (2012) calls "precision nested scales." Behind move and word counts and beyond the borders of genre episodes are chains of hermeneutic components derived from specific historical contexts, which shape future interpretive horizons. Unmasking these hidden components can enable more critical interventions into the work of computational rhetoric, digital humanities, or cultural analytics. Consequently, this chapter shares in the ethos of "hacking" discussed in Cana Uluak Itchuaqiyaq's citation mapping work on multiply marginalized and underrepresented scholars in the field of technical communication (this collection). For Itchuaqiyaq, "hacking" research machines can help reveal voices that have been decentered and obscured by these machines, reminding us that the work of critically reprogramming software can stress existing systems that trade on "friction-free" operations (Gilliard 2018) and invisibility. In the case of the Faciloscope, hacking into development narratives exposes it more fully for critical interrogations.

NOTES

1. The Faciloscope's development was funded by Institute of Museum and Library Services grant LG-25-10-0034-10, with Jeff Grabill as principal investigator.
2. Informal learning differs from formal, classroom-based learning in that relationships and interactions among experts and learner are more flexible, ad-hoc, and situational, and usually lacks traditional pedagogical deliverables such as lectures and graded assignments. For more on the informal learning research that has informed the Faciloscope, see Boud and Middleton (2003, 199); Ellinger and Cseh (2007, 443); Eraut (2004, 247); Falk and Dierking (2002); Germain and Grenier (2015); and Gutwill and Allen (2010).
3. Though less significant in terms of classifier results, the version of the Faciloscope featured in this chapter utilizes a convolutional neural network model provided by the natural language processing suite (see "TextCategorizer SpaCy API Documentation" n.d. for classifier and evaluation code used). For the spaCy natural language processing library, see Honnibal and Johnson (2015) and Montani et al. (2020).
4. Sentences that contain artifacts of the Science Buzz discussion forum infrastructure such as time and date of posts and the name of posters were cleansed of that information in order to reduce the chance of overcounting PERSON and DATE entities when conducting named entity classification.
5. Morris and Hirst (1991, 24) similarly describe an "intentional structure" of communication. For Morris and Hirst, discourse is goal-directed. Consequently, discourse proceeds through the success or failure of discursive segments to complete stated goals.
6. Text-processing steps included the use of the Natural Language Toolkit (Bird, Loper, and Klein 2009). For text preprocessing/normalization practices, see Beveridge (2015).
7. See Lavin's (2019) "Analyzing Documents with TF-IDF" for a comparable tutorial on using TFIDF vectorization for textual data exploration.
8. Scikit-Learn's TF-IDF vectorizer was used in this step (see "Sklearn.Feature_extraction.Text.TfidfVectorizer—Scikit-Learn 0.23.2 Documentation." n.d.)
9. The full term list of Inviting move spans for Ancient Artifacts can be accessed on the book's companion website at https://upcolorado.com/component/k2/item/6219-reprogrammable-rhetoric-supplemental-design-materials-and-programming-scripts.
10. See also John Law's (2004) *After Method: Mess in Social Science Research.*

REFERENCES

"Annotation Specifications SpaCy API Documentation." n.d. Annotation Specifications. Accessed August 5, 2020. https://spacy.io/api/annotation.

Baumer, Eric P. S., David Mimno, Shion Guha, Emily Quan, and Geri K. Gay. 2017. "Comparing Grounded Theory and Topic Modeling: Extreme Divergence or Unlikely Xonvergence?" *Journal of the Association for Information Science and Technology* 68 (6): 1397–410. doi:10.1002/asi.23786.

Bawarshi, Anis S., and Mary Jo Reiff. 2010. *Genre: An Introduction to History, Theory, Research, and Pedagogy.* West Lafayette, IN: Parlor Press.

Beekissed. 2014. "Simulated Natural Nest Incubation~Experiment #1 So it begins. . . ." BackYardChickens.com. February 11, 2014. http://www.backyardchickens.com/t/854946/simulated-natural-nest-incubation-experiment-1-so-it-begins.

Beveridge, Aaron. 2015. "Looking in the Dustbin: Data Janitorial Work, Statistical Reasoning, and Information Rhetorics." *Computers and Composition Online*. http://cconlinejournal.org/fall15/beveridge/.

Biber, Douglas, Ulla Connor, and Thomas A. Upton. 2007. *Discourse on the Move: Using Corpus Analysis to Describe Discourse Structure*. Amsterdam: John Benjamins Publishing Company. https://doi.org/10.1075/scl.28.

Bird, Steven, Edward Loper, and Ewan Klein. 2009. *Natural Language Processing with Python*. Sebastopol, CA: O'Reilly Media Inc.

Blei, David M., Andrew Y. Ng, and Michael I. Jordan. 2003. "Latent Dirichlet Allocation." *The Journal of Machine Learning Research* 3: 993–1022. https://www.jmlr.org/papers/volume3/blei03a/blei03a.pdf.

Boettger, Ryan K., and Laura A. Palmer. 2010. "Quantitative Content Analysis: Its Use in Technical Communication." *IEEE Transactions on Professional Communication* 53 (4): 346–57. doi:10.1109/TPC.2010.2077450.

Boud, David, and Heather Middleton. 2003. "Learning from Others at Work: Communities of Practice and Informal Learning." *Journal of Workplace Learning* 15 (5): 194–202. doi:10.1108/13665620310483895.

Brown, Gilian, and George Yule. 1983. *Discourse Analysis*. Cambridge: Cambridge University Press.

Chen, Nan-Chen, Margaret Drouhard, Rafal Kocielnik, Jina Suh, and Cecilia R. Aragon. 2018. "Using Machine Learning to Support Qualitative Coding in Social Science: Shifting the Focus to Ambiguity." *ACM Transactions on Interactive Intelligent Systems* 8 (2): 1–20. doi:10.1145/3185515.

coops gone crazy. 2013. "Hatching with 2 Broodies." BackYardChickens.com. October 4, 2013. http://www.backyardchickens.com/t/828899/hatching-with-2-broodies.

Crowston, Kevin, Eileen E. Allen, and Robert Heckman. 2012. "Using Natural Language Processing Technology for Qualitative Data Analysis." *International Journal of Social Research Methodology* 15 (6): 523–43. doi:10.1080/13645579.2011.625764.

"Detecting Insults in Social Commentary." n.d. Kaggle. Accessed August 5, 2020. https://kaggle.com/c/detecting-insults-in-social-commentary.

Devitt, Amy J. 2004. *Writing Genres*. Carbondale: Southern Illinois University Press.

Dietterich, Tom. 1995. "Overfitting and Undercomputing in Machine Learning." *ACM Computing Surveys* 27 (3): 326–27. doi:10.1145/212094.212114.

Domingos, Pedro. 2012. "A Few Useful Things to Know About Machine Learning." *Communications of the ACM* 55 (10): 78–87. doi:10.1145/2347736.2347755.

Ellinger, Andrea D., and Maria Cseh. 2007. "Contextual Factors Influencing the Facilitation of Others' Learning through Everyday Work Experiences." *Journal of Workplace Learning* 19 (7): 435–52. doi:10.1108/13665620710819384.

Eraut, Michael. 2004. "Informal Learning in the Workplace." *Studies in Continuing Education* 26 (2): 247–73. doi:10.1080/158037042000225245.

Falk, John H., and Lynn D. Dierking. 2002. *Lessons Without Limit: How Free-Choice Learning Is Transforming Education*. Walnut Creek, CA: AltaMira.

Freedman, Aviva, and Peter Medway. 1994. "Locating Genre Studies: Antecedents and Prospects." In *Genre and the New Rhetoric*, edited by Avia Freedman and Peter Medway. 1–20. London: Taylor and Francis.

Frentz, Thomas S., and Thomas B. Farrell. 1976. "Language-Action: A Paradigm for Communication." *Quarterly Journal of Speech* 62 (4): 333–49. doi:10.1080/00335637609383348.

Geisler, Cheryl. 2018. "Coding for Language Complexity: The Interplay among Methodological Commitments, Tools, and Workflow in Writing Research." *Written Communication* 35 (2): 215–49. doi:10.1177/0741088317748590.

Germain, Marie-Line, and Robin S. Grenier. 2015. "Facilitating Workplace Learning and Change." *Journal of Workplace Learning* 27 (5): 366–86. doi:10.1108/JWL-03-2013-0017.

Gilliard, Chris. 2018. "Friction-Free Racism: Surveillance Capitalism Turns a Profit by Making People More Comfortable with Discrimination" *Real Life*, October 15, 2018. https://reallifemag.com/friction-free-racism/.

Grabill, Jeffrey T., and Stacey Pigg. 2012. "Messy Rhetoric: Identity Performance as Rhetorical Agency in Online Public Forums." *Rhetoric Society Quarterly* 42 (2): 99–119. doi:10.1080/02773945.2012.660369.

Guetterman, Timothy C., Tammy Chang, Melissa DeJonckheere, Tanmay Basu, Elizabeth Scruggs, and V. G. Vinod Vydiswaran. 2018. "Augmenting Qualitative Text Analysis with Natural Language Processing: Methodological Study." *Journal of Medical Internet Research* 20 (6): e231. doi:10.2196/jmir.9702.

Gutwill, Joshua P., and Sue Allen. 2010. "Facilitating Family Group Inquiry at Science Museum Exhibits." *Science Education* 94 (4): 710–42. doi:10.1002/sce.20387.

Halliday, M. A. K., and Ruqaiya Hasan. 1976. *Cohesion in English*. 1st ed. London: Routledge. https://doi.org/10.4324/9781315836010.

Hart-Davidson, William. 2005. "Shaping Texts That Transform: Toward a Rhetoric of Objects, Relationships, and Views." In *Technical Communication and the World Wide Web*, edited by Carol Lipson and Michael Day, 27–42. Mahwah, NJ: Lawrence Erlbaum Associates.

Hart-Davidson, William. 2015. "Genres Are Enacted by Writers and Readers." In *Naming What We Know: Threshold Concepts of Writing Studies*, edited by Linda Adler-Kassner and Elizabeth Wardle, 39–40. Logan: Utah State University Press.

Hearst, Marti A. 1994. "Multi-Paragraph Segmentation of Expository Text." In *Proceedings of the 32nd Annual Meeting on Association for Computational Linguistics*, 9–16. Stroudsburg, PA: Association for Computational Linguistics. https://doi.org/10.3115/981732.981734.

Hearst, Marti A. 1997. "TextTiling: Segmenting Text into Multi-Paragraph Subtopic Passages." *Computational Linguistics* 23 (1): 33–64.

Honnibal, Matthew, and Mark Johnson. 2015. "An Improved Non-Monotonic Transition System for Dependency Parsing." In *Proceedings of the 2015 Conference on Empirical Methods in Natural Language Processing*, 1373–78. Lisbon, Portugal: Association for Computational Linguistics. doi:10.18653/v1/D15-1162.

Jamieson, Kathleen Hall, and Karlyn Kohrs Campbell. 1982. "Rhetorical Hybrids: Fusions of Generic Elements." *Quarterly Journal of Speech* 68 (2): 146–57. doi:10.1080/00335638209983600.

Jockers, Matthew L. 2013. *Macroanalysis: Digital Methods and Literary History*. Urbana: University of Illinois Press.

Latour, Bruno, and Steve Woolgar. 1986. *Laboratory Life: The Construction of Scientific Facts*. Princeton, NJ: Princeton University Press. doi:10.2307/j.ctt32bbxc.

Lauer, Claire, Eva Brumberger, and Aaron Beveridge. 2018. "Hand Collecting and Coding Versus Data-Driven Methods in Technical and Professional Communication Research." *IEEE Transactions on Professional Communication* 61 (4): 389–408. doi:10.1109/TPC.2018.2870632.

Lavin, Matthew. 2019. "Analyzing Documents with TF-IDF | Programming Historian." *The Programming Historian*, May 13, 2019. https://programminghistorian.org/en/lessons/analyzing-documents-with-tfidf.

Law, John. 2004. *After Method: Mess in Social Science Research*. London: Routledge.

Leetaru, Kalev. 2012. *Data Mining Methods for the Content Analyst: An Introduction to the Computational Analysis of Content*. London: Routledge.

Li, Xin, and Dan Roth. 2002. "Learning Question Classifiers." In *COLING 2002: The 19th International Conference on Computational Linguistics*. https://www.aclweb.org/anthology/events/coling-2002/.

Long, Hoyt, and Richard Jean So. 2016. "Literary Pattern Recognition: Modernism Between Close Reading and Machine Learning." *Critical Inquiry* 42 (2): 235–67. doi:10.1086/684353.

Manning, Christopher, Prabhakar Raghavan, and Hinrich Schütze. 2008. *Introduction to Information Retrieval*. Cambridge: Cambridge University Press. https://doi.org/10.1017/CBO9780511809071.

McAuley, Julian, and Jure Leskovec. 2013. "Hidden Factors and Hidden Topics: Understanding Rating Dimensions with Review Text." In *Proceedings of the 7th ACM Conference on Recommender Systems*, 165–72. doi:10.1145/2507157.2507163.

McAuley, Julian, Christopher Targett, Qinfeng Shi, and Anton van den Hengel. 2015. "Image-Based Recommendations on Styles and Substitutes." In *Proceedings of the 38th International ACM SIGIR Conference on Research and Development in Information Retrieval*, 43–52. doi: 10.1145/2766462.2767755.

Miller, Carolyn R. 1984. "Genre as Social Action." *Quarterly Journal of Speech* 70 (2): 151–67. doi:10.1080/00335638409383686.

Montani, Innes, Matthew Honnibal, Matthew Honnibal, Sofie Van Landeghem, Henning Peters, adrianeboyd, Maxim Samsonov, et al. 2020. explosion/spaCy v2.2.4.dev0 (v2.2.4.dev0). Zenodo. March 9, 2020. doi:10.5281/zenodo.3701227.

Morris, Jane, and Graeme Hirst. 1991. "Lexical Cohesion Computed by Thesaural Relations as an Indicator of the Structure of Text." *Computational Linguistics* 17 (1): 21–48.

Muller, Michael, Shion Guha, Eric P. S. Baumer, David Mimno, and N. Sadat Shami. 2016. "Machine Learning and Grounded Theory Method: Convergence, Divergence, and Combination." In *Proceedings of the 19th International Conference on Supporting Group Work*, 3–8. New York: Association for Computing Machinery. doi:10.1145/2957276.2957280.

National Park Service, Department of the Interiors. 2019. "National NAGPRA." https://www.nps.gov/subjects/nagpra/index.htm.

National Park Service, Department of the Interiors. 2020. "Frequently Asked Questions." https://www.nps.gov/subjects/nagpra/frequently-asked-questions.htm.

Omizo, Ryan, William Hart-Davidson, Minh-Tam Nguyen, Ian Clark, Kristi McDuffie, and Jim Ridolfo. 2016. "You Can Read the Comments Section Again: The Faciloscope App and Automated Rhetorical Analysis." *DH Commons Journal*. https://web.archive.org/web/20161110155600/http://dhcommons.org/journal/2016/you-can-read-comments-section-again-faciloscope-app-and-automated-rhetorical-analysis.

Paré, Anthony. 2014. "Rhetorical Genre Theory and Academic Literacy." *Journal of Academic Language and Learning* 8 (1): A83–A94. https://journal.aall.org.au/index.php/jall/article/view/313.

Pedregosa, Fabian, Gaël Varoquaux, Alexandre Gramfort, Vincent Michel, Bertrand Thirion, Olivier Grisel, Mathieu Blondel, et al. 2011. "Scikit-Learn: Machine Learning in Python." *Journal of Machine Learning Research* 12: 2825–30. https://www.jmlr.org/papers/volume12/pedregosa11a/pedregosa11a.pdf.

Pigg, Stacey, William Hart-Davidson, Jeff Grabill, and Kirsten Ellenbogen. 2016. "Why People Care About Chickens and Other Lessons About Rhetoric, Public Science, and Informal Learning Environments." In *Reconceptualizing STEM Education: The Central Role of Practices*, edited by Richard A. Duschl and Amber S. Bismark, 253–70. New York: Routledge.

Piper, Andrew. 2018. *Enumerations: Data and Literary Study*. Chicago: University of Chicago Press.

Ramsay, Stephen. 2011. *Reading Machines: Toward an Algorithmic Criticism*. Urbana: University of Illinois Press.

Russell, David R. 1997. "Rethinking Genre in School and Society: An Activity Theory Analysis." *Written Communication* 14 (4): 504–54. doi:10.1177/0741088397014004004.

Sackey, Donnie Johnson, Minh-Tam Nguyen, and Jeffery T. Grabill. 2015. "Constructing Learning Spaces: What We Can Learn from Studies of Informal Learning Online." *Computers and Composition* 35: 112–24. doi:10.1016/j.compcom.2015.01.004.

Schryer, Catherine F. 1993. "Records as Genre." *Written Communication* 10 (2): 200–34. doi:10.1177/0741088393010002003.

Sam. 2005. "Should Ancient Artifacts Return Home?" Science Buzz. https://www.sciencebuzz.org/blog/should-ancient-artifacts-return-home.

Sinykin, Daniel, Richard Jean So, and Jessica Young. 2019. "Economics, Race, and the Postwar US Novel: A Quantitative Literary History." *American Literary History* 31 (4): 775–804. doi:10.1093/alh/ajz042.

"Sklearn.Feature_extraction.Text.TfidfVectorizer—Scikit-Learn 0.23.2 Documentation." n.d. Accessed August 5, 2020. https://scikit-learn.org/stable/modules/generated/sklearn.feature_extraction.text.TfidfVectorizer.html.

Spinuzzi, Clay. 2003. *Tracing Genres through Organizations*. Cambridge, MA: MIT Press.

Swales, John. 1990. *Genre Analysis: English in Academic and Research Settings*. Cambridge: Cambridge University Press.

"TextCategorizer SpaCy API Documentation." n.d. TextCategorizer. Accessed August 4, 2020. https://spacy.io/api/textcategorizer.

Tsing, Anna Lowenhaupt. 2012. "On Nonscalability: The Living World Is Not Amenable to Precision-Nested Scales." *Common Knowledge* 18 (3): 505–24. doi:10.1215/0961754X-1630424.

Upton, Thomas A., and Mary Ann Cohen. 2009. "An Approach to Corpus-Based Discourse Analysis: The Move Analysis as Example." *Discourse Studies* 11 (5): 585–605. doi:10.1177/1461445609341006.

Van Kuppevelt, Jan. 1995. "Discourse Structure, Topicality and Questioning." *Journal of Linguistics* 31 (1): 109–47.

Van Kuppevelt, Jan. 1996. "Inferring from Topics." *Linguistics and Philosophy* 19 (4): 393–443. doi:10.1007/BF00630897.

Yan, Jasy Liew Suet, Nancy McCracken, and Kevin Crowston. 2014. "Semi-Automatic Content Analysis of Qualitative Data." In *iConference 2014 Proceedings*, 1128–32. doi:10.9776/14399.

Yan, Jasy Liew Suet, Nancy McCracken, Shichun Zhou, and Kevin Crowston. 2014. "Optimizing Features in Active Machine Learning for Complex Qualitative Content Analysis." In *Proceedings of the ACL 2014 Workshop on Language Technologies and Computational Social Science*, 44–8. Baltimore, MD: Association for Computational Linguistics. doi:10.3115/v1/W14–2513.

Yilmazel, Ozgur, Niranjan Balasubramanian, Sarah C. Harwell, Jennifer Bailey, Anne R. Diekema, and Elizabeth D. Liddy. 2007. "Text Categorization for Aligning Educational Standards." In *2007 40th Annual Hawaii International Conference on System Sciences (HICSS'07)*, 73. IEEE. doi:10.1109/HICSS.2007.517.

6

BIG DATA, TINY COMPUTERS
Making Data-Driven Methods Accessible with a Raspberry Pi

Aaron Beveridge
University of North Carolina at Greensboro

Nicholas Van Horn
Capital University

The term "big data" often implies the use of research practices requiring massive computational resources to process datasets too large for more traditional forms of data analysis (spreadsheet software, manual data coding, close reading, etc.).[1] Advertisements from companies like Microsoft and Amazon reinforce big data's inaccessibility, as these companies no doubt prefer that researchers develop data-driven projects using their expensive Azure or Amazon Web Services (AWS) platforms for cloud computing. While these platforms are feasible for researchers with extensive institutional support—either from their home institutions or from grant funding agencies—for many others, the long-term financial costs and steep learning curves associated with such platforms force researchers to look elsewhere for more accessible and affordable options. Of course, making data-driven methods more accessible extends far beyond consumer choices—the work extends all the way down to the computer hardware and to the data itself. In academic research we frequently ask research questions that already-available datasets and ready-made software tools cannot answer, and the production of data-driven research rarely involves plugging a ready-made dataset into an analytics software that returns a publishable result. Rather, data-driven research often requires that researchers employ an inventive *maker* approach when investigating novel questions with previously unexplored data.

Like many other fields, digital rhetoricians have been exploring the value of big data (data science) methods for our own research practices. While this is due in part to the broad influence of the digital humanities (Jockers 2013; Ridolfo and Hart-Davidson 2015), it also stems from the long-standing interdisciplinary nature of the research questions

https://doi.org/10.7330/9781646422586.c006

that emerge in digital rhetoric (Eyman 2015). Rhetoric and writing studies have a long-standing history of employing a *maker* approach when opening new possibilities for the work we pursue (LeBlanc 1993). More recently, projects like the Writing Studies Tree (Miller et al. 2015), Faciloscope (Omizo et al. 2014), Hedge-O-Matic (Omizo and Hart-Davidson 2016), and MassMine (Van Horn and Beveridge 2016) exemplify the vast potential for the type of work emerging from the intersection of data-driven methods and a maker research orientation.

This chapter extends this maker orientation into the realm of physical computing and demonstrates how an inexpensive Raspberry Pi (Rpi) computer (~$35) can address key hardware and workflow issues for long-term data-collection projects. As the co-creators of MassMine, we have been developing tools that reduce the learning curve for scholars who need to collect novel datasets from digital sources. To date, our work has focused primarily on the accessibility of programming practices (software) associated with digital data collection and webscraping. In his chapter for *Circulation, Writing, and Rhetoric*, Aaron Beveridge (2018) explains how the developer access offered by digital networks such as Twitter provide valuable data for understanding social movement trends, but much work remains to make these sources accessible to non-developers and programmers. More recently, Twitter has shown a willingness to work with researchers in making it easier for scholars to acquire developer credentials, but a high level of programming knowledge remains a prerequisite for accessing that data once credentials are acquired. To this end, MassMine began a second round of development in 2019, funded by the National Endowment for the Humanities, to continue supporting access to Twitter and other networks (Wikipedia, Google, Reddit, YouTube, Tumblr, and more) for nonprogrammers. That work is ongoing, focusing on extensive usability testing and interdisciplinary workflows and will result in a new release of MassMine with a full graphical user interface by 2021.

As counterpart to that work, this chapter focuses on the hardware side of data-driven research. From a hardware perspective, long-term data collection activities—those which are needed to create large ("big") research-quality datasets—usually require a computer or cloud server to run for weeks, months, and potentially years. As computing technology continues to get smaller and more affordable, tiny computers and microprocessors—like the Arduino and Rpi—open many possibilities for maker projects that utilize the modularity, portability, and efficiency of these devices. Tiny computers and microprocessors are standard fare in maker labs on university campuses, but usually these devices are used to teach human-computer interaction (such as taking motion-sensor input to control power output to a small LED) or to program a small servo to move a wheel or lever in robotics projects. For example, Matt Ratto's (2011) "Critical Making: Conceptual and Material Studies in Technology and Social Life" discusses the creation of Arduino objects, which he calls "electronic agents," that communicate using infrared and display results using a range of LED lights and artistic creations made by various participants in the project. For Ratto, these projects focus on a collective pedagogical outcome, providing new inroads into engineering and technological development—encouraging participants to bring intuitive and collaborative energies to confront a "deeper disconnect between conceptual understandings of technological objects and our material experiences with them" (253).

While the pedagogical application of tiny computers in makerspaces will no doubt continue to confront this troubling disconnect—work that we hope continues in critical making—we also see the maker orientation as vital to expanding the research possibilities for rhetorical and cultural scholarship more generally. In this collection, Kellie Gray and Steve Holmes's chapter discusses DocNow (short for "Documenting the Now"), an organization that created an archive and provides access to tools for downloading Twitter datasets related to #blacklivesmatter, #amplifywomen, #actionclimate, and many other social and political movements. Similarly, Cana Uluak Itchuaqiyaq's chapter in this collection offers an illustration for critical text mining as part of a decolonial use of data-driven methodologies. In turn, our chapter is in dialogue with these efforts. As researchers in the humanities and social sciences increase their use of data to study and discuss social media trends, it is imperative that our computational and analytical resources meet this demand. However, doing so will require that we confront the rising cost of cloud computing and the sustainability of data-driven research practices.

For many researchers, the issues that affect the efficiency and sustainability of a research project are often more pragmatic in nature. For example, when collecting Twitter data for our *Digital Humanities Quarterly* article, "Attention Ecology: Trend Circulation and the Virality Threshold" (Van Horn, Beveridge, and Morey 2016), we tracked 17,343 unique trends over seventy-four days—accessing Twitter's Application Programming Interface (API) every five minutes (the maximum allowed by Twitter) to collect and archive data. The software used to access Twitter and manage the data collection was an early version of MassMine. While MassMine ran without any problems throughout the entire project, our team ran into multiple pragmatic issues with our hardware setup. When we first started to collect data, MassMine was running in the background on one of our personal laptops. Because MassMine uses limited system resources, MassMine was able to collect data without negatively affecting the performance of other software on the laptop. However, three glaring issues were revealed a couple days into the initial data collection: (1) The laptop could not be reset or turned off without having to restart the data collection, (2) MassMine needed to access the internet every five minutes to collect the maximum data allowed by Twitter, requiring a constant internet connection, and (3) in order to ensure that the data collection was not accidentally shut down, other people were no longer able to borrow or use the computer. While these issues may seem trivial, on a long enough timeline even the most trivial of issues disrupt the feasibility of a research project. Given our intention to track Twitter trends for multiple months, we decided that a dedicated computer was a more feasible solution.

There are other approaches that we could have used to collect data with MassMine, like using a university research computer cluster (supercomputer)—which MassMine supports. However, this option remains narrowly available to only the most advanced researchers who have trained to use clusters. Therefore, we developed a research workflow to use an Rpi as a data collection server, and we have made a new version of MassMine to specifically support this workflow. This is similar, in many ways, to the pragmatic workflow developed for the "Attention Ecology" article mentioned in the previous paragraph, except that it also takes advantage of the resource and space saving efficiencies of the Rpi. While there are many pedagogical benefits to a *maker* research orientation, we hope this work opens new possibilities for applying data-driven methods to well-established rhetorical and cultural scholarship, as well as encouraging new research by enabling the collection of exigent and novel datasets.

The next section discusses various notions of efficiency and compares an Rpi computer to a mid-level laptop—something similar to what a new graduate student in the humanities or social sciences may afford. While it may seem contradictory to suggest using a tiny computer for "big data" projects, as these devices are often purchased for more traditional maker projects, we demonstrate how these capable little machines can help offset the cost of purchasing a dedicated computer or cloud server for data-driven projects. In addition to the comparison that follows, we also provide an example on how to set up and access an Rpi remotely for long-term data collection projects on this collection's companion website at https://upcolorado.com/component/k2/item/6219-reprogrammable

-rhetoric-supplemental-design-materials-and-programming-scripts. We conclude this chapter by discussing ethical issues and the importance of critical making in technology development. To this end, much work remains, but we hope to inspire others to participate in making data-driven methods—an inherently *interdisciplinary* project—more accessible.

COMPARISON

Efficiency, in particular, can have many meanings for big data: computational efficiency (how many steps does it take for a given program to complete its task?), processing efficiency (how many processors are necessary or available for completing the steps of a program?), temporal efficiency (how long will it take for a program to complete its task?), power efficiency (how much electricity is needed to power the computer processors?), and work efficiency (what are the pragmatic restrictions for completing everything involved in a project?). While these are among the more important efficiency considerations for data analytics work, the problems they pose often result in competing and contradictory answers. For example, programs that use multiple processors simultaneously usually produce the most efficient results when processing a large dataset, but these types of programs can be very difficult to design. A programmer could take two weeks to design an advanced program that uses multiple processors and runs 200 percent faster than a more simplified program that already exists. However, if the program that already exists finishes its task in one week's time, then the two weeks required to develop the advanced program would result in once week of wasted effort. Still, other considerations can change how we judge efficiency for this example. If the newly designed program will be used on a regular basis in the future—saving an accumulated amount of time as the new program is reused indefinitely—then the future time savings justify the additional effort. In other words, there are many competing issues involved in trying to efficiently collect and process data for research.

An Rpi computer may not have anywhere near the computational power of a new laptop or desktop computer—and certainly much less than a big data cloud server or supercomputer cluster—but there are other pragmatic considerations beyond computational power. Usually, computational power is only an issue during analysis, when researchers are trying to make sense of a large dataset, but much less of an issue for the data collection aspect of a project—where bandwidth and data storage are among the top considerations. However, for many researchers the problems are even more pragmatic in nature: *Can I dedicate my laptop* or desktop to a long-term data collection project when I also use this computer for writing, emailing, and teaching? What are the costs of purchasing and running a second computer for research? When thinking about the accessibility and efficiency of data-driven research, these issues can be just as limiting as the problems presented by software usability, the difficulties inherent in learning advanced programming methods, and the many issues wrapped up in computational considerations.

While we have already addressed the costs associated with purchasing an Rpi, which are minimal compared to a full laptop or desktop, this section compares the cost (electricity usage) of running an Rpi to that of a mid-range laptop. Simulations were conducted on two separate hardware setups: (1) a recent release of the Rpi and (2) a Lenovo ThinkPad laptop, representative of a typical middle-tier personal computer. The laptop served as a control for comparison to various measurements on the Rpi. We believe the laptop serves as a reasonable baseline as it is neither geared toward power efficiency (like more spartan options such as a Google Chromebook), nor a particularly high-end workstation typical of those laptops or desktops that are built for sound or video processing. As such, it serves its function as a model of the average personal computer. Rather than purchasing a second computer for a long-term data collection project, this comparison shows the Rpi to be a much more cost-effective solution for augmenting a researcher's data-collection workflow.

The hardware of interest was an Rpi 3 Model B+ featuring 1 GB of LPDDR2 SDRAM and an ARMv8 Quad 4 processor with each of four cores running at 1.4 GHz. Collected data was written to an external Seagate hard disk drive for demonstration purposes, and as such the power consumption of the external drive was not considered in the analysis that follows. Data could have just as easily been written directly to the Rpi's onboard USB drive to mitigate this extra power. For compatibility with the typical Rpi user, we installed and used the recommended Rasbian operating system (April 2018 version; kernel 4.14). At the time of writing this chapter, this reflects the most recent and computationally advanced hardware release, and as a consequence, the most energy-intensive model currently available. As such, the chosen model demonstrates the current "worst-case scenario" for energy consumption for the Rpi series of boards. It is worth noting that prior versions of the Rpi can also run MassMine with little effective change in overall run time.

Our comparison laptop featured relatively average hardware for a personal computer, including an Intel i5-4210U processor (four cores at 2.7 GHz) and 8 GB of DDR3 SDRAM. The operating system used during the simulations was the free and open-source Arch Linux distribution (kernel 4.16.12–1-ARCH). Data were written to the computer's internal hard disk drive.

Procedure

Comparisons were made possible through a number of data simulations. MassMine was installed on both computers, and a script was created that subjected both computers to an identical series of routines. For each device, a small amount of data was collected for each of eleven different MassMine "tasks" (each task in MassMine provides access to data from a different online source, like Twitter or Google Trends). This procedure was repeated two hundred times per task for a total of 2,200 data collection iterations per computer. Results reported below reflect the mean outcome (n=200) for each task on each device. Information on each individual simulation was collected using the GNU time command line program, including (1) elapsed real time in seconds and (2) percentage of CPU usage.

Results

While it is difficult to determine precisely how much power is consumed by the Rpi during any given task, we can make some reasonable approximations based on efforts in the maker community (see, for example, Eames 2018 and Geerling n.d.). First, we know that the Rpi requires a 2.5-amp power supply providing 5.2 volts, although only a fraction of this is used to power the Rpi itself. To simplify our estimates, we once again take the worst-case scenario and assume that running MassMine will max out the CPU (although this was not empirically the case, as shown in figure 5.2). The peak amperage for a single CPU core on the Rpi is approximately 0.69. Taken together, this amounts to a minimum of 3.588 watts of power to run MassMine on the Rpi (watts = voltage × amperage = 5.2 × 0.69 = 3.588). Beyond that, it is possible to run multiple instances of MassMine concurrently, with each executing in parallel on a separate CPU core. Thus, it is useful to consider the peak operating amperage across all four cores of the Rpi's processor—approximately 0.98 amps, or 5.096 watts of maximum power consumption. To provide a suitable survey of the possible costs, we take the average national cost (10.27 cents/kWh), as well as the lowest (Louisiana at 7.46 center/kWh) and highest (Hawaii at 23.87 cents/kWh) reported (https://www.eia.gov/electricity/state/).

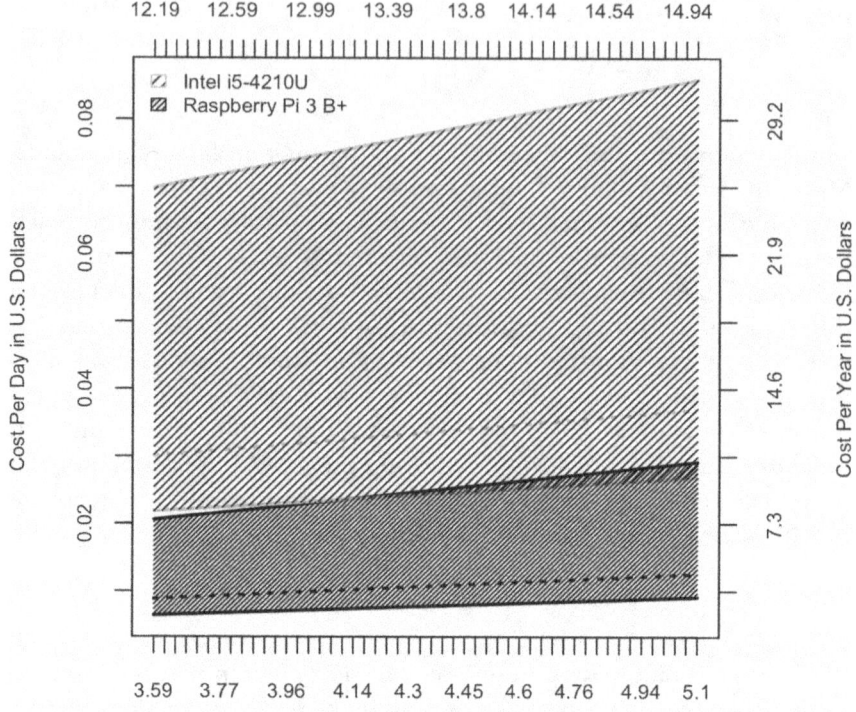

Figure 6.1. Estimated daily and yearly costs associated with nonstop data collection using MassMine on a typical modern laptop computer versus the Raspberry Pi 3 B+. The horizontal axis reflects the range of power consumption associated with lightweight tasks using one processor core (leftmost) to 100 percent of processor cores (rightmost).

Figure 6.1, which depicts cost per day (left vertical axis) and cost per year (right vertical axis) as a function of wattage, reflects estimated operational costs. The abscissa shows the change in wattage from the minimum (that is, our "worst case" minimum of one CPU core at 100%) to maximum (all four cores at 100%) possible while running MassMine. The broken black curve shows the national average cost to run MassMine on the Rpi. The solid black line below reflects the minimum cost (Louisiana), while the upper black line depicts the maximum cost (Hawaii). Thus, the shaded area between these curves reflects the range of possible operating costs for the entire region. The same estimates were generated for the comparison laptop. According

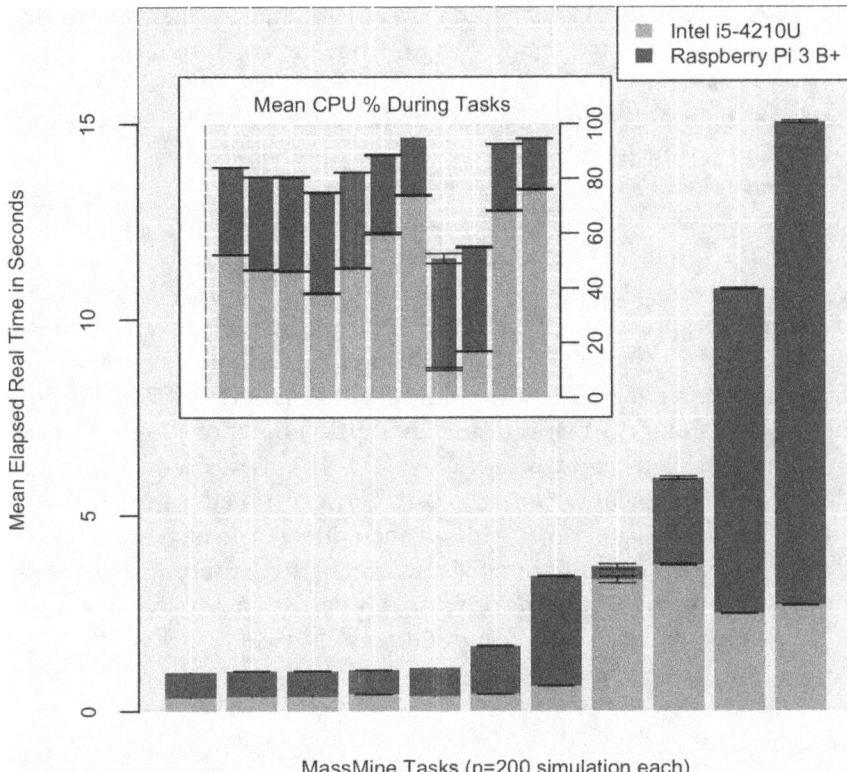

Figure 6.2. Mean (n=200) completion time for eleven MassMine tasks for a typical modern laptop versus a Raspberry Pi 3 B+. Mean CPU usage on each device (inset) is shown for each corresponding task.

to Intel (n.d.) and CPUBoss (n.d.), the Intel Core i5–4210U processor has a thermal design power (peak average power consumption) of 15 watts and typical consumption around 12.19 watts. Combined with our estimated electricity costs, we can generate a similar set of curves for the typical personal computer (figure 6.1, gray curves and lighter shading). Under all comparable conditions, the Rpi has smaller efficiency costs (approximately 30% less). Note that this does not reflect the costs of any attached peripherals. In our setup, the Rpi is not connected to any other power-consuming devices, including a monitor, so the final savings is likely much greater than depicted here.

To assess the role of processing efficiency, we compared both processing time and CPU consumption across devices for a number of data-collection tasks available through the MassMine software. Average completion time and CPU usage are reported for each device (figure

6.2). Error bars, depicting standard error of the mean, demonstrate clear differences between the Rpi and the personal computer across all tasks. However, while there are statistically significant differences between the two devices, the absolute difference, and thus the practical significance, is negligible.

Taken together, these results tell a compelling story: under conditions typical of social media data collection projects, the Rpi is able to handle the same data-acquisition demands as standard personal computers. Further, it is able to accomplish identical tasks with negligible differences in processing time (figure 6.2) while using a fraction of the electrical power (figure 6.1). This efficiency gap can be further widened by setting up a data-collection Rpi in headless mode. In this context, a headless computer is one with no keyboard, mouse, or monitor that is instead controlled remotely over a network. In this way, a dedicated Rpi data-collection machine can be set up and controlled remotely over a shared local Wi-Fi network. Without the added costs of powering connected peripherals—most notably, a monitor—long-running data collections spanning many months or years can persist on a trickle of energy, while freeing up remaining computing resources for other work.

CRITICAL MAKING AND DATA-DRIVEN RESEARCH

As Gray and Holmes's and Itchuaqiyaq's chapters in this section of the collection have made clear, critical making and data-driven research both operate in inherently interdisciplinary and collaborative spaces, they both require that practitioners account for the accessibility and reproducibility of the methods, materials, and/or data utilized in completing a project or in producing results, and they both engage with emerging technologies and toolsets that could have wildly different ethical, social, and material consequences depending on the uses and contexts to which they are applied. For makers who are interested in the critical/data intersection, the consequences stemming from the use of emerging technologies offer contradictory and troubling possibilities—most of which we only learn about by undertaking the work in the first place (in the spirit of critical making's emphasis on making-in-itself as a primary sight of inquiry). For example, on one hand, many researchers who work to archive datasets and protect the historical record of recent social movements—many of which are now digital in nature, in addition to accompanying real-world, in-person protests—also have to account for how they might be contributing to surveillance and disinformation that might equally silence and deform these movements. In part, this step is what Gray and Holmes are trying to do with their text mining project on emoji used by #blacklivesmatter, critical text mining, and a GitHub repository. While many researchers in the critical/data intersection work to anonymize their data, toolsets and workflows also exist to easily de-anonymize those same datasets.

Furthermore, for digital networks like Twitter, which requires that users access Twitter's developer portal each and every time a historical dataset is accessed—rather than allowing for collaborative datasets to be maintained outside of their system—data loss remains a real issue. For example, while DocNow provides many historical datasets that can be accessed for free through Twitter's developer portal, if a user whose tweets are part of those datasets decides to delete their tweets, then that data will not be included in future downloads. For a dataset consisting of millions of tweets, a few individually deleted tweets will likely not affect the reproducibility for future researchers who wish to conduct similar analyses on the same dataset, but the consequences of data loss can vary widely depending on the topic being archived and the amount of data saved. For a forthcoming chapter in *Doing Digital Visual Studies* (Beveridge and Van Horn forthcoming), we produced an analysis of Shepard Fairey's *Obama Hope* image, collecting all tweets related to the artwork from 2008 to 2016. Our analysis looked at the most favorited and retweeted tweets from that time span, and from the time we began conducting the research to acceptance of publication some of the tweets that were among the most favorited or retweeted were deleted by their authors. While we decided to keep those tweets as part of our analysis, other researchers no longer have access to that data.

From a hardware perspective, the low financial cost of an Rpi significantly increases the accessibility of computing projects, but the low financial costs do not come without their own ethical concerns. Low-cost computers and microprocessors like the Rpi rely on rare earth minerals—often known as conflict minerals—to make smaller and faster processing chipsets and components. Rare metals such as gold, tantalum, tin, and tungsten are used in many electronics and microchips, and these resources are regularly acquired from international locations where intense conflict and warring factions use them to finance a wide range of violent and abusive activities. While there are no easy answers for solving the complex supply chain issues that, in part, enable the atrocities related to conflict minerals, it is imperative that we continue to press computing companies to take an active role in addressing these issues. In general, these issues stem from broader technological trends that favor the newest hardware and the fastest processors, and we ought to balance our enthusiasm for innovating with a continued eye toward sustainability and avoiding unnecessary e-waste.

Much of the work involved in the critical/data intersection must focus on ethical issues, but we cannot confront them on purely ethical grounds. In other words, we will likely have to *make* our way out of these dilemmas. In the case of data loss and deleted tweets, the question of *the right to be forgotten* is a real issue without an easy answer. On one hand, users deserve the right to control their data and delete their tweets and accounts should they decide to do so, but on the other hand researchers and journalists have an equal right to analyze public discourse and work to understand public opinion in all its emerging forms—and issues surrounding the anonymization of data are just as contradictory in nature.

Tools like MassMine or maker workflows like the Rpi approach we share in this chapter—that attempt to deal with some of the problems in the critical/data intersection—are developed not only because of how quickly new technologies emerge, but also because of how the ethical issues are constantly evolving. Given the ethical orientation of the humanities and social sciences, this offers immense collaborative opportunities between these disciplines and computer science. As new technologies in cyber security and data management are developed to deal with these and other similar issues, students and scholars who are comfortable with the critical/data intersection should be well positioned to help developers address these important ethical questions. Of course, this will require that we continue to expand our own data-driven research capabilities should we hope to be valued contributors in these endeavors. The more comfortable and capable we are with collecting data and analyzing it, in addition to making our own tools and workflows to do so, the more effective our voices will be in shaping the ways these technologies emerge in the future.

NOTE

1. Both authors contributed equally to the writing and research for this chapter.

REFERENCES

Beveridge, Aaron. 2018. "Circulation Analytics: Software Development and Social Network Data." In *Circulation, Writing, and Rhetoric*, edited by Laurie E. Gries and Collin Gifford Brooke, 243–61. Logan: Utah State University Press.

Beveridge, Aaron, and Nicholas Van Horn. Forthcoming. "Mining Hope: Preserving and Exploring Twitter Data for Digital Visual Studies." In *Doing Digital Visual Studies*, edited by Laurie Gries and Blake Hallinan. N.p.: Computers and Composition Digital Press.

CPUBoss. n.d. "Intel Core i5 4210U." CPUBoss. http://cpuboss.com/cpu/Intel-Core-i5-4210U.
Eames, Alex. 2018. "How Much Power Does Raspberry Pi 3B+ Use? Power Measurements." *RasPi.TV*, March 18, 2018. http://raspi.tv/2018/how-much-power-does-raspberry-pi-3b-use-power-measurements.
Eyman, Douglas. 2015. *Digital Rhetoric: Theory, Method, Practice*. Ann Arbor: University of Michigan Press. doi:10.3998/dh.13030181.0001.001.
Geerling, Jeff. n.d. "Power Consumption Benchmarks." *Raspberry Pi Dramble*. https://www.pidramble.com/wiki/benchmarks/power-consumption.
Intel. n.d. "Intel® Core™ i5–4210U Processor." Intel. Accessed August 1, 2014. https://ark.intel.com/content/www/us/en/ark/products/81016/intel-core-i5-4210u-processor-3m-cache-up-to-2-70-ghz.html.
Jockers, Matthew L. 2013. *Macroanalysis: Digital Methods and Literary History*. Champaign: University of Illinois Press.
LeBlanc, Paul. 1993. *Writing Teachers Writing Software: Creating Our Place in the Electronic Age*. Urbana, IL: National Council of Teachers of English.
Miller, Ben, Amanda Licastro, and Jill Belli. 2015. *Writing Studies Tree*. https://www.writingstudiestree.org.
Omizo, Ryan, Bill Hart-Davidson, Ian Clark, and Minh-Tam Nguyen. 2014. *Faciloscope*. http://faciloscope.cal.msu.edu/facilitation/.
Omizo, Ryan, and Bill Hart-Davidson. 2016. "Hedge-O-Matic." *Enculturation: A Journal of Rhetoric, Writing, and Culture* 22. http://hedgeomatic.cal.msu.edu/hedgeomatic/.
Ratto, Matt. 2011. "Critical Making: Conceptual and Material Studies in Technology and Social Life." *The Information Society* 27 (4): 252–60. doi:10.1080/01972243.2011.583819.
Ridolfo, Jim, and William Hart-Davidson, eds. 2015. *Rhetoric and the Digital Humanities*. Chicago: University of Chicago Press.
Van Horn, Nicholas M., and Aaron Beveridge. 2016. "MassMine: Your Access to Data." *The Journal of Open Source Software* 1 (8): 50. doi:10.21105/joss.00050.
Van Horn, Nicholas M., Aaron Beveridge, and Sean Morey. 2016. "Attention Ecology: Trend Circulation and the Virality Threshold." *Digital Humanities Quarterly* 10 (4). http://www.digitalhumanities.org/dhq/vol/10/4/000271/000271.html.

SECTION 3

Eversion and Critical Making

7
TOUCH-INTERACTIVE RHETORICS
Exploring Our "First Sense" as a Rhetorical Act of Eversion

Matthew Halm
Georgia Institute of Technology

David M. Rieder
North Carolina State University

Touch, perceived and a perception, is always relational, bringing bodies in contact and creating a new space. Touch, at its most productive, depends on more than one body or a sense of "the body" that exceeds singularity.

–Shannon Walters, *Rhetorical Touch*

INTRODUCTION

In today's post-PC era of physical computing, audience interactions with rhetorically engaging, digital projects are no longer trapped behind the looking glass of the screen, nor are they limited to the mouse and keyboard—or, in what can be thought of as an extension of the mouse's capabilities, the relatively limited actions of tapping and swiping on a trackpad or touchscreen. In the new era, due in large part to the wide range of sensors that are available, the virtual[1] and the real are combined. In other words, reality is increasingly a mixed or hybrid experience. To borrow a term from William Gibson (2010), which has been picked up by digital humanists like Steven E. Jones (2013), popular computing today leads to *eversion*.

In an essay in which he helped popularize the term eversion, Gibson (2010) described our early understandings of cyberspace as "a specific elsewhere, one we visited periodically, peering into it from the familiar physical world" (para. 6). It isn't difficult to recognize how this definition of cyberspace, a term he coined decades earlier, is expressive of the personal computing (PC) era during which he wrote novels like *Neuromancer*. Behind a screen, made all the more distal with as limited a set of sensors as are the keyboard and mouse, the computational realm of the computer can seem like an elsewhere. But unlike the cyberspace of the 1980s, Gibson explains that cyberspace today "has everted. Turned itself inside out. Colonized the physical" (2010, para. 6). Sidestepping the question of whether the era of physical computing is a new form of colonization, Gibson's description of eversion reflects the changes that we are experiencing in the post-PC era. In an era during which physical computing technologies are smart, ubiquitous, and pervasive, oftentimes designed to fit seamlessly into our everyday routines, the distinction between our interactions in the "real" world and the computational realm of the computer are increasingly blurred. In our estimate, this blurred or everted reality is what makes the new era of physical computing compelling for digital rhetors. The reason is that eversion is, by definition, a moving experience, and that experience can be designed rhetorically to lead toward a suasive end (Rieder 2017, 5).

When we think of the process of eversion as a folding back upon the real of some of the affordances of the virtual realm, thereby leading to a hybrid and moving experience, sensors play an essential role. A primary inventional step for digital rhetors is to choose one or more sensors capable of detecting the kinds of physical energy with which they want to work. Sensors capture physical energy from the real world and transduce it into usable data. With that transduced data in hand, a digital rhetor can generate multimodal feedback directed at their audience that leads to a stylized experience. Static and animated images, text, and sound can be fed back to the audience in ways that are meant to change the way they think and feel about their interactions with the surfaces, materials, objects, and environment with which they are engaged, thus moving them. Our interests in expanding digital rhetoric and writing studies toward sensory-based approaches to eversion resonate with this general interest in an expanded, procedural approach to writing.

There are a wide range of inexpensive sensors available on the market. US-based companies like SparkFun Electronics and Adafruit Industries sell many different kinds, including some of the following: temperature, humidity, and moisture sensors; light sensors; capacitive-touch sensors; static and dynamic gravity sensors (accelerometers); motion and distance sensors; gyrometers; magnetometers; altimeters; barometers; force, pressure, and flex sensors; and pulse (heart rate) sensors. These sensors can be built into the surfaces and objects with which an audience interacts, integrated into the walls, ceiling, or floors of a physical space, and some can be sewn into clothing.

In this chapter, we focus on one particular kind of sensor technology, capacitive touch. Touch is a ubiquitous and important way in which the post-PC era achieves the goal of blurring the boundaries between the virtual and the real. From the prevalence of touch screens on phones, kiosks at supermarkets and banks, touch-interactive displays in museums, touch-interactive hand and fingerprint sensors, and more experimental applications associated with digital media arts, touch is a basic interface technology. The challenge and opportunity for digital rhetors is to begin from the premise that touch opens up new spaces and experiences for audiences. As Shannon Walters (2014) explains, touch "is always relational, bringing bodies in contact and creating a new space. Touch, at its most productive, depends on more than one body or a sense of 'the body' that exceeds singularity" (4). Touch, in other words, is more than an instrumental, individualist action. Borrowing from Sean Morey and M.Bawar Khan's argument in this collection that AR is a medium, not "simply a technology," touch can be valued as a metamedium, too. It is an event that opens up a body to something beyond it. For digital rhetors, touch and the context of eversion work well together to generate a moving, rhetorically meaningful experience.

Capacitive touch sensors are a relatively easy way to explore the possibilities of touch-interactive interface design. The specific sensor that is used by both of the boards that we reference is the twelve-input, MPR 121 sensor. Although NXP Semiconductor has stopped producing it, the sensor is still widely available and is well supported with open-source project code, open-source libraries, and documentation.

The MPR 121 offers a digital rhetor twelve inputs with which to work, and each of those inputs can be extended with conductive thread, wire, tape, or paint to make almost any surface, material, or object an interactive part of a novel interface. With capacitive touch inputs, a digital rhetor can transform almost any aspect of an audience's "real world" interactions into an interactive interface, leading to a rhetorically-engaging, everted experience.

In what follows, we begin with excerpts from Walters, Matthew Fulkerson, and Erin Manning, to explain how touch ultimately leads to a deeper engagement with our relational bodies. As a way of further exploring this perspective, we also include a review of the way in which German media theory helps frame the cultural value of touch interactivity. After those theoretical sections, we devote the rest of the chapter to applied and pedagogical examples of touch-interactive work. We conclude with a reminder that all bodies are in movement. We may not experience that fact often enough in our routinized and habituated realities, but movement is there. When an audience is caught in the wonder of an everted experience—a relational act—the potential for movement, change, and exploration is there. And in that moment of

Figure 7.1a. Bare Conductive's touch board. MPR 121 sensor is circled.

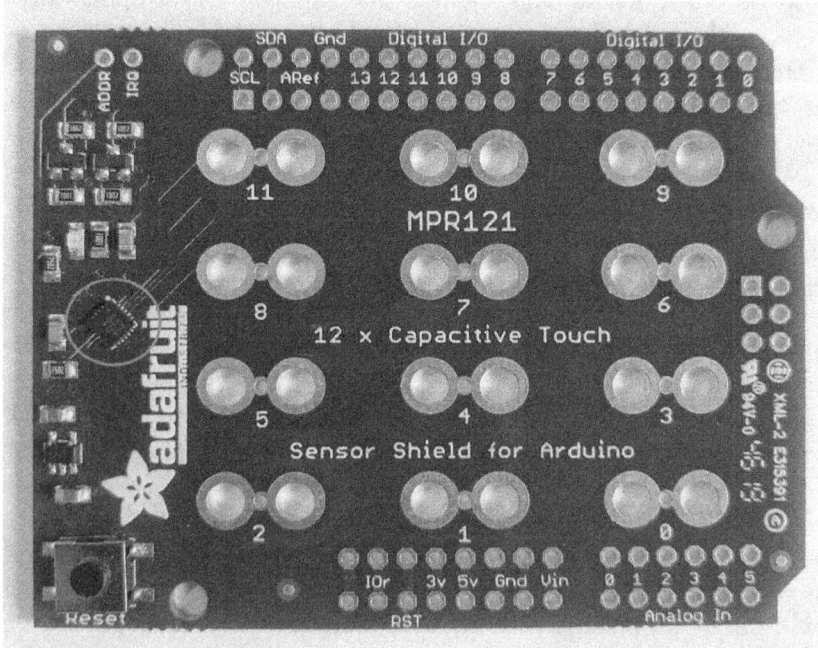

Figure 7.1b. AdaFruit's capacitive touch shield. MPR 121 sensor is circled.

movement, change, and eversion, the digital rhetor is presented with an opportunity to move their audience toward a rhetorically informed end.

TOUCH, MOVEMENT, AND TECHNOLOGY

Of the five senses traditionally demarcated in the West, touch has not been as well studied or explored as sight and hearing have been. Matthew Fulkerson (2014) offers an explanation for why this oversight has persisted. One important reason is that touch cannot be reduced to a single sensory modality. Touch implicates a wide range of systems in our bodies: "It is difficult to say exactly what counts as the organ of touch, or what sensible features, subsystems, actions, and abilities define it as a separate and unique sensory modality" (2–3). Similarly, Walters (2014) explains that, unlike the other four senses, touch is linked to the largest organ in our bodies, which complicates attempts to locate and understand it. There are a wide range of sensations that seem to be ultimately related to touch, but because they are implicated in other systems of sensation, they are not so easily categorized. Can our sensation of touch include the use of objects with which we explore our environment? Fulkerson (2014) provides an example: "Using only a pen or pencil, you write upon two sheets of paper laid out side by side in front of you (assume your eyes are closed). You experience one paper as rough; the other is felt to be very smooth" (4–5). Fulkerson's in-depth study of touch foregrounds the extent to which touch is not just complicated and co-implicated in many of our experiences, but that it is comprehensive. As digital rhetors, the opportunity to work with touch-based interactive technologies is an opportunity to work even more closely (and broadly) with the human experience and sensations of the world.

In addition to these physiological complexities, touch is both rhetorically and culturally situated, which is one of Walters's important contributions to rhetoric theory in *Rhetorical Touch*. In her explorations of Empedocles's and Aristotle's rhetorics, among others, Walters elucidates the extent to which touch is at the center of how rhetorical appeals and persuasion occur. Since touch is such a fundamental aspect of our humanity, and exceeds the experience of any one person, it is a sense that opens up bodies to each other and the rest of the world. Walters writes, "Touch is a sense that transcends bodily boundaries; it demands an approach that also transcends boundaries" (8). In the following sections we explore these dimensions to demonstrate understanding touch as an everted and transcendent experience.

Rhetorics of the Relational Body

The act of touch itself is a complex rhetorical movement. In addition to Fulkerson's and Walters's contributions to the topic, Erin Manning (2007) describes touch as a process through which bodies become something other than what they have been. She rethinks the binary oppositions of mind/body and subject/object by suggesting that the body is generative of its environment. She argues that "the body does not move into space and time, it creates space and time: there is no space and time before movement" (xiii). Touch is a particularly important aspect of this generation—it is a way of relating bodies to each other, and of creating bodies that "will-have been" (xiii). Bodies in contact create each other and their environment by creating a relation between them that has not existed before. Manning writes that touch is not "a stable concept," the effects of which can be understood in advance; instead, "The only thing we can grasp, momentarily, are touch's inventions" (xiv). Touch produces new ways of being. The concept of identity is relational and enabled by touch: "I reach out to touch you in order to invent a relation that will, in turn, invent me" (xv). Under this description of touch, subjectivity is an embodied process of relation that is always changing.

A fundamental consequence of Manning's depiction of the touching body is the mutual construction of subjectivity produced by touch. The body is composed of components of other bodies, and those other bodies are further composed of other bodies, and so on. To articulate this complex interconnectedness and blurring of boundaries between bodies, Manning draws on Deleuze and Guattari's description of the Body without Organs (BwO), which is invoked to demonstrate the insufficiency of describing bodies and the notion of subjectivity in terms of components (organs) possessed by those bodies. To know what a body can be is not as simple as knowing what it contains. Instead of organs, bodies "can be occupied, populated only by intensities" (Deleuze and Guattari 1987, 153). For Manning, this is important because it means that identity is not stable enough to be located in a single body, and by focusing on touch as a movement that reciprocally produces new modes of being, she demonstrates that subjectivity is "a phenomenon of accumulation, coagulation, and sedimentation" of strata of the BwO rather than an individual organism distinct from its surroundings (Deleuze and Guattari 1987, 159; quoted in Manning 2007, 139). Manning argues that "if my body is created through my movement toward you, there is no 'self' to refer back to, only a proliferation of vectors of intensity that emerge through contact" (136–37). Ultimately, "There is no identity to the BwO. There is only movement" (137). What we observe as identity is a relational movement between bodies. Manning's arguments in favor of an expanded and relational view of embodiment and subjectivity augment Walters's study of rhetorical touch in the context of disabilities studies by demonstrating how bodies engage one another to produce the act of touch.

Walters, Fulkerson, and Manning help explain that a rhetoric of touch describes the ongoing production of subjectivity that results from intensive contact between bodies as a persuasive event. By engaging in physical contact with an entity other than oneself, the act of touch invokes a rhetoricity in the world that depends upon relationality and turns it toward persuasive ends. Echoing some of what Andrew Pilsch argues in his chapter for this collection, touch can be valued as an important actor in a process of invention-from-the-middle, as one of several eloquence adjuncts working in a moving and complex collaboration. This persuasive movement everts the lived environment of an audience and reveals it to be a component of their subjectivity. The posthuman depiction of identity provided by Manning suggests that this eversion is ongoing. Eversion in this case is not a question of opening the body to the world because it is already fundamentally open. Instead, eversion makes connections between already-open bodies in the world sense-able; exploiting eversion in a rhetorical sense is less an active inverting than a process of highlighting latent connections among subjects and their environment to reveal their having been everted. Because touch forms a material connection between bodies, both technological and organic, a rhetoric of touch can reveal the originary relationality and ongoing processes of mediation operating around those bodies.

In the act of touching, bodies are revealed to have been moving in ways that generate certain subject positions. Eversion reveals that these movements and subject positions always exceed a stable notion of an individual self. For rhetoricians, eversion is an opportunity to move an audience, to take feedback from their interactions with an environment and reveal it to have exceeded their expectations of their own subject positions. Everting movements, especially habitual or otherwise unnoticed ones, shows how they are a generative force of an audience's subjectivity.

Touch as Cultural Technique

The concept of cultural techniques, prominent within German media theory, is useful for elaborating on the ways certain kinds of touch are made possible by media and vice versa. The term "highlights the operations or sequences of operations that historically and logically precede the media concepts generated by them" (Siegert 2011, 15; quoted in Geoghegan 2013, 69). Though he did not use the term "cultural technique" directly, this approach is a primary conclusion of Friedrich Kittler's work as exemplified in *Gramophone, Film, Typewriter*, in which he famously (and cryptically) declares that "media determine our situation" (1999, xxxix). Kittler shows how the titular technologies of the gramophone, film, and typewriter changed the ways cultures think: the gramophone, for example, brought with it the notion that sound waves exist in the world in a way that can be recorded or observed separately from organic means of hearing. But in order for the gramophone to be thinkable and therefore able to be invented, the concept of sounds existing in such a recordable way must have also preceded the technology.

Research on cultural techniques demonstrates an "interest in describing and analyzing how signs, instruments, and human practices consolidate into durable symbolic systems capable of articulating distinctions within and between cultures" (Geoghegan 2013, 67). This research "commences with an inchoate mixture of techniques, practices, instruments, and institutional procedures that give rise to a technological set-up" (Geoghegan 2013, 70). In the case of touch, a number of technologies and practices have developed over the past several centuries to give rise to what David Parisi (2018) calls the "haptic subject" (4). Parisi documents a history of touch interfaces, from eighteenth-century electric shock devices to contemporary capacitive touch technologies. The haptic subject, Parisi explains, "served to both mark and steer the drastic changes touch underwent as it became increasingly an 'object-target' of scientific knowledge, engineering and design practice, bureaucratic management, therapeutic discourses, and commercial investment" (4). These processes transformed touch into haptics by "revealing, laying bare, and imposing structure on the messy assemblage of human sensations designated by the term 'touch'" (19). For Parisi, haptics is the routinization and codification of the biological sense of touch into processable data.

New technologies must be culturally situated, and certain kinds of touch are made more or less thinkable by these technologies. Parisi recounts numerous examples of touch-based technologies and the ways they are introduced to the public via advertisements. As Parisi explains in the case of the iPhone, early advertisements for the device used the phrase "touching is believing" and visually recalled Michelangelo's painting of Adam and God reaching toward each other with fingers outstretched.[2] This explicit reference to knowledge and creation "simultaneously attempted a retraining of interfacing habits and a revaluation in touch's position in the cultural hierarchy of the senses" (Parisi 2018, 280). By placing touch in such a privileged position, the iPhone "promise[d] to awaken and recall an alternative epistemic orientation, where manual and digital rather than visual inspection would provide the most reliable pathway to knowledge and belief" (280). But, as Parisi notes, touchscreens such as that of the iPhone do not actually provide any haptic feedback. The device registers a user's touch in various ways depending on the gesture performed, but the finger of the user only ever feels the glass of the touchscreen. Instead of the traditional understanding of haptics used to refer to touch-based feedback created by the device (such as a vibration), Parisi argues that the routinized gestures associated with touchscreen devices themselves create a form of embodied feedback, even if the device alone is not the source of the feedback (283). The relationship between the device and the body, as well as the sociocultural environment within which the device exists, functions to provide a sense of material interaction with the device. When a gesture is performed on a touchscreen, the user's body feels itself performing that gesture differently than it might feel in another context or with another gesture, and that feeling is associated with the particular gesture and with the function it performs in concert with the device. The ways that humans touch influence what technologies we use, and the technologies we use influence the ways we touch things and each other (think of stories of young children attempting to "swipe" non-touch-enabled screens like televisions).

Extending Parisi's insights on technologies of touch, positioning touch as a cultural technique suggests that the embodied experience of a user of a touch interface is the result not only of the physical interaction between the user and the interface but also of the social ecology of technologies and practices within which the user performs various gestures and touches. A certain swipe might gain significance for a user after repeated interactions with a device and as a result of the way they understand that swipe to exist culturally. As these techniques of touch gain cultural currency, we "must ask how, and under what conditions, cultural techniques strategically and temporarily consolidate these forces into coherent technologies" (Geoghegan 2013, 79). There is no predetermined outcome of this consolidation, only a complex array of practices, which constantly increase and decrease in significance. Everting such a cultural technique for rhetorical ends would reveal the extent to which an individual's practices and habits connect them to a wide array of other subjects, human and technological. Related to Morey and Khan's "monumental" app about which they've written for this collection, touch interactivity could add to their MEmorialist approach to digital rhetoric and writing. In particular, it could introduce an explicit engagement with eversion, which could lead to a stronger, punctum-like connection with the events associated with the Deepwater Horizon disaster.

EXPLORING TOUCH WITH THE MPR 121 SENSOR

With the understanding in mind that touch-interactive projects can be developed in order to evert and ultimately move an audience rhetorically, and that this eversion relies on a complex array of cultural and technological relations, we now turn to a series of examples and pedagogical approaches. All of the following rely on the twelve-input MPR 121 sensor mentioned above (figures 7.1a and 7.1b). The digital interactive artwork that is discussed first was developed with Bare Conductive's Touch Board and conductive paint. The first of the two pedagogical approaches to touch was also based on that board. The second pedagogical approach (and final example) was based on AdaFruit's capacitive touch shield for Arduino (Adafruit Industries n.d.).

In 2009, Bare Conductive began distributing their nontoxic, water-soluble conductive paint (figure 7.2). Originally conceived of as a kind of wire for flexible circuit designs, the conductive paint has more commonly been used as a proximity or touch sensor. Several years later, they released an Arduino-compatible Touch Board. These two technologies work well together as a basis for exploring the possibilities of touch-interactive digital rhetorics.

To help explain the serendipitous similarities between a painted wire and a sensor, let's take a closer look at the circuit in figure 7.2. The words written in cursive, "draw a" and "circuit," are essentially two continuous lines—two painted wires—that conduct electricity to and from the negative and positive terminals of the battery (on the right). As the electricity flows from the positive to the negative terminals, it energizes an LED light (on the left). The paint is used in the same way that a length of wire would be used to complete an electrical circuit. But the painted wire isn't insulated, so it's susceptible to energy draws from other conductive bodies. If a user touches a section of the painted wire, they will draw electricity from that circuit because their body is conductive, too. When this happens, the LED light will dim. Not only has their body become part of the circuit, but the circuit has become part of their body, and in doing so everts its nature as a circuit.

Due to the painted wire's susceptibility to energy draws, it can be used as a touch-sensitive technology. When a surface of conductive paint is

Touch-Interactive Rhetorics 155

Figure 7.2. Electrical circuit written with Bare Conductive's electrically conductive paint.

connected to one of the inputs on a capacitive touch sensor, the surface becomes a touch-sensitive interface, and as friends and others experimented with Bare Conductive's paint, they demonstrated to the co-founders its use as a platform for a wide range of innovative, touch- and proximity-based interactive projects. And more: since the paint could be applied to most any surface, objects and surfaces of all kinds, including walls, floors, and even human skin could be transformed into a point of interaction—and this relates to the everting tendency of these technologies. When data can be generated from a conventional, interactive experience, and that data can become the basis for a novel experience, then the paint becomes a platform for innovation—for making everyday surfaces into a moment in which eversion can be generated.

The digital interactive "soundscape" *Contours* demonstrates the range of possibilities for the touch board. In 2013, Bare Conductive collaborated with artist Fabio Antinori and designer Alicja Pytlewska to develop an interactive art piece for the MAK Museum in Vienna, Austria (figure 7.3). Bare describes the project on their blog as follows:

> What we created was an interactive installation which invited visitors to approach and even touch three interactive screenprints. The output of the interaction was a generative soundscape which will continue to evolve as

Figure 7.3. Contours at MAK Museum (2013).

the piece is interacted with. . . . The entire installation is run from a Max/MSP patch . . . [that] uses inputs from the Touch Boards to both trigger and manipulate a series of sounds derived from machinery which is associated with examining the human body such as EKG machines and various scanners. These sounds are combined with synthesizers and manipulated to create a landscape of sound which visitors navigate through and change. Any interaction with the piece is logged and will affect the nature of the sound from that point forward. (Bare Conductive n.d.)

In videos on Vimeo and YouTube, participants walk between the sheets of Tyvek (figure 7.3), moving their hands across the vertically painted columns, which are connected to the inputs on several Touch Boards. The sounds they produce are complex and overlapping. There is a stylized parallel between the undulating movement of the sheets and the undulating waves of sound. The software "remembers" past engagements with the project, so the sounds that are produced are continually changing and evolving from one participant to the next. Borrowing from some of Manning's discussions of touch and wonder, *Contours* can be appreciated as a wondrous project that elicits from its audience's interactions a new experience of and connection with the world. And with the notion of wonder in mind, perhaps it could also be valued as another kind of eloquence adjunct, based on what Pilsch writes in his chapter. Some of the computational "magic" of the virtual realm bleeds through into the "real" world as touch is transformed stylistically into a subtle but powerful activation of movement along the length of fabric and sound. Like the ripples introduced across a watery surface, when a participant touches the fabric, they create a ripple across the fabric and the soundscape.

For digital rhetors, this touch-interactive project and the technologies on which it is based demonstrate both the versatility of the paint, which allows a rhetor to make virtually any surface into a basis for eversion, and how space gives way to its potential for movement. There are a wide range of possibilities for generating a stylistic experience that leads not just to a moment of eversion but especially to one that is then directed toward a rhetorical goal.

PEDAGOGIES OF TOUCH

In addition to the digital interactive work detailed above, the following pedagogical approaches to touch-interactive project design demonstrate the educational potential of highlighting eversion with touch. Compared to works of art, which focus on the user or audience, what stands out in the following two approaches are the ways in which students began to see the world from the standpoint of eversion. In other words, both groups of students began to experience the world around them as something that could be transformed into an interactive and virtualized experience.

Colegio Atalaya

In the fall of 2018, while Rieder was on research leave in Spain, Bare Conductive's Touch Board was introduced to a group of eighteen 5th and 6th grade students in a course on robotics at Colegio Atalaya, a K–12 school in the city of Santander in the northern region of Cantabria. During that fall semester, the students were offered a basic introduction to creativity via touch interactivity. First, the students were introduced to the Touch Board and its capacities. Students were introduced to the idea that computer-based creativity and interactivity had broken free of the "looking glass" of the computer screen, mouse, and keyboard. Examples of touch-based systems in movies were listed, followed by examples from their own lives in Spain and Europe. These ten- and eleven-year-old students understood that computing involved more than a keyboard and mouse, but they could now appreciate how it was connected to a new post-PC paradigm of computing.

Once the class recognized that the Touch Board was part of a new post-PC paradigm of popular computing, students were introduced to

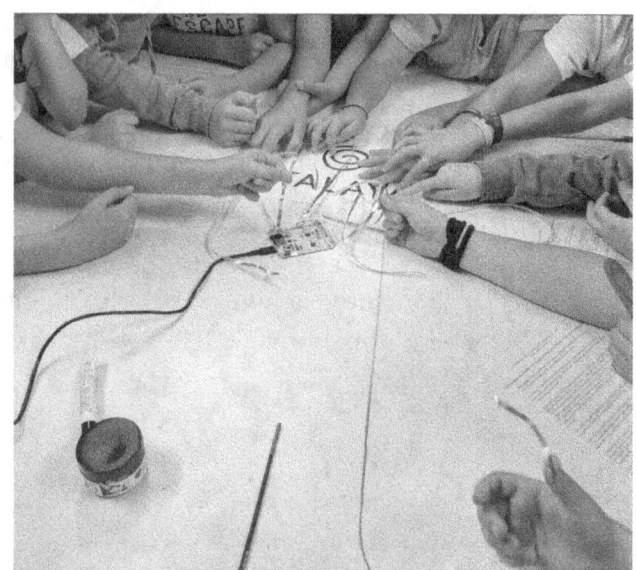

Figure 7.4a. Students at Atalaya playing with a touch interface drawn with conductive paint that is connected to a touch board.

Figure 7.4b. A simple visualization of the changing values from the twelve inputs on the touch board.

the focus of their work that fall: creating original touch-interactive interfaces. Copper tape and conductive paint were provided as the creative materials with which they would be developing their original interfaces. And as a way of appealing to their age, it was explained that they were like magicians (which some students excitedly associated with Harry Potter) when they worked with these technologies. Like a magician, they had the power to transform any surface or object into something that would speak, sing, or play music. Like a magician, they could make more of the world come to life in new ways, surprising their friends and family, making them laugh or dance or stop and listen to an important message. They understood that the surfaces of the tables around which they were seated as well as in the lunchroom, the walls and floors throughout the school, the words in their books, any doorknob, button, or handle, playground equipment, even the leaves and bark on plants and trees outside could be made to talk or sing or play music when touched. Without using the terminology on which this chapter has relied, the students understood quickly and intuitively that their world could be everted; moreover, that eversion was, like magic, a way of being creative with computers today.

A simple visualization of the constantly changing values from the twelve-input sensor was written in the open-source programming language, Processing, so that the students could begin to understand the relationship between their interactions with tape or paint and the way a software application can be written to respond to them (see figure 7.4a). Related to this, a basic introduction to the sine wave of electrical voltage was presented in order to help them see the connection between electricity and the numerical values with which a programmer would work at the level of code.

While the art projects discussed above expose an audience to an experience that leads them to an awareness of their body's relational capacities with the world and with technology by everting their lived experience, pedagogical exploration of physical computing demonstrates this relationality in an applied, and in some ways more interactive, context. By constructing and tinkering with technologies of touch, the underlying mechanisms that enable touch as it has been culturally and technologically constructed are laid bare for experimentation.

Studies in Rhetoric and Digital Media at North Carolina State

For several semesters at NC State, including the spring semester of 2018, Rieder taught an undergraduate special topics course, ENG 395, Studies in Digital Rhetoric. In the spring of 2018, the special topic was "Touch

Interfaces with Arduino and Processing." The technologies assigned in that class were an MPR 121 sensor installed on a "shield" for an Arduino microcontroller (see figure 7.5a), an Arduino starter kit, a spool of copper tape, and a pen of Bare's conductive paint.

The course started with an introduction that was similar to the one offered at Atalaya: The many opportunities for developing digital rhetoric projects with post-PC technologies like touch-interactivity, albeit with more emphasis on the history and technical background of the MPR 121 sensor. After those introductions, students began learning how to write simple, multimodal software programs in the Processing language, which is a simpler dialect of the Java language designed to help teach nonprogrammers how to transition their work to a computational medium. The language was originally designed for visual artists, but it is well-suited for teaching students in both the humanities and communication studies how to think computationally, in order to create original, digital interactive projects.

Very few of the students enrolled in sections of ENG 395 have a technical background. Sometimes two or three engineering majors will enroll, but the majority of the twenty-two-student class are humanities or communications majors. They do not know how to write a software program or how to solder, but they do have a basic computational mindset for the creative work in which they will be engaged, so learning the basics of computer programming and how to design an original touch interface can take up most of the focus of the class. Echoing something that Morey and Khan mention in their chapter, this class is meant to help humanities students see the possibilities for bridging the persistent gap between the 'two cultures' of humanities- and science-based approaches to research and engagement. To this end, the class is ultimately focused on the creation of two touch-based, rhetorically engaging, digital interactive projects for which students must work technically and collaboratively to complete.

To help guide the students toward a rhetorical focus on their technical work, a first step is for students to recognize that their projects are the supporting argument for an implicit or explicit claim. Reminding students through the technical learning process that there needs to be an answer to the "so what?" question is helpful. A second step is to help students understand what kinds of supporting arguments they are designing. One significant, argumentative difference between the supporting arguments they have been accustomed to developing in educational settings and those toward which their projects tend is that there is far more of a focus on emotional appeals. Eversion is an experience that

Figure 7.5a. AdaFruit's capacitive touch shield based on the MPR 121. See also figure 7.1b.

Figure 7.5b. Early draft of a final project in ENG 395, spring 2018, at NC State University.

is not particularly "heady" or logos-oriented in its appeal. Movement, change, and wonder are oftentimes grounded in experiences that are pathetic in their appeal. Especially for touch-interactive projects, the supporting arguments are inextricably tied to the feeling of the surfaces and materials with which audience members will engage physically, and these sensory experiences are oftentimes personal, subjective, and pathetic in orientation.

One of the final projects in the spring 2018 class was meant to generate a sense of identification with NC State's history and "brand." Figure 7.5b is a photograph of an early draft of the project. The final version was a large-scale map of NC State's campus. On the map, the group used conductive paint to make twelve locations on the campus the basis for a multimodal experience. Their claim was conjectural as well as qualitative: They wanted students new to the campus to use the map to discover more about their new campus. When a user touched one of the locations on the map that had been linked to their Processing sketch, videos, music, and text would play. The screen that is in figure 7.5b was eventually supposed to be replaced with a screen large enough (from floor to ceiling) to create a far more immersive experience, and the map would have been on display as an interactive kiosk.

CONCLUSION

The rhetorical force of touch is a fundamental part of a body's capacity to relate to other bodies and to the world. Touching is suasive, and the ways cultures construct practices of touch create an intricate relational network that forms a ground for rhetoric. If, as Fulkerson (2014) argues, touch is our first sense, it is in some ways our first rhetoric. As a way of perceiving the world, touch is also a means for making things matter in the world according to their ability to touch or be touched. As emerging technologies continue to push past the PC era, digital rhetors have new opportunities to engage audiences and evert the digital world in tangible ways. Physical computing offers a tactile mode for digital rhetoric and reveals complexities of the post-PC era.

As Fulkerson and Walters have argued, touch is a complex sense because it engages so many subsystems in both the mind and body. In a post-PC era, this complexity is augmented by the ways in which touch can be associated with an even wider range of materials, surfaces, and space-times. And while Manning has developed compelling and convincing theoretical arguments in favor of a posthuman understanding of touch, digital rhetors can communicate a similar idea at the level of practice. In the moment of eversion and wonder, touch is transformed into a becoming that is all the more complex for the ways in which it opens up a body to others.

Related to this transformative moment is a final point: By demonstrating a body's fundamental openness and capacity for change, eversion offers possibilities for an individual's subjectivity that may cause them to imagine things they had not previously thought possible. Wonder moves bodies—physically (in the sense that they are brought into contact with other bodies, generating new subject positions) and rhetorically (in the sense that they are influenced to act based on that movement). A rhetoric of touch accesses this movement directly. Echoing some of the arguments present in this part of this collection, a rhetoric of touch is one that is best explored inventionally from the middle as well as one that leads to a more choric (and eloquent) engagement with the worlds of which a body is expressive.

NOTES

1. As Alex Reid (2007) explains in *The Two Virtuals*, there is an important distinction between the "familiar" virtual "produced by modern computing" and the "philosophical" virtual of philosophers like Spinoza, Bergson, and Deleuze and Guattari (3). Our use of the term "virtual" remains within the more colloquial understanding, but it is important to note that the "virtual" can have very tangible effects on the "real" world under both uses of the term.
2. This image is also discussed extensively by Manning (2007, 53ff).

REFERENCES

Adafruit Industries. n.d. "Adafruit 12 × Capacitive Touch Shield for Arduino—MPR121." Adafruit Industries. Accessed February 18, 2021. https://www.adafruit.com/product/2024.

Bare Conductive. n.d. "BLOG | MAK Fashion Lab #2: Scientific Skin." Bare Conductive. Accessed February 18, 2019. https://www.bareconductive.com/news/mak-fashion-lab-2-scientific-skin/.

Deleuze, Gilles, and Félix Guattari. 1987. *A Thousand Plateaus: Capitalism and Schizophrenia*. Translated by Brian Massumi. Minneapolis: University of Minnesota Press.

Fulkerson, Matthew. 2014. *The First Sense: A Philosophical Study of Human Touch*. Cambridge, MA: MIT Press.

Geoghegan, Bernard Dionysius. 2013. "After Kittler: On the Cultural Techniques of Recent German Media Theory." *Theory, Culture & Society* 30 (6): 66–82. doi:10.1177/0263276413488962.

Gibson, William. 2010. "Google's Earth." *The New York Times*, August 31, 2010, https://www.nytimes.com/2010/09/01/opinion/01gibson.html.

Jones, Steven E. 2013. *The Emergence of the Digital Humanities*. New York: Routledge.

Kittler, Friedrich A. 1999. *Gramophone, Film, Typewriter*. Translated by Geoffrey Winthrop-Young and Michael Wutz. Stanford, CA: Stanford University Press.

Manning, Erin. 2007. *Politics of Touch: Sense, Movement, Sovereignty.* Minneapolis: University of Minnesota Press.

Parisi, David. 2018. *Archaeologies of Touch.* Minneapolis: University of Minnesota Press.

Reid, Alexander. 2007. *The Two Virtuals: New Media and Composition.* West Lafayette, IN: Parlor Press.

Rieder, David M. 2017. *Suasive Iterations: Rhetoric, Writing, and Physical Computing.* Anderson, SC: Parlor Press.

Siegert, Bernhard. 2011. "The Map *Is* the Territory." *Radical Philosophy*, no. 169: 13–16. https://www.radicalphilosophy.com/article/the-map-is-the-territory.

Walters, Shannon. 2014. *Rhetorical Touch: Disability, Identification, Haptics.* Columbia: The University of South Carolina Press.

8
WHAT THE COMPUTER SAID
Poetic Machines, Rhetorical Adjuncts, and the Circuits of Eloquence

Andrew Pilsch
Texas A&M University

In the university humanities classroom, computers are often incorporated to increase the marketable skills of humanities majors. This goal has a long history, dating to efforts in the mid-twentieth century to teach computer programming as part of a liberal arts education. In this chapter, I survey this history in the context of the older rhetorical tradition of using programmatic exercises to produce eloquence. By considering computer poetry programs as rhetorical tools tuned to the production of eloquence, I show how programming in the humanities classroom can also be used to, and perhaps is most important for, enforcing basic tenets of critical making, namely a rearticulation of our relationship with technology.

First, I consider the computer as a co-producer of eloquence, what, in a different context, Richard Doyle (2011) has called an "eloquence adjunct" (103). For Doyle, eloquence adjuncts are compounds that allow humans to fully experience their individual capacity for speaking, flexing their linguistic muscles to their limits. I connect Doyle's concept to James J. Brown Jr.'s (2014) work on Erasmus's *De Copia*, a text that offers "conceptual machines for the generation and maintenance of eloquence" that help "us reconsider the distinctions we draw between humans and computational mechanisms" (499).

To combine these two ideas, I trace the history of using what Richard Lanham (1989) calls "aleatory poetry"—generated by randomly substituting words or phrases into predefined templates—to teach computer programming to humanities students in the university (275). I then discuss my experiments with aleatory poetry. The computer and I, in the process of working together, produced something neither of us could have produced on our own. By imagining the computer as an eloquence adjunct, we can further think through digital rhetoric as a product of intimate relationships with our devices and, as Matthew Halm and David M. Rieder argued in the previous chapter, something that cuts across a clean break between analog and digital. In doing so, we can see that when considered as part of a critical making program and in the context of a rhetorical education focused on the production of eloquence, computer programming reprograms our understanding of the limits of language.

ELOQUENCE ADJUNCTS

Richard Doyle explains eloquence in terms of its etymology: "to speak out" (qtd. in Doyle 1997, 101). He uses speaking out to develop his thesis that ecodelics, the perception-altering substances that are the focus of *Darwin's Pharmacy*, are "adjuncts to eloquence" whose role is to enable "us to speak out about the ineffable until we exhaust the capacities of our language" (Doyle 2011, 103). For Doyle, training in rhetoric—"the practice of learning and teaching eloquence, persuasion, and information architecture by revealing the choices of expression or interpretation open to any given rhetor, viewer, listener, or reader"—involves the ability to speak out unto the very limits of language (Doyle 2011, 8). Doyle traces the rich tradition of using chemical compounds to facilitate the speaking out about the nature of the world. As an example, Doyle cites Walter Benjamin's writings on hashish: "You follow the same paths of thought as before. Only, they appear strewn with roses" (qtd. in Doyle 2011, 107). For Benjamin, hashish renders his thoughts more elegantly.

From Doyle, I want to develop a more general idea of eloquence adjuncts because work in digital rhetoric figures the computer as another substance-mediated speaking out "until we exhaust the capacities of our language." In "The Machine That I Therefore Am," Brown (2014) argues that Erasmus's *De Copia* provides "conceptual machines for the generation and maintenance of eloquence" (499). *De Copia*, Desiderius Erasmus's 1512 textbook, is a particularly zany node in the history of rhetoric, promising to teach students to wield the power of an "abundant style." The most famous moment in the text, which Brown explores in his essay, is chapter 33 of Book I, in which Erasmus demonstrates his methods by presenting 195 variations on the sentence "Your letter pleased me greatly."

Brown (2014) argues that Erasmus is training his readers—schoolboys learning rhetoric for the first time—to become eloquence machines. He suggests Erasmus is comparatively neglected in the rhetorical tradition because such demonstrations highlight the mechanical dimensions of rhetoric and eloquence, the machinations that rhetoricians might shy away from since it lands us in the realm of rulebooks, robots, and 'mere' rhetoric. If eloquence is merely the robotic use of an abundant style, the superficial use of different words, then rhetoric is merely cosmetic, a way to dress up ideas. (501)

Tapping into anxieties motivated by the history of rhetoric, Erasmus's set of instructions seems to play into derisive views of rhetoric as ornamentation rather than a substantive set of methodologies for, as Doyle figures, exploring the edges of our capacity to speak and our ability to know the world.

However, Brown sees in Erasmus's machines a meditation on the nonhuman adjuncts that accompany all writing, whether my MacBook Pro or Walter Benjamin's hashish. As Brown (2014) writes, "To become more human and more able to spontaneously respond to the kairotic situation, the student must exaggerate the robotic machinations of rhetorical study" (504). Neither the product of human action nor passively received from the computer, eloquence is a product of interweaving or, as Doyle (2011) argues, "It is perhaps this capacity of some rhetorical practices to induce and manage the breakdown of borders . . . that deserves the name 'eloquence'" (101). In breaking down the border between self and world, we practice eloquence.

Does all writing work this way? Doyle's writing about eloquence might suggest a rarified experience, within the ecodelic, encountering the ineffable. However, Brown's conception of machined eloquence suggests a more quotidian understanding. Summarizing Levi Bryant's machine-oriented ontology in the context of Erasmus, Brown (2014) claims that "machines as inscribed and dynamic—we can see all relations, human and extrahuman, as dynamic machines, shifting their procedures to address exigencies" (509). From this characterization—where "all relations" are "dynamic machines" that require eloquence—we might conclude a more everyday model at work.

Such a model is also called into view by Doyle's revision of Aristotle's definition of rhetoric, which I quoted above but repeat here: "Rhetoric is the practice of learning and teaching eloquence, persuasion, and information architecture by revealing the choices of expression or interpretation open to any given rhetor, viewer, listener, or reader" (Doyle 2011, 8). Doyle's use of "information architecture" is particularly suggestive. For Doyle, the intersection between ecodelic and rhetorical eloquence emerges from the need to navigate the complex tangle of information that is the world:

> Having introduced this baroque vision whose sheer number of activities already challenges the mind attempting to imagine the scene—contemplating, singing, flitting about, crawling, tangling—Darwin also introduces the scalar and spatial differences across which this entangling bank now spreads into his suggested contemplation. The bird sings in the symbolic, aesthetic, informational realm of birdsong and its habitat, sky, while Darwin's implicit exploration of the space of visible ecosystems continues down through which insects, perhaps grasshoppers, vaguely "flit," and ending finally, as it must, with the worms and the earth beneath the feet of each and every mortal reader. (131)

This comes in a reading of Charles Darwin's attempt to observe a riverbank that borders a jungle. As Darwin attempts to catalog the systems he encounters, he finds these overlapping information systems begin to infect his thinking about the systems he is observing. For Doyle, this is an example of a rhetorical practice that tends toward eloquence. Darwin's abundant prose mirrors the abundant information system of the riverbank and responds to the experience of being entangled by the forest itself. His sense of stable, observing self is disrupted in a way that is rhetorically productive.

We can synthesize a theory of ubiquitous computational eloquence, based on the tension between self and system in Doyle and Brown. When Darwin's consciousness drifts into contemplation of the riverbank, he loses the boundaries of himself and is moved to speak out about this ineffable experience. However, as Doyle invokes by mentioning "the worms and the earth beneath the feet of each and every mortal reader," the possibility for these experiences is all around us whenever we encounter the dense information ecologies that surround us.

An aspect of what is happening to Darwin, developed briefly by Doyle in *Darwin's Pharmacy*, is the twin processes of "symmetry breaking and rhythmic entrainment," which philosophers John Collier and Mark Burch (1998) claim are "the two processes . . . responsible for much (if not all) of the complexity and organization in the Universe" (n.p.). Rhythmic entrainment, "the formation of regular, predictable patterns in time and/or space through interactions within or between systems that manifest potential symmetries," first emerges as an idea in musicology explaining how animals learn to speak by sampling and repeating noise before whittling that noise into repeatable patterns of meaning (n.p.). Collier and Burch's article is notable for providing a powerful explanation of not only language acquisition but a variety of scientific and social phenomena, suggesting that measurements and memes are nothing more than the creation of new patterns of symmetry by populations. New symmetries grasped through rhythmic entrainment break up old patterns of organization and, Collier and Burch argue, allow a system to dissipate greater and greater quantities of excess energy.

Rhythmic entrainment in human psychology suggests that pattern recognition and rhythm habituation are key to individual human learning. However, as Collier and Burch show (and Doyle draws out by connecting to rhetoric), novel information is acquired when humans are jolted out of collective, habituated ways. Such a need for a jolt has long been central to the study of rhetorical eloquence. As Cicero wrote, "eloquence which does not startle I don't consider eloquence" (Cicero 2002, 331).

TEACHING HUMANITIES COMPUTING

The exploration of computer programming as a means of generating eloquence and producing the means to startle rhythmic entrainment has a history nearly as long as the concept of programming itself, especially as a pedagogical tool. There is a large body of literature from the 1960s and 1970s considering the use of programming to generate poetry in the humanities classroom. Given the focus on literary writing, rather than composition, this material has not been widely considered within the field of digital rhetoric. Under the banner of eloquence, however, this material makes interesting contributions about the use of creative writing in support of critical making.

In a 1970 conference presentation, R. J. Collens—reporting on his use of YORICK (a poetry robot) at York University to introduce humanities students to computers—outlines the stakes for integrating computer programming into the general university curriculum:

> Computer science curricula traditionally have ignored the humanities student and have left him to graduate with little or no understanding of computers and computing. This has led to beliefs, too often expressed publicly, concerning giant brains about to take over the world or worse yet starting a war of annihilation. (Collens 1970, 103)

For Collens, the increasing prevalence of doomsday computer narratives (1970 saw the release of *Colossus: The Forbin Project*, for instance) necessitated the creation of a course in, as he calls it, "computer appreciation" (103).

Such a move was fairly common during this period, as evidenced by the series of articles Joseph Rudman published detailing work using computer-generated poetry to teach computer programming to humanities students. Covering work in the 1960s, '70s, and '80s, Rudman's essays during this period[1] provide a valuable resource for understanding the state of digital pedagogy in the early days of computing. Many efforts to teach programming to humanities students do not start with domain knowledge familiar to its audiences:

> Problems and exercises for an introductory computer programming course for humanities students should be taken from the humanities rather than the sciences. I specifically suggested the area of computer generated literature. It is an area that is understandable to the humanities student who therefore can spend his time on the programming aspects of the course. (Rudman 1986, 116)

Rudman suggests that meeting humanities students where they are with literary texts and their composition works better than many programming for poets texts that focus on mathematics. As an example of this problem, he offers the following:

> In 1976, William Ralph Bennett Jr., a professor of Physics and Engineering and Applied Science at Yale University, published . . . *Introduction to Computer Applications for Non-Science Students (BASIC)* (11). The first three chapters cover introductory programming, more advanced programming, and plotting. The fifth problem in chapter 1 is:
>
> WRITE a program that compares values of N! with those obtained from Stirling's approximation,
>
> $$N \approx \sqrt{2\pi N} e^{-N} N^N$$
>
> for increasing integral values of N. How large a value of N! can your computer handle? HINT: One easy way to evaluate N! consists of setting up a loop for another variable M which goes from 1 to N in integral steps and within which we repeatedly let F=F*M (where F=1 initially).
>
> And the problems get more complicated. . . . Only in the final chapter does Bennett get to computer generated literature. (Rudman 1986, 117)

I cannot imagine a humanities student being able to make much of this exercise. I have a background in computer science at the undergraduate level and I cannot understand what this exercise is asking.

Rudman's point is computing must be made into a field with which humanities students can interact. By clinging to the origins of computing in mathematics, as Rudman points out, "technology shock" is a prominent feature of these types of classes. Instead, focusing on a knowledge domain with which fledgling programmers are already familiar "allows the student to concentrate on programming and not the problem to be programmed" (Rudman 1986, 121).

However, for Rudman, and for most of the literature on the topic, the payoff for creating computer poetry is programming. But what of eloquence? Can a computer programming exercise facilitate a faculty for language itself?

To return to Collens's (1970) article about YORICK, after revealing the secrets of how YORICK works to his students, Collens reports that

> most of the students lost interest in YORICK at that point. Some remained curious and obtained copies of the programme and tried to understand why the programme did what it did. These people became intensely involved in programming. Others simply modified the pattern of the lines and changed the vocabulary in an effort to discover what poetry was all about. One student, who had taken a poetry appreciation course previously, discovered finally that she really liked poetry and understood it much better after using YORICK. (111)

For me, this one student is perhaps the most interesting outcome from Collens's experiment. She, like the ecstatic rhetoricians of Doyle's work on ecodelics, cannot recall, Collens informs us, what it is about YORICK that so inspires her: "Somehow YORICK made poetry more fun," is all Collens can offer by way of a conclusion (112). And yet, she is insistent that something important, even transformational, has happened to her facility with language.

In a 1970 piece that critiques one of the earliest articles on computer-generated poetry, Richard W. Bailey, a professor of English at the University of Michigan, articulates the limits of an early attempt to generate poetry using a computer. Bailey praises an anonymously authored piece ("by a professor of operations research who has asked that his name not be mentioned in connection with essay") from the previous year for "drawing attention to this interesting question [of generating poems with computers]" but suggests it "may misdirect people seriously interested in the problem" due to the low quality of the resulting poetry and "blunders in explaining the background of such work" (Bailey 1970, 10). Suggesting that the poems "may have been picked from a mass of randomized texts by someone with a taste for cliché," he offers the opinions of an astute reader of poetry, in contrast to the anonymous author's "hasty and sometimes careless treatment of both linguistic and poetic problems" (10). Moreover, he suggests that the difficulty in generating poetry on the computer may lie in "the many 'structures' of poetry—grammatical, thematic, metrical, and so on" (13). Further, he suggests that

> to attack all of them at once results in the kind of failure just described. Surely the best approach is to leave most of them invariant in a given experiment while the programmer explores one or two at a time. A fixed grammatical schema, like the one described above, may offer a beginning and allow the programmer-poet to explore dimensions of thematic organization. (13)

The problem, as Bailey notes, lies in underestimating the difficulty of poetic composition and a focus on the programming problems themselves. Like Collens's student, Bailey suggests that teaching poetry to a computer can teach the programmer about the nature of language alongside teaching computer programming.

Further, Bailey suggests a more gradual experience of poetry generation, experimenting with the many structures of poetry by varying a small number in each iteration. This rhythmic entrainment would take apart and lay bare individual aspects of the process of poetic composition, breaking up previous symmetries and building a more nuanced understanding of poetry. Poetry programs, such as Collens's YORICK, would play the role of an eloquence adjunct, disrupting previously certain rhythms about poems and replacing them with a newer and richer engagement with language itself.

COMPUTING POETICS, A PERSONAL REFLECTION

I have found this to be the case in my own work with computer poetry. Starting with a collaborative project with Margaret Konkol[2] at the Digital Humanities Summer Institute[3] where we produced a poetry box that would, when a button was pressed, display a randomly generated poem for only so long as the viewer held the button down. Konkol is a talented poet, while I run toward Bailey's "someone with a taste for cliché," but the process was fruitful because I could program our Arduino controller and allow her to fill in the bank of poetic lines that our device chose from.

After this initial experience, I began a series of experiments exploring particular aspects of the poetic process as Bailey identified, while experiencing Rudman's point of seeing the work as a series of programming problems. Inspired by Brown's Twitter bot (discussed in "The Machine That Therefore I Am"), I first produced @HomeworkCopia, a Twitter bot that generates variations on the classic homework excuse "my dog ate my homework." This project was much like YORICK, in which a particular template—<A|My|Your|Our|The> <Sometimes an Adjective> <Animal|Synonym For Homework> <Synomnym for Ate|Synonym for Was Eaten By> <A|My|Your|Our|The> <Sometimes an Adjective> <Animal|Synonym For Homework>—was filled in from banks of words, including lists of articles, adjectives, animals, synonyms for homework, and active and passive forms of "to eat." To generate the tweets, I used George Buckenham's *Cheap Bots, Done Quick* service, which uses Kate Compton's Tracery library to generate bots of this type (Compton, Kybartas, and Mateas 2015).

From that project, I next created @InfiniteQuintilian. I was teaching a history of rhetoric course and observed how *long* Quintilian's *Institutes of Oratory* is (the Loeb Classical Library translation spreads across five volumes and runs to more than two thousand pages). I was also interested in powering a bot using Markov chains, a simple artificial intelligence algorithm that models the likelihood of a future outcome based on previously observed behavior. Markov chains are particularly interesting to bot makers to generate random sentences that are statistically similar to a provided set of texts. So, I trained a bot on *Institutes of Oratory* and now make Quintilian's work even longer.

Where the first bot used a fixed grammatical structure to experiment with the effects of word choice and the second used a fixed corpus to generate novel sentences, I started a third project to experiment with form and content. Specifically, I wanted to make *haiku* using the text of *Moby Dick*, for reasons I can no longer recall but do recall making sense at the time.

Rather than use the more dynamic forms of actual Japanese *haiku*, I chose to work in the *5-syllable / 7-syllable / 5-syllable* form often taught in elementary schools. My choice here was primarily due to it being easier to generate fixed syllabic lines with a computer. This project required a particular dataset: the text of *Moby Dick* broken into words and sorted by syllable count. Syllable count is a bit difficult to measure on a computer, but Daniel C. Howe's RiTa library can extract this information with reasonable accuracy.[4]

Thinking about this as a programmer rather than a poet, I assumed I could just combine words into five- and seven-syllable units (2-3, 3-2, 1-4, 4-1, 1-3-1, 3-1-1, etc.) to generate poems. However, choosing words at random produced gibberish. I began to read Basho's *haiku*, noticing the structures of poetry discussed by Bailey that I had missed by approaching this as a programmer.

RiTa can also extract part-of-speech information. After some computing, I now had a dataset that was sorted by syllable count and part-of-speech information. At this moment, I should have been close to generating *haiku*, but I wasn't.

RiTa tags words with part-of-speech information derived from the Penn Treebank,[5] which defines part-of-speech using a two- to three-letter code. For instance, an adjective would be tagged jj while the word "to" is tagged to. While meaningful to linguists, the Penn Treebank is an esoteric grammar for describing words and has some rather opaque tokens. Keeping straight nnp (singular proper noun) and nns (plural noun) was not particularly inviting me into the linguistic eloquence I hoped to express.

To address this issue, I wrote a development utility that would let me key in a set of POS tags and syllable numbers to generate poems. So, for instance, I could input

The 'word(1'JJ')' 'word(2,'NN')'.

and it would make sentences of the form The <1-syllable-adjective> <2-syllable-noun> (so "The red cabin" or "The big masthead"). This let me input and verify in real-time the *haiku*-ness of a particular scheme.

It was while using this testing utility that I had what Gregory L. Ulmer (1994) describes as a "eureka" moment (7). While I was experimenting with certain patterns, I started being able to judge which ones would create poetic outputs *before* I generated poems. This moment was particularly important for me, because, as I've mentioned, I am a *terrible* poet. And, in fact, I still cannot write a decent *haiku* to save my life, but as I worked with my utility, I found I could generate templates that let my computer generate decent poems.

Working in this way did not change my ability to write a poem without the system. I am left to conclude that the particular configuration of human and machine, in this moment, produced something that neither would be capable of producing on their own. My *haiku* project provides a more fleshed out example of Collens's student: Poetry became fun by turning an activity I do not understand (poetic invention) into one I do (generating and extracting patterns from input).

STUCK IN THE MIDDLE

My experience writing *haiku* programs suggests eloquence adjuncts facilitate the Greek understanding of "invention-in-the-middle," as Debra Hawhee (2003) explains in "Kairotic Encounters" (17). As she shows, "Occasionally the verb ['to invent'] takes on what philologists refer to as the middle voice, a reflexive grammatical construction which conflates the active and passive meanings: the subject at once also becomes the object" (17). She suggests that the production of eloquence is a process by which the rhetor and language collaboratively produce discourse. Further, this collaboration "marks the opportunity for a subject to produce discourse, even as it marks the other side of subjection—i.e., one is *called upon* to produce discourse" (18).

As Hawhee establishes, invention-in-the-middle disrupts the normal way we think about ideas, specifically that they come from a particular someone. In the moment of working with my poetry bot, I could not claim that I was writing the poems. I could, however, neither claim that it was the program. We were producing something beyond either of our capacities for eloquence.

When I first encountered Marie Borroff's 1971 article detailing her experiments in computer-generated poetry, I was optimistic that I had found someone else who shared my experience based on her title, "Computer as Poet." However, this article was not by a kindred spirit. Borroff opens by explaining:

> The poetry on these pages was produced during the series of "Random Stanza" programs which I composed and conducted at the Yale Computer Center last spring. In these programs an IBM 7094–7040 DeS computer was made to generate "poetic imagery" arranged in stanzaic free-verse patterns.... I was able to instruct the computer to make an unpredictable selection from among any set of numbered alternatives and to proceed in accordance with the result.
>
> I designed the programs to obtain a maximum of vividness and "originality" in a grammatically if not rationally coherent language clearly affiliated with poetry rather than with prose. (Borroff 1971, 22)

Borroff's claims that the "computer was made to generate" its output bothers me. Borroff consistently emphasizes her own agency throughout the piece, as can be seen in this selection. She "designed the program"—she made the computer work, she is the poet.

However, parts of the essay suggest otherwise. I can excerpt a confession that "the computer ... has in some sense a poetic style of its own" by removing her insertion "so programmed" from between "computer" and "has" (Borroff 1971, 24). Later she asks, "Is the IBM 7094–7040 DeS in this series, just a poet's mind that happens to be made out of electronic circuitry?" (25). In both cases, there is a grudging acknowledgment that the poetry her program produced is not the same as the poetry she might produce with a pencil or a typewriter.

If I considered myself a professional poet, perhaps I would be quicker to circumscribe the agency of my *haiku*-writing computer program? I take from Borroff's account that to experience the effects of an eloquence adjunct, one has to first be open to it. But what would it mean to be open to the linguistic lessons programming a computer poet might teach us?

In "Abandoned to Writing," Victor J. Vitanza (2003) answers the question, "What is the place of rhetoric in composition today?" by meditating instead on what it is that writing wants. As Vitanza argues, "'We' have yet to write 'writing'! 'We' may never write 'writing,' but 'we' must start letting writing write" (n.p.). Writing is the process of unfurling a particular linguistic space beyond the stable, expected arguments of the *topoi*. In the writing Vitanza imagines, writing is not a question of what the author wants: "'Writing' does not want what either the university thinks it needs nor what 'we' think we want" (n.p.).

While Vitanza is not particularly discussing the possibilities of writing with a computer, his model of writing writing the writer speaks to the idea of eloquence adjuncts, as I have been discussing in this chapter:

> What writing or composition wants is a writer! To invite someone to become a writer! What rhetoric wants is a body that comes to expressing itselphs [*sic*]. A writer. A body filled with tics that cannot but (not) write! (Vitanza 2003, n.p.)

Writing inviting a person to become a writer aptly expresses the broader implications of Doyle's idea of eloquence adjuncts. In my *haiku* project, neither I nor the computer are poets on our own, but together we write something we otherwise would not. Our linguistic abilities extend. We invent together, in the middle, something new. This middle is the space of eloquence and the territory of rhetoric: to disrupt rhythmic entrainment to produce new patterns of thinking.

CRITICAL DATA CAPTURE

I wish to conclude this chapter by considering the critical affordances of bot-making and eloquence adjuncts. Where Mark Sample (2014) highlights content as the oppositional quality of bots, extending a tradition of protest literature into a digital, data-driven world, my project focuses more on method as a site for critical intervention. As a model for this method of criticality, Edmond Y. Chang (2018) suggests "play" as a term in the middle between theory and practice in the digital humanities. For Chang, play is a mode of engagement that "is ultimately about learning the rules, the affordances and limitations of the platform, interface, and program, about understanding not the code itself but sensing and manipulating the contours, the structures of the code" (363). By thinking about play as a way of engaging with technology that maps its limits, we can think further about the role of eloquence adjuncts and rhetorical machines in the history of rhetorical education: Erasmus's invocation of so many variations of "your letter pleased me greatly" in *De Copia* is ultimately an invitation to play a game.

Lanham (1989) also argues for play as a central method in digital literary studies. Imagining a still, as-yet-unrealized future in which all English majors will be, on enrolling in college, presented with a CD-ROM containing "all the texts you were asked to read—or ever could read—in your undergraduate career," Lanham asks, "Wouldn't you begin to play games with it? A weapon in your hands after 2500 years of pompous pedantry about the Great Books . . . ?" (269). Play such as taking *Moby Dick* apart to see how it works and putting the pieces back together differently.

Understanding digital practice in the humanities as Chang's idea of play—exploring the edges of systems and seeing what else might be possible—is possible only absent the view that computers provide truth. In a 2011 essay, Johanna Drucker, highlighting the need for humanistic engagement with computational tools, suggests that programs used in humanities work "are a kind of intellectual Trojan horse, a vehicle through which assumptions about what constitutes information swarm with potent force" (n.p.). Instead of subjecting these artifacts to "the humanistic tenets of constructed-ness and interpretation," we engage computational artifacts as if they provide "mere descriptions of *a priori* conditions" (Drucker 2011, n.p.). This lack of sophistication "[renders] *observation* . . . as if it were *the same as the phenomena observed*" and this move "collapses the critical distance between the phenomenal world and its interpretation" (Drucker 2011, n.p.).

To counteract this tendency, Drucker suggests a substitution for the word "data" in digital humanities practice: "Capta is 'taken' actively while data is assumed to be a 'given' able to be recorded and observed" (Drucker 2011, n.p.). Unlike data, a term suggesting facts lie around in the world waiting to be discovered by researchers, capta highlights the mangling of all facts in the process of their becoming: We go into the world and slice off segments that become facts. As data, computers are sources of truth, passively reflecting the world around us. Capta, focusing on the violence of fact creation, affirms criticality.

Matt Ratto (2011) defines critical making as "[Using] material forms of engagement with technologies to supplement and extend critical reflection and . . . to reconnect our lived experiences with technologies to social and conceptual critique" (253). The maker movement has an inexorable tie to the digital side of technology, as it emerged in response to the impersonality of the digital age. However, bot-making placed in contexts of humanities computing pedagogy and rhetorical education highlights language as a technology that can also be reconnected through supplemented critical reflection. The startling effects of novel rhetorical formations suggest that eloquence has a strong affinity with the aims of critical making, despite predating that concept and the emergence of digital technology. With the affordances of the computer, however, we can more easily visualize, conceptualize, and deploy the programmatic prospects of rhetorical language.

NOTES

1. There are three essays Joseph Rudman wrote surveying computers for humanities courses. See Rudman 1978; 1986; 1987.
2. Konkol's own work with physical computing resulted in (Konkol 2018).
3. In Jentery Sayers's "Physical Computing" course
4. Originally developed for use in Processing, the library has since been ported by the original author for use in JavaScript, the language this project was developed in.
5. For more information on the history of this important computational linguistics project, see Marcus, Marcinkiewicz, and Santorini (1993).

REFERENCES

Bailey, Richard W. 1970. "Automating Poetry." *Computers and Automation* 19 (4): 13–14.
Borroff, Marie. 1971. "Computer as Poet." *Yale Alumni Magazine*, January 1971, 22–25.
Brown, James J., Jr. 2014. "The Machine That Therefore I Am." *Philosophy and Rhetoric* 47 (4): 494–514. https://muse.jhu.edu/article/562412.
Chang, Edmond Y. 2018. "Playing as Making." In *Disrupting the Digital Humanities*, edited by Dorothy Kim and Jesse Stommel, 357–68. New York: Punctum. https://doi.org/10.21983/P3.0230.1.00.
Cicero. 2002. *Letters to Quintus and Brutus. Letter Fragments. Letter to Octavian. Invectives. Handbook of Electioneering; D. Letters*, 1st ed. Translated by D. R. Shackleton Bailey. Cambridge, MA: Harvard University Press.
Collens, R. J. 1970. "Computer Generated Poetry as a Pedagogical Tool." In *Proceedings of the Conference on Interdisciplinary Research in Computer Science: University of Manitoba, Winnipeg, June 8–10, 1970*, vol. 2, edited by R. G. Stanton and M. G. Saunders, 99–122. Winnipeg, Canada: University of Manitoba.
Collier, John, and Mark Burch. 1998. "Order from Rhythmic Entrainment and the Origin of Levels Through Dissipation." *Symmetry: Culture and Science* 9 (2–4): 165–78.
Compton, Kate, Ben Kybartas, and Michael Mateas. 2015. "Tracery: An Author-Focused Generative Text Tool." In *Interactive Storytelling*, edited by Henrik Schoenau-Fog, Luis Emilio Bruni, Sandy Louchart, and Sarune Baceviciute, 154–61. Lecture Notes in Computer Science. Cham, Switzerland: Springer International Publishing. https://doi.org/10.1007/978-3-319-27036-4_14.
Doyle, Richard. 1997. *On Beyond Living: Rhetorical Transformations of the Life Sciences*. Stanford, CA: Stanford University Press.
Doyle, Richard. 2011. *Darwin's Pharmacy: Sex, Plants, and the Evolution of the Noösphere*. Seattle: Washington University Press.
Drucker, Johanna. 2011. "Humanities Approaches to Graphical Display." *Digital Humanities Quarterly* 5 (1). http://www.digitalhumanities.org/dhq/vol/5/1/000091/000091.html.
Hawhee, Debra. 2003. "Kairotic Encounters." In *Perspectives on Rhetorical Invention*, edited by Janet M. Atwill and Janice M. Lauer, 16–35. Knoxville: Tennessee University Press.
Konkol, Margaret. 2018. "Prototyping Mina Loy's Alphabet." *Feminist Modernist Studies* 1 (3): 294–317. https://doi.org/10.1080/24692921.2018.1505273.
Lanham, Richard A. 1989. "The Electronic Word: Literary Study and the Digital Revolution." *New Literary History* 20 (2): 265–90. https://doi.org/10.2307/469101.
Marcus, Mitchell P., Mary Ann Marcinkiewicz, and Beatrice Santorini. 1993. "Building a Large Annotated Corpus of English: The Penn Treebank." *Computational Linguistics* 19 (2): 313–30. http://dl.acm.org/citation.cfm?id=972470.972475.
Ratto, Matt. 2011. "Critical Making: Conceptual and Material Studies in Technology and Social Life." *The Information Society* 27 (4): 252–60. doi:10.1080/01972243.2011.583819.

Rudman, Joseph. 1978. "Computer Courses for Humanists: A Survey." *Computers and the Humanities* 12 (3): 253–79. doi:10.1007/BF02400087.

Rudman, Joseph. 1986. "The Use of the Rational and Techniques of Computer Generated Literature in an Introductory Programming Course for Students in the Humanities." In *Proceedings of the Second Annual Eastern Small College Computing Conference*, 112–36. Scranton, PA: University of Scranton.

Rudman, Joseph. 1987. "Selected Bibliography for Computer Courses in the Humanities." *Computers and the Humanities* 21 (4): 245–54. http://www.jstor.org/stable/30207395.

Sample, Mark. 2014. "A Protest Bot Is a Bot So Specific You Can't Mistake It for Bullshit." *Medium*, May 30, 2014. https://medium.com/@samplereality/a-protest-bot-is-a-bot-so-specific-you-cant-mistake-it-for-bullshit-90fe10b7fbaa.

Ulmer, Gregory L. 1994. *Heuretics: The Logic of Invention*. Baltimore, MD: Johns Hopkins University Press.

Vitanza, Victor J. 2003. "Abandoned to Writing: Notes Toward Several Provocations." *Enculturation* 5 (1). http://enculturation.net/5_1/vitanza.html.

9
ACTIONABLE MONUMENTS
Making Critical Augmented Reality Activism

Sean Morey
 University of Tennessee

M. Bawar Khan

This project demonstrates critical making through an augmented reality-based MEmorial mobile application. The app helps its user to intervene in discussions of oil by MEmorializing the BP oil spill, pointing to the assumed value of unabated driving, but also the abject sacrifices needed to sustain this value. To demonstrate this intervention as interactive monumentality—toward what we will define as an "actionable monument"—we will model an augmented reality (AR) project that includes the following components: an AR application created with Unity software, corresponding image and video assets, and C# scripts that will link this app with the user's automobile via a Bluetooth onboard diagnostic scanner. In total, these elements provide real-time data on the user's relationship to the petrol economy and the animal sacrifices that result from this participation. However, the project files could be emptied of this particular topic and adapted to other pedagogical uses, which we explain at the end of the chapter.

AR technologies provide a metamedium through which site-specific information can be delivered, requiring location-based sensitivity when designing texts that rhetorically engage with a particular environment. While some scholars in rhetoric and writing studies have begun to produce AR texts—most prominently, researchers in the University of Florida's TRACE Innovation Initiative—artists have provided the prototypes for how this emerging technology might be used for activism, public critique, and other types of critical making. For example, the artist Conor McGarrigle (2017) has used AR to bring attention to the bailout of Irish banks; Mark Skwarek and John Craig Freeman (Freeman 2017) have created a digital memorial for the Mexican immigrants who have died while crossing the Mexico–US border. In these cases, the artists

https://doi.org/10.7330/9781646422586.c009

attempt to intervene in conversations about the worldwide banking system and immigration, respectively.

As Alan Craig (2013) has explained, AR should be understood as its own medium rather than simply a technology; it goes beyond being a tool and instead becomes a human-cultural interface like any other medium (xvii). However, we might better understand AR as a metamedium—that is, a medium similar to a web browser that arranges or transmits other media. While a variety of genres across a variety of media can be displayed with AR—including games, film, audio, and/or text—this project builds on Skwarek and Freeman's work toward incorporating the genre of Gregory L. Ulmer's MEmorial (also Electronic Monument) into the AR environment (Ulmer 2005), which expresses hidden or overlooked values that society holds, but doesn't recognize as unofficial, latent values. While critical making is often focused on hacking and building generalizable technology, we're in need of particular pedagogical applications that help rhetoric and writing scholars shift from critical making's genuine spirit of "invention as tinkering" to "what do I do?" David M. Rieder (2017) offers some examples in his work *Suasive Iterations*, particularly critical physical computing projects that are meant to "disturb the peace," or make the inner workings of technology visible to its creator and audience. Of course, other kinds of inner workings may become revealed, such as the values that such technology supports.

Although not developed from physical computing directly, Ulmer's genre of the MEmorial turns out to be infinitely suitable to critical making and applicable to classroom use. The point of building a MEmorial, then, is to make these values present so that if they do not align with those that society or an individual overtly expresses, the user can invent new practices. Since, at its own core, AR technologies are computer-based, an AR MEmorial can actualize these values through associative virtual monuments, allowing the user to digitally install them in public spaces to cultivate discussion and awareness of whatever value the AR interaction produces.

MAIN CONTEXT

On April 20, 2010, an explosion shook the Deepwater Horizon drilling platform in the Gulf of Mexico, a platform owned by Transocean, and leased by British Petroleum. This explosion precipitated what would become America's worst accidental oil spill (United States Department of Homeland Security 2011). Officially, the sacrifices of the rig workers have been honored in many ways, but they were honored monumentally in September of 2016 by artist Jason Kimes, who created and installed eleven 500-pound statues on the Elysian Fields Avenue neutral ground in New Orleans, Louisiana—one statue for each fatality (MacCash 2016).

Of course, the final death toll of biological organisms far exceeded those eleven humans, and so, how should they be memorized? This critical making project—which we call the Roadkill Tollbooth Actionable Monument (RTAM)—uses Ulmer's MEmorial as a point of departure, demonstrating the practicality of this theory of monumentality while shedding light on new ways to understand our relationship to the BP disaster (figure 9.1). The RTAM demonstrates an application of these concepts, scholarship that tests out these ideas to offer how they work and how they might be transferable across disasters. In addition, this project offers a model for future research and extends Ulmer's original work, building upon the MEmorial as a critical making theory/prototype/methodology to refine and make new prototypes.

Ulmer (2005) notes that a "MEmorial witnesses (monitors) a disaster in progress" (xxvii). But unlike their physical, official counterparts, MEmorials exist in the virtual. As such, they usually exist at a distance from the official monuments they augment, usually housed on a remote web server, and accessible via a desktop computer, even if accessed from a mobile device on site (which we'll return to later). As such, most previous forms of MEmorials maintained a critical and affective distance from the object of investigation and critique. The process of critical making reduces this distance.

The MEmorial upon which this project builds exists online as a proposed monument that suffers from these restrictions of time and distance. Memorializing the abject sacrifices of the animals killed in Deepwater Horizon oil spill of 2010, the project maps the user's automobile trek south along the Florida Turnpike from its northern boundary near Ocala, Florida, to its southern terminus at Florida City, Florida. The RTAM makes motorists conscious of the relationship of an official memorial (the tollway itself, commissioned in 1998 as the Ronald Reagan Memorial Turnpike) and their daily uses of petroleum needed to access that memorial, allowing them to fulfill their "right to drive," which is part of the statute that created the Ronald Reagan Turnpike (State Library and Archives of Florida 1998). The RTAM, then, shifts the focus from the official values and sacrifices to unconscious values and sacrifices abject.

Actionable Monuments 183

Figure 9.1. Screenshot of the Roadkill Tollbooth Actionable Monument while passing under a tollbooth gantry.

ABJECT CONTEXT

Official memorials usually honor humans. However, images of animals, saturated with oil, saturated the post-spill media in turn. Much of the conversation shifted toward the spill's effect upon the environment and wildlife. News reports continually returned to discussions of oil-covered birds, deceased dolphins washed upon the shore, massive fish kills—thus, mostly the charismatic megafauna of the gulf region. While some reports focused on invertebrates and microorganisms, most news reporting covered these larger vertebrates. Eventually, as might be predicted, the coverage shifted back to BP, the humans responsible, and the humans affected, including the eleven who lost their lives, the many injured rig workers, and those whose water-based and tourist-based livelihoods were now threatened.

The RTAM monument could focus on these human casualties simply through proper spatial positioning. For example, it could be appended as a virtual monument to Kimes's statues. The memorial's placement within a wide roadway medium that separates the six-lane street—as well as its juxtaposition next to traffic aids such as street signs, stop lights, bus stops, parked cars, and other markers of a petrocentric culture—provide ample reference to the end point of the oil flow that had begun from the very oil well that would lead to the eleven statues. In this way, the monument does provide a witness to the process for which these eleven oil rig workers sacrificed their lives.

However, this monument only looks at the human cost and does so at the margin, the buffer, between literal and figurative paths in the larger petrochemical network. Instead of holding focus on the human victims, we want to mourn the unofficial sacrifices. Most monuments link to their supported value in some explicit way. War memorials usually indicate the war in which the soldiers were killed. Monuments such as Mount Rushmore explicitly depict its revered figures who, by past position, are easily historically linked to an ideal (i.e., freedom). Kimes's memorial is simply titled "ELEVEN" with no clear connection back to the Deepwater Horizon disaster. And after two years in the Elysian Fields location, the statue was moved to a private sculpture garden, one piece of artwork among many, the sign only loosely connected to its referent. If the RTAM were attached to ELEVEN, the connection to the oil flows that link them would be severely restricted, if not severed.

ACTIONABLE MONUMENTS

While an electronic monument can provide this sort of psychic interaction, helping the creator/user better understand their role within an event (disaster or not), these monuments are usually remote, either in terms of physical proximity (they only exist online) or practically (they could be built on-site, but no official organization would probably greenlight the construction). While a visitor to a virtual, online monument could bring the digital experience to the official site via a mobile device, this interaction still causes problems that remove the user—to some degree—from the juxtaposition needed for an optimized experience. To exceed such limits of typical mobile use, John Tinnell (2017) provides an updated metaphor to describe mobile media interfaces as actionable media, and through this updated interface, we propose that we extend Ulmer's electronic monuments to mobile computing in the subgenre of actionable monuments.

Unlike its overarching category of electronic monument, an actionable monument requires this kind of place-based integration, just as participating in a traditional, physical monument requires corporeal visitation. However, an actionable monument incorporates the digital functionality of the web, particularly ATLAS, to increase the user's participation in public monumentality. In this instance, the RTAM transforms the virtual monument into a memorial both present and actionable for the user.

Tinnell's concept of actionable media extends Ulmer's electrate theories of monumentality, providing a theoretical underpinning for an AR-based electronic monument: an actionable monument. The RTAM takes the platform of the MEmorial and adds Tinnell's ubicomp interface of ATLAS to provide a genre that allows for the dissemination of scholarship to the general public, scholarship written through ATLAS (apps, tags, layers, actuators, and sensors) rather than modes native to the desktop. Tinnell's argument mirrors Matthew Halm and Rieder's claims that "digital projects are no longer trapped behind the looking glass of the screen, nor are they limited to the mouse and keyboard—or in what can be thought of as an extension of the mouse's capabilities, the relatively limited actions of tapping and swiping on a trackpad or touchscreen" (see their chapter in this volume). Instead, as they argue—and what is showcased by AR—"reality is increasingly a mixed or hybrid experience." Moreover, while they focus on capacitive touch rather than driving a car, the objective of these critical making projects is the same: "This persuasive movement everts the lived environment of an audience and reveals it to be a component of their subjectivity." The RTAM provides a pedagogical tool for embodied invention, or what Andrew Pilsch in this volume identifies as Ulmer's writing of "eureka."

DIGITAL PRECEDENTS

The RTAM thematically and conceptually builds off several art projects developed by artists working with AR, artists who enact the modes of actionable media identified by Tinnell. In general, these artists deploy the mobile and emersive aspects of AR to emplace art in the very locations they seek to criticize, or the powers that control those locations, either explicitly or implicitly. Thematically, the RTAM links to artist Mark Skwarek's (2014) "The Leak in Your Hometown." Skwarek developed an app that, through a device's camera view, will recognize the BP logo and overlay it with an animated pipe that gushes out oil. Tamiko Thiel (2017) makes an environmental statement through her work *Clouding Green,* which "floats a massive augmented reality cloud—in colors ranging from sooty carbon black to brilliant renewable green—over eight major Silicon Valley cloud computing providers" as indicated by Greenpeace's report "How Clean is Your Cloud?" (Thiel 2017, 272). For instance, viewing the Apple headquarters through the app, the user will see a "15.3% Clean Energy Index" cloud over the building.

The RTAM extends these projects, in some ways combining them, to demonstrate the occurrence of animal deaths and linking the cause of death to oil. However, although "The Leak in Your Hometown" and *Clouding Green* might implicitly cast blame, the RTAM directly implicates the user as one node of causation within the larger petroleum network, attempting to draw out an introspective, affective, and critical response. More than the examples above, the user participates in the artwork rather than just viewing it. Skwarek and Thiel never directly implicate the audience, instead pointing—implicitly or explicitly—to more localized actors or causes, such oil companies or tech industries. Although the RTAM app looks outward at the surrounding environment through the camera, accelerometer, GPS, gyroscope, and other sensors, the goal of such sensing is to relay users' actions back to their own internal awareness of how their immediate actions affect their environment—that may be invisible to them at the time—through their physiological senses. Beyond the limited technological concept of augmentation that AR provides, the app provides a broader psychic augmentation.

In doing so, the app provides what Rieder (2017) might identify as a "transducer" within the context of physical computing. Rieder explains that transduction converts one form of energy into another, often when "the virtual and real are folded together in some novel way" (11). Software programs convert electricity into the movement of analog objects, for instance, such as when a programming script signals a light to turn on. Since transduction is methodological, it can apply to a host of processes whereby some form of energy must be transformed into another. In the case of the RTAM, we might argue that a series of transductive moments occur, as signals from the car engine, GPS sensor, and a driver's behavior all create different transductive outputs, which then act further upon the driver as they receive the feedback about their own driving behavior, oil consumption, and the effect on deceased (and potentially deceased) animals they will most likely never directly encounter. From the burning of oil converted into movement to that movement providing a reflective feedback system about the process of burning that oil, a cascade of critical transductive moments occur within this particular application, making the mundane act of driving a critical rhetorical activity.

TECHNICAL SPECIFICATIONS

This project required several platforms and elements to create a mobile AR application that was location-aware, interacted with the user's car, and provided an AR experience. As such, the pieces included an AR application composed with the game-building platform Unity software and an onboard diagnostic sensor (OBD-II). The gameobjects in the Unity3D scene are driven by five primary scripts that are attached to them: (1) a script to indicate where, geographically, overlays should appear; (2 and 3) two scripts that integrate tweets into the scene; (4) a script that enables tweets to scroll across the screen; and (5) a script that adds the car's diagnostic data from the OBD-II. In the next section, we have included a general tutorial for how to change the variables for each of these scripts.

GPS Script

This project uses location-based, markerless AR. This method processes location input from the device's GPS sensors and displays overlays according to predetermined locations—in this case, the coordinates of each toll gantry. As the user approaches a toll gantry, an overlay appears on the user's device—ideally mounted to a heads-up bracket, allowing the user to view the overlay hands-free. The overlay provides a reminder of the disaster and imports current Twitter conversations related to #BPoilspill. Without having to remove their hands from the wheel, the user is able to participate in this public mourning, all while (hands)freely participating in the official value supported by the Ronald Reagan Memorial Turnpike.

The C# file "GPSController.cs" sets the location-based functionality for the app. The script includes values for GPS coordinates and a value for accuracy (see below). This distance provides some buffer in case of GPS inaccuracies to allow the information to appear onscreen entering and exiting the toll gantry area. For projects requiring more accuracy—since GPS reliability can change depending on a variety of factors—other programmers might need to greatly improve upon the app's location reliability.

Twitter Scripts

The basic functionality of the Twitter script (Twitter.cs) was programmed by Craig Tinney (2018). The script uses Twitter's API (which must be set up through a Twitter developer account) to log into the social media platform and pull tweets depending on several variables. Using his tutorials, we wrote a second script (tweet.cs) that will show tweets when attached to a TextMeshPro object in Unity.

The script tweet.cs includes two main classes: "tweets," which uses the method "SearchForTweets" and can pass a search term, the number of tweets that the user wishes to return, and the type of tweet (recent, popular, mixed). In the code we share on this collection's companion website at https://upcolorado.com/component/k2/item/6219-reprogrammable-rhetoric-supplemental-design-materials-and-programming-scripts, we've set the variables to "#BPoilspill" for the search term, and "1" as the number to return. The results must then be passed through to the TextMeshPro method so that it will display on the TextMesh gameobject. However, because "tweets" returns as an array, a for loop converts them into a string that can then be displayed as text.

Twitter Scroll Script

Because a single tweet might fill up too much of the user's screen, we added a script that directs the text of the tweet to scroll horizontally across the screen, much like a sports or stock ticker. The main variable for this script is the scroll speed, which can be adjusted depending on the developer's preference.

OBD-II Script

In addition to these networking features, the app also connects with the automobile itself. If equipped with a diagnostic port, the user can purchase and install an Onboard Diagnostics II sensor (OBD-II) that will transmit vehicle information (such as speed, temperature, and other diagnostic readings) to the user's device. The app performs calculations based on these readings to determine the user's current fuel flow rate in US gallons per hour, and then multiplies this figure by 4,859, a rough estimate of how many animals died per gallon of oil spilled during the event. This number is then displayed for the user as they approach the tollbooth gantry to allow them to see the rate of virtual animals sacrificed per hour. As an example, if the fuel flow rate returned a value of 3.5 gallons per hour, the user would simultaneously be sacrificing 17,006.5 animals per hour.

We have based this multiplier on some incalculable numbers but believe it is conservative and not an inflated value. According to the Center for Biological Diversity's (2011) estimates, "We found that the oil spill has likely harmed or killed approximately 82,000 birds of 102 species, approximately 6,165 sea turtles, and up to 25,900 marine mammals." However, new estimates—revealed from the federal government's lawsuit against BP—indicate that billions of animals most likely died, including between 4 billion and 8.3 billion oysters. The government based these numbers on a "combination of on-site sampling, previously known populations of key species, laboratory studies of the toxicity of the oil and the dispersants used to break it up and the mathematical extrapolation of those results" (Smith 2015). While one billion is too low, as the oil spill still affects the gulf regions, this number also doesn't overcount the deaths. This figure of one billion animals is then divided by the estimated gallons of oil spilled during the disaster—205.8 million US gallons—to provide the number of animals sacrificed per gallon spilled, or 4,859 A^SPG.

For the individual user, their personal A^SPG number is displayed on screen as they pass through the toll gantry, as the GPS signal initializes the action. As an example, a driver passing through the gantry, whose fuel flow reads 4.3 gallons per hour, would see a reminder that they are relationally responsible for the deaths of 19,672 A^SPG [4.3GPH × 4,859 = 20,893.7].

GENERAL TUTORIAL

This section provides details for how to adapt the RTAM files for individualized projects or implement it within a classroom for students to use as a template.

Required Software

To edit the files for this app, first download and install the following software:

- The Roadkill Tollbooth Actionable Monument source files available on this book's companion site at https://upcolorado.com/component/k2/item/6219-reprogrammable-rhetoric-supplemental-design-materials-and-programming-scripts.
- Unity software version 2018.1.3f1, located at https://unity3d.com/unity/whatsnew/unity-2018.1.3. Unity provides extensive documentation and a support forum to help with installing their software. Depending on the computer's particular settings and software,

additional companion software may be needed (Unity will prompt the user to download it if necessary). Below, we refer to different elements of the Unity interface. For a more detailed description, visit: https://docs.unity3d.com/Manual/UsingTheEditor.html.
- To edit the C# files, necessary for changing the Twitter hashtags that are displayed, a source code editor is required. Some examples include Microsoft Visual Studio, Notepad++, or MonoDevelop.

Basic Installation

To open the project's source files, follow these steps:

- **Step 1:** Download and install Unity software (as listed above).
- **Step 2:** Download the file Roadkill Tollbooth Actionable Monument source files (a .zip file).
- **Step 3:** Unzip the file.
- **Step 4:** Open Unity.
- **Step 5:** From the startup screen, click on "Open."
- **Step 6:** Select the unzipped file folder for the Roadkill_Tollbooth_Actionable_Monument.

Once the folder is opened, the app should be visible within Unity's interface, comprised of several windows (figure 9.2). The most basic changes the user will probably want to make include changing the amount and locations for GPS coordinates, as well as the Twitter hashtags. The user may also want to adjust the graphic design. The following sections provide information about how to adjust these variables.

Changing GPS Settings

Changing the GPS settings can be done through Unity's WYSIWYG interface. With the file open in Unity, follow these steps:

- **Step 1:** In the Hierarchy window (on the left side of the interface; see figure 9.3), click on "Controller."
- **Step 2:** In the Inspector window (on the right side of the interface), locate the section titled "GPS Controller (Script)."
- **Step 3:** Click on the dropdown arrow next to "Gps Coordinates." You'll see that this project includes nine locations (Element 0 through Element 8).
- **Step 4:** In the field labeled "Size," enter the desired number of locations.
- **Step 5:** Click on the dropdown arrow for the first element.
- **Step 6:** Enter the latitude and longitude using decimal degrees.
- **Step 7:** Repeat steps 5 and 6 for each additional element.
- **Step 8:** Under "File" in the main menu, select "Save Project."

Actionable Monuments 191

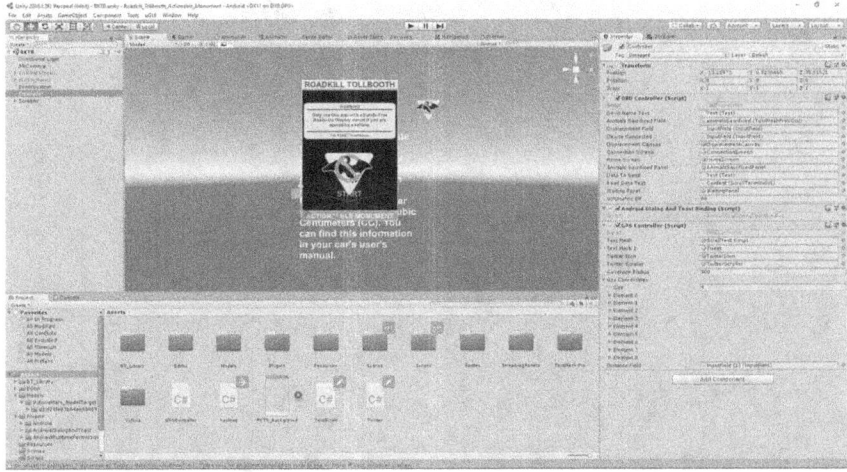

Figure 9.2. Unity interface showing all of the app's programming components.

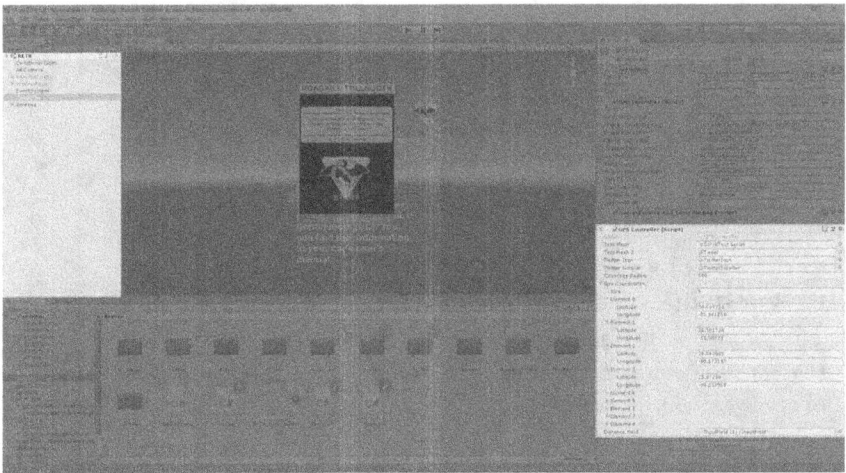

Figure 9.3. Unity interface highlighting the GPS controller.

Changing the Twitter Scroll Hashtag

To change the hashtag that scrolls along the bottom of the screen, the script "Hashtag" must be edited. Follow these steps to make this change.

- **Step 1:** In Unity's Assets window (located at the bottom of the interface), open the file titled "Hashtag."
- **Step 2:** Locate line 50, which should read (see figure 9.4):
  - yield return StartCoroutine(Twitter.API.
    SearchForTweets("#BPoilspill", accesstoken, (tw) => { tweets
    = tw; }, 1, Twitter.API.SearchResultType.recent));

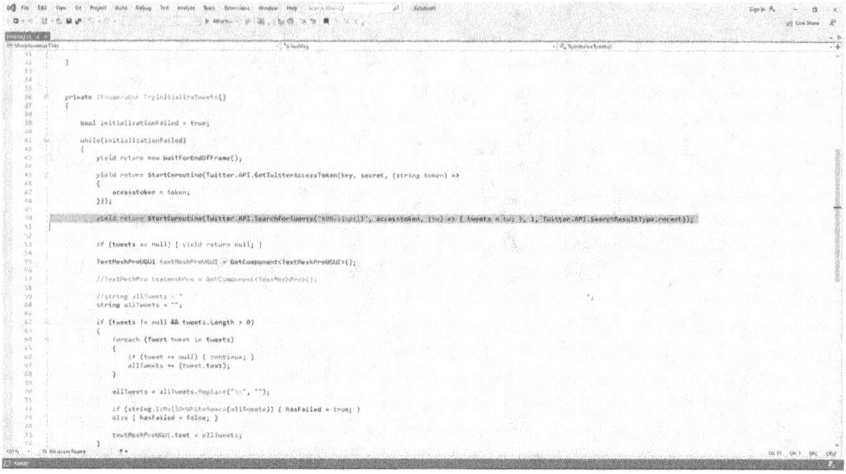

Figure 9.4. Line 50 contains the code that should be edited to change the Twitter hashtags.

- **Step 3:** Change #BPoilspill to another hashtag to display tweets tagged with the new hashtag.
- **Step 4:** Save and close the hashtag.cs file.
- **Step 5:** Return to Unity and save the project.

EDITING TEXT IN THE GRAPHIC INTERFACE

The text in the app's user interface can be changed by clicking on an element and adjusting the properties in the Inspector window (see figure 9.5). Below are the steps to change the title at the top of the window, but the general principles apply to all the text elements.

- **Step 1:** Click on the title "Roadkill Tollbooth" in the #Scene window.
- **Step 2:** In the Inspector window, locate the Text field with the text "ROADKILL TOLLBOOTH."
- **Step 3:** Change this text to your own title.
- **Step 4:** Under "File" in the main menu, select "Save Project."

Editing Images in the Graphic Interface

The images in the app's user interface can be changed by clicking on an element and adjusting the properties in the Inspector window (figure 9.6). Below are the steps to change the icon in the center of the window, but the general principles apply to all the image elements.

Actionable Monuments 193

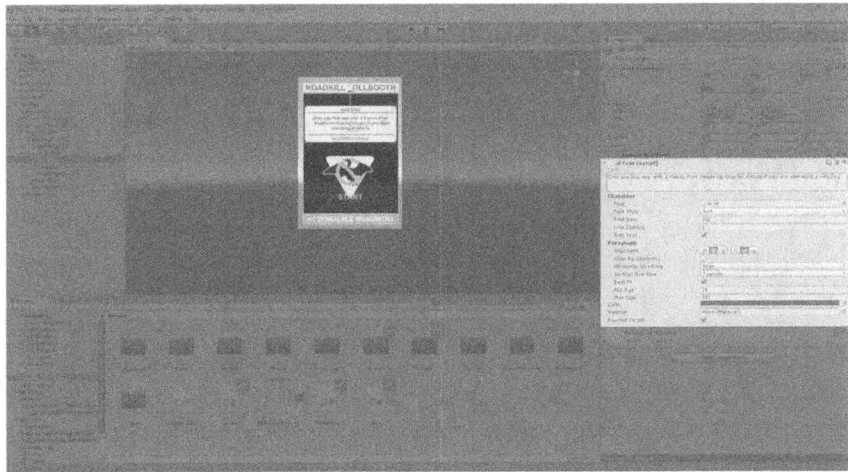

Figure 9.5. Clicking on the text in the #Scene window will make it editable in the Inspector window.

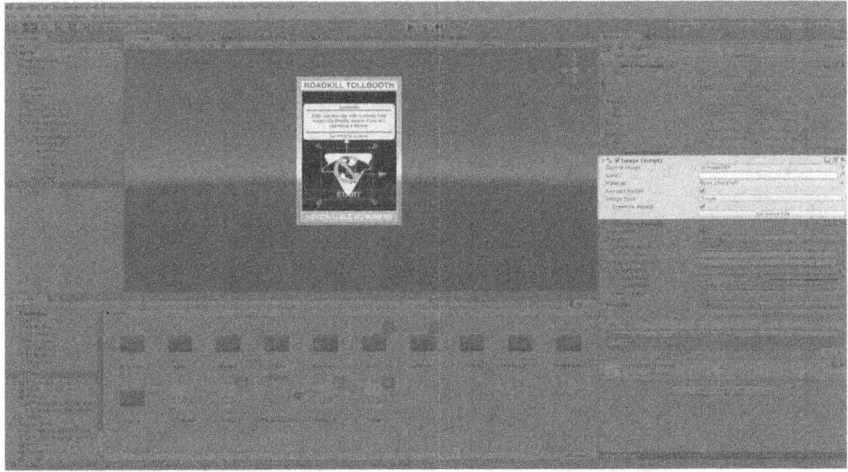

Figure 9.6. Clicking on an image in the #Scene window will make it editable in the Inspector window.

- **Step 1:** Drag and drop an image into the Assets window.
- **Step 2:** Click on the app's icon in the center of the #Scene window.
- **Step 3:** In the Inspector window, locate the Image field with the file name "imag845."
- **Step 4:** Replace this image file with your own image by dragging your image from the Assets window into the field.
- **Step 5:** In the Tool panel below the main menu, select the Scale tool.

- **Step 6:** In the #Scene window, adjust the image scale as necessary by dragging the corners of the image.
- **Step 7:** Under "File" in the main menu, select "Save Project."

ACTIONABLE PEDAGOGIES

As noted in the introduction, this project can be emptied of its oil-based content and adapted to other critical making projects. Using the RTAM in a writing classroom—whether or not it is focused on critical making—could help students explore rhetorical aspects of digital writing, especially related to programming. For a class assignment, Ulmer's (2005) *Electronic Monuments* could be paired with general tutorials about using Unity before then incorporating the code detailed above. *Electronic Monuments* and Tinnell's (2017) *Actionable Media* would provide the theoretical template for students to map their own MEmorials, and the code could be tinkered with to showcase other disasters. For instance, a student might research the Flint water crisis through the lens of MEmorials and then construct an AR project that worked with the local environment, removing the OBD-II car interface and instead just placing geolocation-based AR overlays in rhetorically appropriate places. As Halm and Rieder point out in this volume, "Sensors play an essential role" in the creation of a rhetorically engaging, everted experience. Thus, "A primary inventional step for digital rhetors is to choose one or more sensors capable of detecting the kinds of physical energy with which they want to work." Adapting the RTAM, or even investigating its use of sensors, becomes a lesson in rhetorical invention. As Rieder (2017) also makes clear, part of tinkering with such critical making projects includes moments for invention, and we hope students are able to use the RTAM to invent beyond the scope of this project.

CONCLUSIONS

Meshing the interfaces of automobile, highway infrastructure, and actionable media, the RTAM becomes an actionable monument, one that must be experienced though action within a specific place and specific time. The app provides both the critical awareness of Ulmer's electronic monuments with the *in situ* aspects of AR artwork such as those by Skwarek, Freeman, and Thiel. Moreover, future features would allow users to participate in other ways besides just through a gas pedal.

The app does have a few problems. Although somewhat hands-free, the app could pose potential hazards while driving. The safest use would be to dock the phone in a third-party heads-up display mount, turning the app on before shifting the car into drive, or using the app with AR eyewear. A more technical problem arises from smartphone and OBD-II compatibility. This app has only been successfully tested on Android 4.4 and higher, and it would need a different OBD-II script for iOS or other devices. In addition, not all cars provide the same OBD-II data to a sensor.

Despite these constraints, we hope others can extend the application to develop their own critical making projects, whether they are actionable monuments or new genres and forms of actionable media. Working across both Ulmer and Tinnell's humanistic (and theoretical) approaches to technology—in tandem with a variety of devices and programming languages—gives students an educational experience that bridges C. P. Snow's "two cultures" of the sciences and humanities. Such an experience follows Annette Vee's (2017) argument that programming "is not bound by any one discipline" and is "more available to fields that transcend those historical divisions" (42), to which we would add many forms of critical making.

As Matt Ratto and Megan Boler (2014) note, "Critical making invites reflection on the relationship of the maker to the thing produced, reflection on how elements (whether nuts and bolts, bits, and bytes, or breath, blood, flesh, brain, and neurons) work together—in short, consideration and awareness of the mediated and direct experiences of interacting with the material world" (3). While the connection between the BP oil spill, animal deaths, and individual oil consumption had been established through the MEmorial prior to constructing the AR app, the process of making the app taught us more about our relationship with our vehicles as well. We learned about engine mechanics that relied upon the oil we sought to critique, but also that many automobile manufacturers limit access to the kinds of data a typical owner can acquire through aftermarket sensor technologies.

And while other AR authoring platforms exist that could provide similar (though not all) interactions without the need for extensive coding, we share Vee's (2017) perspective that a "concept of coding literacy can help us to understand the stakes of who codes and how" (224) and feel that authoring through the metamedium of AR with images, sounds, and social media must also include some degree of making with code to learn how programming serves as another of Rieder's (2017) transducers in a larger ecology of human, nonhuman, environmental, and media interactions. Ultimately, the genre of the MEmorial is itself a transducer already and precisely adaptable to the ethos of critical making—and echoing Tinnell, Rieder, Halm, and others—we should develop modes of rhetoric that are more dynamic and attuned to our surrounding environments.

Acknowledgments. We would like to recognize and thank two other programmers who contributed their code to this project: Craig Tinney, who programmed the original version of the Twitter script and provided help to implement it, and Shalev G., who helped to program the Twitter Scroll script. We also thank Miranda Campbell for her help in researching the technical requirements of this project. Finally, Sean thanks the University of Tennessee Humanities Center for awarding him their Digital Humanities Faculty Fellowship to complete the RTAM application.

REFERENCES

Center for Biological Diversity. 2011. "A Deadly Toll: The Gulf Oil Spill and the Unfolding Wildlife Disaster." *Center for Biological Diversity.* https://www.biologicaldiversity.org/programs/public_lands/energy/dirty_energy_development/oil_and_gas/gulf_oil_spill/a_deadly_toll.html.

Craig, Alan B. 2013. *Understanding Augmented Reality: Concepts and Applications.* Burlington, MA: Morgan Kaufmann.

Freeman, John Craig. 2017. "*Border Memorial: Frontera De Los Muertos* [Arizona, 2012]." In *Augmented Reality: Innovative Perspectives Across Art, Industry, and Academia*, edited by Sean Morey and John Tinnell, 288. Anderson, SC: Parlor Press.

MacCash, Doug. 2016. "Deepwater Horizon Memorial Sculpture 'ELEVEN' Arrives in New Orleans." *NOLA.com*, September 10, 2016. https://www.nola.com/arts/index.ssf/2016/09/deepwater_horizon_sculpture_ne.html.

McGarrigle, Conor. 2017. "*NAMAland* [Dublin, 2010–2012]." In *Augmented Reality: Innovative Perspectives Across Art, Industry, and Academia*, edited by Sean Morey and John Tinnell, 276–77. Anderson, SC: Parlor Press.

Ratto, Matt, and Megan Boler, eds. 2014. *DIY Citizenship: Critical Making and Social Media.* Cambridge, MA: MIT Press.

Rieder, David M. 2017. *Suasive Iterations: Rhetoric, Writing, and Physical Computing.* Anderson, SC: Parlor Press.

Skwarek, Mark. 2014. "Augmented Reality Activism." In *Augmented Reality Art: From an Emerging Technology to a Novel Creative Medium*, edited by Vladimir Geroimenko, 3–29. New York: Springer.

Smith, Matt. 2015. "The Federal Government Says Billions of Animals Died from the BP Disaster." *Vice*, October 15, 2015. https://news.vice.com/en_us/article/ywjwdj/the-federal-government-says-billions-of-animals-died-from-the-bp-disaster.

State Library and Archives of Florida. 1998. *Designation of Buildings/Facilities/Programs/RoadsCong: Senate Bill 1014.* http://laws.flrules.org/node/1033.

Thiel, Tamiko. 2017. "*Clouding Green* [Singapore, 2012–2013]." In *Augmented Reality: Innovative Perspectives Across Art, Industry, and Academia*, edited by Sean Morey and John Tinnell, 272. Anderson, SC: Parlor Press.

Tinnell, John. 2017. *Actionable Media: Digital Communication Beyond the Desktop.* Oxford, UK: Oxford University Press.

Tinney, Craig. 2018. "Twitter REST API tools for Unity (Version 1.0) [Source code]." Craig Tinney. https://ctinney94.itch.io/unity-twitter-tools.

Ulmer, Gregory L. 2005. *Electronic Monuments.* Cambridge, MA: MIT Press.

United States Department of Homeland Security, United States Coast Guard, National Response Team. 2011. *On Scene Coordinator Report: Deepwater Horizon Oil Spill.* https://https://repository.library.noaa.gov/view/noaa/283.

Vee, Annette. 2017. *Coding Literacy: How Computer Programming Is Changing Writing.* Cambridge, MA: MIT Press.

SECTION 4

Critical Play as Critical Making

10

REPARATIVE MAKING
Re-Orienting Critical Making for Queer Worldmaking

Michael J. Faris
Texas Tech University

In recent years, the maker movement has begun to receive attention because of how the movement values personal making, collaboration, problem solving, play and tinkering, discovering and invention, and the potential to put into practice materialist theories of rhetoric (Breaux 2017; Rieder 2017; Sayers 2017; Sheridan 2016; Shivers-McNair 2018). However, makerspaces and the maker movement, despite the revolutionary and utopian rhetoric from their advocates, may not be the idealized models for composing that proponents claim. Unreflexively turning toward recent maker movements as models of making in rhetoric and composition risks ignoring issues of difference, identity, access, and power, reducing access to mere material access instead of what Adam J. Banks (2006) calls "meaningful access" (41).

In this chapter, I turn to queer making practices in order to reconsider Matt Ratto's concept of critical making for queer purposes. In Ratto's formulation, critical making is designed as a pedagogical practice to assist participants in developing embodied and material understandings of sociotechnical theories and critique (Ratto 2011; Ratto and Hertz 2015; Ratto and Hockema 2009). Drawing on Eve Kosofsky Sedgwick (2003), I repurpose critical making as *reparative making*, or restorative making for individuals and groups that can assist in queer worldmaking practices, opening up the world to new ways of thinking and being. In doing so, I am seeking out different paradigms, models, and traditions to think through critical making. Like Steven Hammer (this volume) and the contributors to Jentery Sayers's (2017) *Making Things and Drawing Boundaries*, I am suspicious of "the normative assumptions and effects of popular maker cultures—usually white, cisgender, straight, male, and able-bodied" and seek alternative "histories and paradigms of making" (Sayers 2017, 7).

My argument in this chapter draws on queer indie video game design and practice in order to conceptualize what it might mean and look like to practice reparative making as a queer practice, one that opens up the world to thinking and practicing "otherwise" (Muñoz 2009, 22). I first explain what I mean by *queer* and then argue that recent enthusiasm for the maker movement risks eliding concerns about difference, identity, access, and power. From there, this chapter conceptualizes *reparative making* as a form of critical making and provides examples from queer indie video game designers, exploring how queer makers use video games to create sustenance, reparative feelings, and the restoration of selves and communities in a hostile world. As a form of conclusion, I explore some of the complexities of access, queerness, and power regarding making. In many ways, my discussion in this chapter is motivated by a desire for the survival (and hopefully further—thriving) of those marginalized by sexuality, race, ability, gender identity, income, and other axes of difference. As Daniel Charny (2012) suggests in the zine *Critical Making: Manifestos*, "For many people, making is critical for survival" (5). Reparative making often makes that survival possible.

QUEERNESS AS DESIRING OTHERWISE

By queerness, I mean an orientation toward the world that believes the world could be "otherwise" (Muñoz 2009, 22). For Sara Ahmed (2006), queerness is an orientation toward the world and others that is "'oblique' or 'off line'" (161), that sees "the world 'slantwise'" (107). Thus, queerness is disorientation or ill-fitting in the world, and "an effect of being 'out of place' is also to create disorientation in others" (160) to, in effect, "disturb the order of things" (161). Because queerness involves such an ill-fitting or disorientation toward the world, it produces a "shared critical dissatisfaction" with the world as it is, as José Esteban Muñoz (2009, 189) explains. In this understanding, "Queerness is primarily about futurity and hope" (11). Further, "Queerness is not yet here; thus, we must always be future bound in our desires and designs" (185).

As I explore what reparative making might look like, I have in mind how queerness's disorientation toward the world involves inventing new ways of being in the world, often with and through new media and technologies. This inventional work is both critical and reparative. As Michael Warner (1993) suggests, queerness is a form of criticality: "Queers do a kind of practical social reflection just in finding ways of being queer" (xiii). But queerness is also reparative, about finding ways to survive and thrive and invent ways of being in an otherwise hostile world. Kara Keeling (2014) explains in her theorizing of Queer OS, "*Queer* offers a way of making perceptible presently uncommon senses in the interest of producing a/new commons and/or proliferating the senses of commons already in the making" (153).

Queerness as an orientation toward the world, an attempt to imagine it otherwise, is ideally also an intersectional orientation, one that takes into account how various axes of power and difference are mutually constitutive. Cathy Cohen (1997) argues that we should understand queer politics as based on "shared marginal relationship to dominant power," not solely on identity, in order to build coalitions (458). Such an approach, feminists of color and queer theorists have argued, involves studying power through an intersectional lens that understands that sexuality and gender are not isolated identity categories but co-constitutive with race, class, ability, citizenship status, region/location, nationality, religion, and other axes of difference/power (Cohen 1997; Combahee River Collective 1983; Crenshaw 1991; Ferguson 2004; Johnson 2001; Muñoz 1999).

ENTHUSIASM FOR AND EXCLUSIONS WITHIN THE MAKER MOVEMENT

In the 1990s, Gail Hawisher and Cynthia Selfe (1991) argued that "uncritical enthusiasm" for electronic writing classes risks reproducing the status quo by ignoring the social and political dynamics of culture, technology, and pedagogy (56). Likewise, we should be concerned about uncritical enthusiasm for making and the maker movement. Turning uncritically to the maker movement as a model for making in rhetoric and composition risks eliding issues of identity, difference, access, and power. Chet Breaux (2017), for example, in his argument for a maker pedagogy, draws on key figures in the maker movement to argue that making involves a do-it-yourself mentality, expands our notions of literacy, and can promote commons-based collaboration. While his history of making and craftwork is somewhat inclusive (going back to craftwork in the nineteenth century and including domestic spaces), his argument is overly optimistic about the maker movement, praising it as an inclusive site of collaboration (27).

But the maker movement, in its various instantiations, has been a site of exclusion, even if that exclusion is unintentional. Due to a variety of factors, makerspaces and similar movements are often utilized by majority men, and women and girls may experience barriers to entry, including feeling unwelcome or unsafe in male-centric spaces; lacking mentorship; coming from nontechnical backgrounds and thus having their ways of knowing, making, and inventing not valued in these spaces; discouragement by gender norms and stereotypes; and not having childcare (Bean, Farmer, and Kerr 2015; Faulkner and McClard 2014; Nafus 2011). Even for groups that have relatively egalitarian ethics, gendered divisions of labor may arise. For example, in her study of a "geek group" that tinkered with FM radio technologies, Christina Dunbar-Hester (2008) found that most of the tinkering fell to men, in part because a lack of resources and time in a volunteer group made teaching technical skills inefficient, and in part because some women in the group "felt that the dynamics of the group were masculine and competitive" (217). Even within feminist spaces, "Economic and cultural assumptions" about making as production—including that one should have "the time, resources, abilities, and support to become cultural producers"—can make these spaces exclusive (Chidgey 2014, 106). As Dunbar-Hester (2014) concludes elsewhere,

> An unforeseen and unintended consequence of maker identity as an activist strategy was the persistence of social structure and identity categories that have historically limited access to technical artifacts and expertise: the proposition that technical participation can be a route to social or civic egalitarianism tends to elide a history in which the codes of masculinity and whiteness have been codified around dominance, over both technology and other social groups. (84)

While I've mostly focused on gender here, makerspaces can also be exclusionary based on race, social class, ability, location, and a variety of other factors. In short, as Phoebe Sengers observes, the maker movement seems to privilege the cultural elite, whose middle class and urban status affords them opportunities to engage in makerspaces and gain attention for that work (Sengers and Hertz 2015, 16). We should, then, be cautious of too quickly extolling the virtues of the maker movement.

One way to resist too much enthusiasm is by complicating our notions of access. For example, Breaux (2017) focuses primarily on material access to technologies, observing that Raspberry Pis and Arduino kits are relatively inexpensive (30). But, as Banks (2006) argues, conceiving of access as simply material access reduces technologies to mere tools, ignoring how those technologies work within networks of power, information, and economics (40). Banks proposes that we understand access more broadly. In addition to material access, we need to conceive of access as including *functional access* (access to skills and knowledges), *experiential access* ("access that makes the tools a relevant part of their lives"), *critical access* (access to abilities to critique and challenge technologies within systems of use), and *transformative access* (41–45). Importantly, transformative access involves not mere inclusion but the ability to change both the technologies and "the codes that determine how the system works" (45). In the maker movement, access is not automatically transformative because cultural codes around difference are often unaddressed. As we turn to making and critical making in rhetoric and composition, then, we might turn to a wider array of histories and practices as models for making and critical making than the maker movement.

TOWARD REPARATIVE MAKING

Ratto (2011) developed the concept of critical making with the goal of using "material forms of engagement with technologies to supplement and extend critical reflection and, in doing so, to reconnect our lived experiences with technologies to social and conceptual critique" (253). As he explains, critical making attempts to combine two intellectual traditions: critical thinking, especially sociotechnical critique, which, as an "abstract, explicit, linguistically-based" way of thinking can inhibit embodied understandings of critical concepts; and physical making with objects, which involves embodied and tacit knowledge (Ratto and Hockema 2009, 52).

My goal in this chapter is to suggest that we might explore a different type of critical making—reparative making. Reading Ratto's work, I am struck by the rather narrow range of sociotechnical theorists he draws upon (e.g., Heidegger, the Frankfurt School, Latour). My hope is that by turning toward reparative making, we might open the histories and ways of making that we consider critical making, moving beyond traditions in critical social theory that, as Warner (1993) observes, has a history of neglecting issues of sexuality and even of marginalizing queer ways of being in the world (viii–ix).

In the spirit of arguing for more capacious and promiscuous traditions, histories, and models for critical making, I suggest that we consider how making is also reparative and how reparative making should be valued in rhetoric and composition. In "Paranoid Reading and Reparative Reading," Sedgwick (2003) argues that paranoia has become "coextensive" with critical practice in the humanities and queer theory (126). By paranoia, Sedgwick means those critical readings driven by a hermeneutics of suspicion that provide a strong theory, setting out to "prove the very same assumptions with which it began" (135) and hoping to expose the truth of a phenomenon through demystifying the ideological structures underpinning it. As Sedgwick argues, paranoia makes "it less rather than more possible to unpack the local, contingent relations between any given piece of knowledge and its narrative/epistemological entailments for the seeker, knower, or teller" (124)—or, we might add, maker.

In place of paranoid readings, Sedgwick (2003) offers what she calls *reparative readings*, or readings that attend to pleasures, ameliorations, everyday theories, hope, and localized, contingent practices. As Sedgwick explains, reading reparatively means attending to "contingent developments" (147), tracing affects and paying attention to how individuals and collectives survive and thrive in a world we already know to be threatening to them. Thus, as reparative reading is a project in learning from "the many ways selves and communities succeed in extracting sustenance from the objects of culture—even a culture whose avowed desire has often been not to sustain them" (150–51), reparative making entails the *making of those objects that create sustenance, reparative feelings, and restoration of selves and communities.*

I am not the first to connect Sedgwick's concept of reparative reading to making and craftwork. Video game designer and theorist Kara Stone (2018) shares her experiences making video games and argues that "reparative game design" involves a "slow and sensual process of crafting [that] can be a healing experience." The experiences of creating and playing queer games, Stone suggests, can reorient makers and creators to time, creating a queer or crip time that rejects the normative logics of productivity and reproduction (see Halberstam 2005). Stone's game *Ritual of the Moon*, she argues, through both production and play, challenges both normative imperatives to be happy and therapeutic solutions that end depression: "There is no end point because there is always something that is draining" (Stone 2018). For Stone, "Reparative game design is a method to work through difficult feelings, but it is also a means to stay in them as long as they need to be felt. . . . Sometimes the work is to make yourself and others feel *worse*." Thus, reparative making becomes about the mundane rather than the exceptional, "a daily ritual" for surviving and potentially healing and thriving in a world that is anti-queer, anti-crip, exploitative financially and emotionally, and racist.

Projects like Stone's and studies of the literacy and performance practices of queers of color (Muñoz 1999; Pritchard 2017) make clear that marginalized folks turn to literacy and making for acts of self-care, wellness, and worldmaking. Consequently, I understand reparative making as a mode of critical making that is both process- and product-based, attending to how making can be a restorative practice for selves and collectives and how critical making can open up possibilities for queer worldmaking, opening up the world to new ways of thinking and being. In this way, reparative making carries forward Ratto's vision that "critical making names a mode of engagement in the world that is about seeing and making a world that has somewhat different characteristics from the world that we live in now" (Ratto and Hertz 2015, 53). But rather than focus on learning sociotechnical critique, participants in reparative making engage in practices that assist in reparative and worldmaking projects for both makers and users. I turn, then, to examples of reparative making in which marginalized subjects who, through their mundane, everyday practices, invent new ways of being in the world and in relation with each other.

QUEER INDIE VIDEO GAMES AS REPARATIVE MAKING

What does it mean to make games when we're unhealthy? When we're under threat of violence? When we're hungry? When we have no money?
—Porpentine (2013)

I really hate this world and so when I make something I'm making something I can bear to live in.
—Porpentine (quoted in Iadarola 2017)

These two remarks by indie video game maker Porpentine suggest the reparative potential of queer indie video games. Identifying as a "queer tranarchafeminist" (Petit 2013), Porpentine (2013) frames making video games as an act of self-care: They are "radical acts of joy to combat terror." Making video games is about survival and thriving and about communicating that survival in evocative, sometimes disruptive, disorienting, or estranging ways. As Porpentine explains, "I don't have time to explain to 1000 people about why I deserve to live and why being trans is beautiful, but I can make 1 game about it that 1000 people play."

Porpentine's games often communicate trans experiences through poetic language that disorients players from typical game play or typical understandings of reality. Like many other indie designers, Porpentine uses Twine (http://twinery.org/), an open-source WYSIWYG developer tool that exports a project into HTML and CSS, to create her video games. (See Kendall Gerdes's chapter in this collection, where she argues that Twine games provide alternatives to the norms of mainstream video games.) In her browser-based game *Parasite* (http://slimedaughter.com/games/twine/parasite/#b), players follow the experiences of a narrator who decides to try to sell some possessions because "i don't have enough money to buy the hormones i need to survive." However, these "possessions" turn out to be dreamscapes that later will be sold to clients for entertainment or distraction. The game is entirely text-based, and players select hyperlinked text (in the mode of choose your own

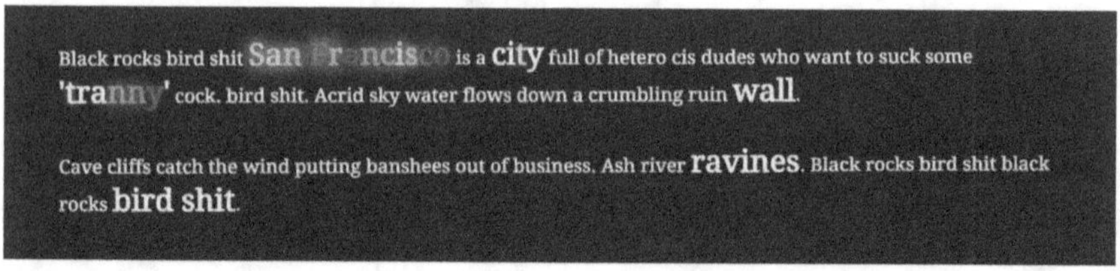

> Black rocks bird shit San Francisco is a city full of hetero cis dudes who want to suck some 'tranny' cock. bird shit. Acrid sky water flows down a crumbling ruin wall.
>
> Cave cliffs catch the wind putting banshees out of business. Ash river ravines. Black rocks bird shit black rocks bird shit.

Figure 10.1. Scene from the dreamscapes in Porpentine's Parasite. *Larger text represents links, which, when hovered over, fade away and then slowly return. When clicked, some text changes (e.g., from "San Francisco" to "America" to "the world"), and others proceed through the game. Screenshot by author.*

adventure games) and are taken through a disorienting world in which the narrator and the buyer traverse her dreamscapes together. At one point, the dream-buyer tells the narrator of her dreamscape, "I don't think our studio will be interested in this region. Many of our customers are cis males," but the art department can edit the dreamscape for customers, and "we can even leave the trauma behind." The dreamscapes are haunted with the traumas of living in a cis hetero world—visualized through disappearing text, changing text, and text presented at disorienting angles (see figure 10.1)—and the narrator knows that just like the rest of the city, country, and world, this dream-selling industry is made for cisgender heterosexuals, whose "heterosexuality perfectly incorporates everything they want to fuck, staining everyone but themselves."

Queer indie video games like *Parasite* are often made as survival or reparative tactics in a world quite hostile to queerness, gender nonconformity, and mental disability and illness. In *Videogames for Humans*, game designer and theorist merritt kopas (2015) explains, "Twine changed my life" (5). Kopas discovered Twine games while in graduate school, and playing and producing those games became "the main outlet through which I processed my emotions, working through personal and political struggles by making something out of them" (6–7). In contrast to her experiences in graduate school—where she divorced her emotional self from her work—kopas was able to use video game design to make "works about family, love, sex, bodies: things I'd never been able to examine directly before" (7). For kopas, the accessibility of Twine—both as a designer and a player—was central to the queer indie video game scene. Most Twine games are simple and text-based, and players use a mouse (and sometimes a keyboard) to interface with the game. For designers, a simple WYSIWYG developer tool allows easier entry into game development than a program that requires coding. Twine's ease of entry made inventional work possible for kopas: "When I opened up Twine, I felt free to just start writing fragments, each in their own passage. The connections could come later" (7).

With the development of accessible video game creation software like Twine, indie video games have proliferated in recent years, especially indie games made by queer and transgender individuals and small teams. Indeed, queer orientations toward video games are garnering increasing attention for their experimental approaches that reimagine game play and logics (Harvey 2014; Keogh 2013; Ruberg 2018; Ruberg and Phillips 2018; Ruberg and Shaw 2017; Shaw 2015). These games are often short (sometimes as short as ten seconds, in the case of Anna Anthropy's *Queers in Love at the End of the World*, discussed in Gerdes's chapter in this collection) and rather simple in design. As Naomi Clark (2017) observes, "It's the refusal to obey orthodox conventions about games, and a willingness to embrace bare systems, that makes it easier for queers to achieve striking new forms of interplay and consonance between the experiences and aspects of queer existence they represent and the structures of interaction that players encounter" (9).

In addition to being sites of reparative making for game designers, these games are also reparative sites for players as well. What queer indie video games do is reorient players to their environments, to others, and to their own bodies. Many of these games are cases of reparative making: rather than provide an explicit critique of sexual, gender, and able-bodied norms, they provide an operant theory of embodied rhetoric, remaking experiences with the world. As Lisa Nakamura (2016) argues, these games are accessible to nongamers, providing "new types of access: access to the pleasure of games, and to new types of feelings" (39). These games provide "alternative ways of being" that "purposefully complicate rather than distribute representations of marginalized people's lives" (Ruberg 2018, 426). (See Wendi Sierra's chapter in this collection, where she discusses the limitations of representations of Indigenous characters in games and shares experiences creating video games with Indigenous participants.) Let me provide a few more examples of queer reparative making through queer indie video game design, making, and play: another game that queerly (dis)orients players toward the world, the possibilities of queering physical game interfaces, and an example of making spaces for queer and feminist inclusion.

Queering Play, Queering Feelings: No World Dreamers
Porpentine's *No World Dreamers: Sticky Zeitgeist* (http://slimedaughter.com/sticky_zeitgeist/) provides another example of queer reparative

making, as it refigures gameplay as not tied to skill and success but rather to depression and estrangement. As Porpentine's game and others exhibit, queer indie games provide "ways of being and knowing that stand outside of conventional understandings of success," like "failing, losing, forgetting, unmaking, undoing, unbecoming, not knowing," as Jack Halberstam (2011, 2) suggests of queer modes of being in the world. In this way, queer games allow players to play with "failure as a way of life" that doesn't attempt to avoid disappointment or incorporate failure into logics of success (186; see also Ruberg 2015).

Episode 1: Hyperslime of *No World Dreamers* opens in a futuristic world with the main character, Ever—a transgender humanoid with animal facial and ear features—masturbating and getting "high on girlchunks." However, this scene is quickly interrupted by a text from "M.O.M." informing Ever that she needs to attend orientation at work. But before navigating Ever out of her home and off to work, the player spends some time interacting with items in Ever's bedroom and the rest of her home. One option a player is presented with is rereading her old blog posts, and in one of the few moments where the player is directly addressed, they are asked, "Why are you reading this. Don't you have something better to do than seek out other people's sickness?" When the player navigates to the living room, Ever thinks, "I can't imagine taking up this space." After leaving the house and waiting for the bus, Ever is reminded via text that she's running late, and she experiences a drug-induced panic attack as text moves about the screen: "IT'S TOO MUCH TO HANDLE." At orientation, Ever takes a urine test and has a psychological exam in which we learn more about her depression: She can barely remember a time when she was happy. But Ever passes the exam, and Episode 1 ends as Ever learns she has her first mission—to collect debris that has fallen from space.

Unlike mainstream video games, Episode 1 of *No World Dreamers* isn't a game designed for winning, and player choices don't affect the outcome of the game. Instead, the game narrates Ever's day and journey in depression, but not in a way to resolve that depression. As Brendan Keogh (2013) writes, queer indie games "are frequently less concerned with being beaten or mastered, and more concerned with being participated with in order to communicate an idea." Rather, players sit with Ever through the rather estranging world of being depressed in a post-apocalyptic future. And players sit with these feelings of failure, disappointment, and forgetting. The game further resists empathizing with Ever or with Porpentine—we are presented with a futurescape that is just estranging enough that we aren't meant to "feel with" Ever, but rather to observe and be with her.

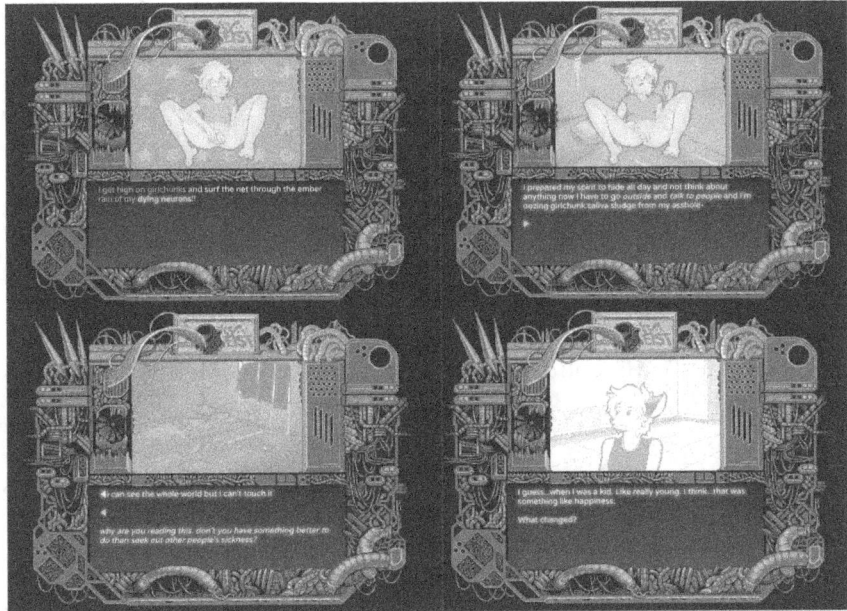

Figure 10.2. Scenes from Episode 1 of Porpentine's No World Dreamers. *Top Left: Ever gets high on girlchunks and masturbates. Top Right: Ever is interrupted by a text from M.O.M. requesting she attend orientation. Bottom Left: While reading Ever's blog, the player is chastised, "why are you reading this? don't you have something better to do than seek out other people's sickness?" Bottom right: Ever explains that she once experienced "something like happiness" when she was "really young." Screenshots by author.*

Players are challenged to ask why they are reading a depressed woman's blog, follow along as Ever struggles to leave the house and get on the bus, and witness how Ever lives with depression. Thus, players aren't meant to identify with Ever and inhabit her subject position; instead, they become witnesses who leave behind traditional goals of winning and sit alongside Every on a journey in depression. (On how queer indie games challenge the notion of empathy, see Pozo 2018; Gerdes, this collection.)

Porpentine's games are often disorienting and confusing. She has observed, "People don't really know how to talk about my work. . . . If people could talk about it more easily it would mean it wasn't about the unspoken things that I care about, or the slimy, decomposed things. As long as people struggle with it I know I'm good" (quoted in Iadarola 2017). *No World Dreamers* acts as reparative making—both in the process of making, as I've discussed above, and through the critical act of play as it opens up new ways of thinking about depression, trans experiences, and one's position in the world.

Queering Physical Interfaces

While I have focused so far on indie video games played through browsers, physical interfaces, like game controllers, are also sites of queer making. When we consider video games, we often think about the visual aspects of the game on screen but rarely about the physical interface—the joystick, keyboard, mouse, or console controller (though touch screens have gained some attention because of their relative newness). Game controllers are both physical interfaces with a game and tools for controlling actions and choices in the game. They are also sites of exclusion, often assuming "a universal body . . . with two mobile hands able to act independently" (Marcotte 2018, quoting Boluk and LeMeiux) and requiring intensive time and practice to develop proficiency (Anthropy 2012, 15)—what Jess Marcotte (2018) calls control literacy, a habituated ability to manipulate controllers that becomes naturalized to players.

Marcotte (2018), Gregory L. Bagnall (2017), and Miguel Sicart (2017) have argued for reimagining game controllers in order to make gameplay more accessible, to potentially estrange players from environments, and to reimagine gameplay and embodied interactions with games themselves. Marcotte (2018) argues for queer games controllers so that "we can access more ways to question, transform, resist, imagine, and bring difference to game design more broadly." She suggests, "Alternative controllers represent a desire to see, make, and play with something that does not exist within the standard set of interfaces, and there is queer, intersectional feminist, political potential in that speculative possibility space." And, Sicart (2017) argues, "We need controllers for the alternative emotions, alternative bodies, and alternative experiences that games now foster." Sicart suggests both that we redesign already-existing controllers to allow for new inputs and outputs and that we design "queer controllers from scratch."

Because of the relatively low cost of physical computing and other making equipment, prototyping and making alternative controllers is now more of a possibility for game players and designers. Take Marcotte and Dietrich "Squinky" Squinkifer's game *Rustle Your Leaves to Me Softly*, developed out of the Technoculture, Arts and Games Research Centre at Concordia University. This game is "an ASMR Plant Dating Simulator" in which players simulate "dating" a plant, connected to Arduino sensors, by touching and stroking it. As Marcotte (2017) explains, she researched autonomous sensory meridian response (ASMR) communities, and she and Squinkifer designed *Rustle* so that "the human inputs caresses, touches, and other inputs (such as blowing wind across the

Figure 10.3. Still from promotional video for Rustle Your Leaves to Me Softly *(Squinkifer 2017).*

plant's surface, or speech) and the plant, in this narrative, responds to the touches that it appreciates by outputting an ASMR soundscape that it hopes will be pleasing to the human." In order to dissuade players from seeing the plants as merely "controllers," Marcotte designed and wrote dating profiles for the four plants, and the soundscape produced by the game included remixed poetry and a variety of pleasing soundscapes. Marcotte (2017) was surprised by the intimacy created with the game and how frequently players, after their initial laughter, became "quiet, contemplative." By considering different materials for construction, different senses and organs to appeal to, and different ways of interfacing with video games, we might develop more accessible game play and create "new ways to use our bodies in video games" (Marcotte 2018).

Reparative making of physical interfaces can range from the whimsical, to the ameliorative, estranging, and affective, to the assistive and social. And they could go further, becoming what Sara Hendren (2011) calls "tools for estrangement" (59), or "speculative and practical technologies that upend all our expectations about what 'assistive aids' should do, who they are for, and how mysterious and often invisible the economy of human needs really is" (62).

Queering Spaces and Inclusivity

Another mode of queer reparative making in video game cultures is rethinking and practicing anew the codes of working together in makerspaces, hackerspaces, and similar initiatives. As I noted above, even those spaces that attempt to be egalitarian often reproduce normative power differentials along lines of access, difference, and identity. Stephanie Fisher and Alison Harvey (2013) have shown that interventionist approaches can reproduce power structures when they start with assumptions about participants—constructing those participants' subject positions as naive, mystified, intimidated, or lacking access—and view education as a unidirectional process rather than a reciprocal one in which both organizers and participants learn from each other (29). That is, even spaces and initiatives that are developed to address gender disparity can reproduce gendered norms and power structures if they don't take into account how invisible power differences are at play for participants (31).

Some spaces and organizations, however, engage in what I am calling reparative making by focusing more on reciprocal education and challenging the codes of culture in order to create what Banks (2006) calls transformative access. These spaces are akin to the Women's Studies Multimedia Studio at the University of Maryland that Melissa Rogers (2017) discusses: "Rather than producing 'things,' the activists . . . produce *space* for modes of thought" that can challenge our typical conceptions of making and "the digital" (234). One such organization is Dames Making Games (DMG, https://dmg.to/), "A self-organized hub for female-organized developers" located in Toronto and founded in 2012 (Fisher and Harvey 2013, 37). In contrast to makerspaces and similar sites that attempt to be inclusive of women without considering structural codes of inequality, DMG is explicitly feminist and intersectional in its practice and design. DMG makes it a priority to serve those most marginalized in society, taking an antiracist, antioppression approach that "prioritize[s] people who are traditionally marginalized in tech and game spaces" along axes of gender, race, and disability (DMG Toronto 2019). Through development workshops, game design jam sessions, speaker socials, roundtables, mentorship programs, and collaboration with other social justice–oriented organizations, DMG creates a reparative and transformative space in Toronto for cisgender and transgender women, genderqueer and nonbinary individuals, two-spirit individuals, and others marginalized by race, gender, sexuality, ability, and access.

DMG provides space (both a physical space and an intellectual/social space through a variety of engagements) to transform the codes of gender, race, economics, and video game design and play. While DMG provides material access to video game making, organizers also understand that focusing solely on material access is a limited approach: it leaves in place social structures and assumes that what is already available is enough. As co-founder Jennie Faber explains, "We are dissatisfied with technology available for making games" because that technology can encourage traditional game design; new representations and game play require rethinking the technologies used to design games (quoted in Kaszor 2014). Through their programming, DMG promotes functional access through teaching skills and knowledges; experiential access in which video game making is relevant to participants' lives; critical access in which participants can critique video games and the social, cultural, and economic structures of the video game industry; and transformative access in which participants and organizers can create "alternative forms of economic power grounded in solidarity, openness and collective values" (Dames Making Games n.d.).

We might consider DMG, then, as a site of reparatively making culture, as worldmaking. This approach entails challenging and reimagining cultural codes around gender, race, access, and even making itself, as DMG "challenges the solo auteur vision of the indie designer in general by emphasizing the important role of colleagues, collaborators and support groups for those seeking to engage in independent game design" (Fisher and Harvey 2013, 37). As DMG director Amy Leitch explains, one goal of DMG and its participants is "to make a thing that's good enough to convince ourselves that we can change and have some say in how we live our lives and what we do with them" (quoted in Westecott 2013, 85).

OTHER POTENTIAL SITES OF QUEER REPARATIVE MAKING

I have, thus far, focused mostly on video games as sites of queer reparative making, but reparative making is not limited to video games. Indeed, queers, feminists, transgender folks, folks with disabilities, and people of color have turned to a wide variety of media and practices for reparative making, including zines, which can "operate as critical making tools capable of fusing vernacular theorizing and material production" (Chidgey 2014, 105); indie comics from feminist, disabled, and queer perspectives (e.g., Walters 2015); craftwork and craftivism (e.g., Orton-Johnson 2014; Rogers 2017); and both nondigital and online queer and trans archival making, where individuals and collectives work together to make "a world where trans lives count, a world where everyone makes history, and a world of shared experience" (Rawson 2014, 39).

Other sites of potential queer reparative making include new media art, physical computing, and coding that reimagine computing and embodiment, like Micha Cárdenas and Elle Mehrmand's *technésexual*. *Technésexual* combined biometric sensors to track heart rates and body temperatures in a live performance of erotic acts to create an "'organic interface,' or a physical interface that responds to one's organs" (Cárdenas et al. 2011, 92). These biometric sensors, connected to an Arduino kit, translated heart rates and body temperature into an aural experience for the audience: the sounds were broadcast from a visual project of Cárdenas's and Mehrmand's avatars (also engaged in erotic encounters) in *Second Life* (98–100). Consequently, "bodies become instruments" (99) and organic interfaces created new forms of intimacy—both between performers and with the audience (101–2).

By focusing on video games, I have hoped to be suggestive and not limit an understanding of the potentials for reparative making, which I see as a concept that can broaden our understanding of critical making to include both processes of making and the final products that can help create sustenance, repair selves and feelings, and challenge and estrange norms.

CONCLUSION: ACCESS, QUEERNESS, AND POWER

Earlier in this chapter, I warned against too readily turning to maker rhetoric as revolutionary, inclusive, and accessible. I want to continue that warning with queer reparative making as well. Queerness may be utopian in that it desires a world otherwise, but in practice, queerness is constantly on the horizon, never actually achieved (Muñoz 2009, 185). Likewise, the queer video game cultures I discuss in this chapter are not fully utopian in practice. Kopas (2015) explains that despite the popularity of some games and designers, "A very small number of authors gained visibility . . . and almost all of them still struggle with material insecurity" (8). Many queer indie video game makers are critical of capitalist modes of circulation, as they allow conditions in which "some people get a little famous, nobody gets rich, and years later, people who have more resources than you steal your ideas and use them to get richer and more famous than they already were" (8). Likewise, in her acceptance speech at the 2012 Arse Elektronika conference on games and sex, Anthropy (2013) pushed back against a focus on technology: "There's something more important to games about sex than technology, and that's *access to technology*" (163). As she observes, "Almost all of the queer people I know, including myself, live in poverty" (163).

I want to close with some self-reflexivity—to attend to my "own practices of knowing and experiencing" (Sedgwick 2003, 145)—and concerns about the maker movement in rhetoric and composition and about queerness in making. First, who benefits by our turn to maker movements? How does our work as scholars and teachers perpetuate a focus on leading figures in maker movements and critical theory and ignore those who are most marginalized? If we understand rhetoric as a matter of circulation (Porter 2010), what does our circulation of certain ideas over others contribute to the actual lives of those who make, both inside and outside of the academy? What does it mean that I, a casual and inconsistent video game player, have turned to queer indie video games as a model of reparative making without contributing anything (other than a few dollars to buy games and books) to that community?

And what risks do I make in promoting queerness for critical making in rhetoric and composition? Jonathan Alexander and Jacqueline Rhodes (2011) have argued that queerness may be antithetical to composition because of its attention to rupture, excessiveness, and decomposition. Though perhaps there is room to queer composition in individual classes, especially as we turn to new media and their medial logics that challenge us to think differently about composing than print literacies do (Alexander and Rhodes 2014). However, as rhetoric and composition teachers attempt to queer their assignments and classrooms, how do we prevent queerness from becoming a mere "generic badge of subversiveness, a more trendy version of 'liberal'" that is "abstracted from the quotidian realities of lesbian and gay male life," as David M. Halperin (2003, 341, 343) warns. How might attending to queerness in making "help us think what has not yet been thought" (343) in rhetoric and composition? And further, when we do turn to queerness, how do we think through identities and issues of power as intersectional, so that queerness doesn't continue to be, as it has been in some theoretical circles, a largely white, middle-class, urban enterprise (Cohen 1997; Ferguson 2004; Halberstam 2005; Johnson 2001; Muñoz 1999).

I don't have easy answers to these questions, but I can gesture toward the work of my former colleague Kendall Gerdes, whose teaching and scholarship has animated some of my thinking in this chapter, and who introduced me to some of the video games I discuss here. In her upper-division composition course at Texas Tech, Kendall themed the course on "weird video games," asking students to play a variety of games that challenge their conceptions of what constitutes typical gameplay and how rhetoric and writing could be conceived. Toward the end of the course, after reflecting on play, analyzing games, and engaging with a variety of interfaces, students created their own weird video games. There was no explicit injunction in this course to perform queerness, but Kendall opened spaces for queer and antinormative approaches to making in this course. I think Kendall's course worked so well for students (it was a quite popular course) because Kendall opened the world to new ways of thinking, of being, and of being in relation with each other. This ultimately should be a what a maker pedagogy is about: How can we imagine new possibilities for being together—with other humans, with objects, and with designed environments and objects—in a world that often doesn't want many of us to even exist. As Stone (2018) asks, "What can academia do to heal, to comfort, and to reform dominant understandings of the world into something that enables queer people to keep living"? If making for some people is simply about survival, how can we enable such reparative making in our classes and curriculum?

Acknowledgments. I would like to acknowledge that this chapter was written within the historical territories of the Teya, Jumano, Apache, and Comanche peoples. I would also like to thank Kendall Gerdes for introducing me to many of the video games I discuss (and many I don't discuss) in this chapter and helping me be a better colleague, teacher, scholar, and human. I am indebted to Kendall for much of my thinking about video games and affect and feelings in this chapter.

REFERENCES

Ahmed, Sara. 2006. *Queer Phenomenology: Orientations, Objects, Others.* Durham, NC: Duke University Press.

Alexander, Jonathan, and Jacqueline Rhodes. 2011. "Queer: An Impossible Subject for Composition." *JAC* 31 (1–2): 177–206.

Alexander, Jonathan, and Jacqueline Rhodes. 2014. *On Multimodality: New Media in Composition Studies.* Urbana, IL: National Council of Teachers of English.

Anthropy, Anna. 2012. *Rise of the Videogame Zinesters: How Freaks, Normals, Amateurs, Artists, Dreamers, Dropouts, Queers, Housewives, and People Like You Are Taking Back an Art Form.* New York: Seven Stories Press.

Anthropy, Anna. 2013. "Acceptance Speech for Prixxx Arse Elektronika 2012: Play." In *Screw the System: Explorations of Spaces, Games and Politics through Sexuality and Technology,* edited by Johannes Grenzfurthner, Guenther Friesinger, and Daniel Fabry, 162–63. San Francisco, CA: RE/SEARCH.

Bagnall, Gregory L. 2017. "Queer(ing) Gaming Technologies: Thinking on Constructions of Normativity Inscribed in Digital Gaming Hardware." In *Queer Game Studies,* edited by Bonnie Ruberg and Adrienne Shaw, 134–43. Minneapolis: University of Minnesota Press.

Banks, Adam J. 2006. *Race, Rhetoric, and Technology: Searching for Higher Ground.* Mahwah, NJ: Lawrence Earlbaum and National Council of Teachers of English.

Bean, Vanessa, Nicole M. Farmer, and Barbara A. Kerr. 2015. "An Exploration of Women's Engagement in Makerspaces." *Gifted and Talented International* 30 (1–2): 61–67. doi:10.1080/15332276.2015.1137456.

Breaux, Chet. 2017. "Why Making?" *Computers and Composition* 44: 27–35. doi:10.1016/j.compcom.2017.03.005.

Cárdenas, Micha, Elle Mehrmand, Amy Sara Carroll, Ricardo Dominguez, Brian Holmes, James Morgan, Allucquére Rosanne Stone, and Stelarc. 2011. *The Transreal: Political Aesthetics of Crossing Realities.* New York: Atropos Press.

Charny, Daniel. 2012. "Power of Making." In *Critical Making: Manifestos,* edited by Garnet Hertz, 1–6. Hollywood, CA: Telharmonium Press. http://www.conceptlab.com/criticalmaking/.

Chidgey, Red. 2014. "Developing Communities of Resistance? Maker Pedagogies, Do-It-Yourself Feminism, and DIY Citizenship." In *DIY Citizenship: Critical Making and Social Media,* edited by Matt Ratto and Megan Boler, 101–13. Cambridge, MA: MIT Press.

Clark, Naomi. 2017. "What *Is* Queerness in Games, Anyway?" In *Queer Game Studies,* edited by Bonnie Ruberg and Adrienne Shaw, 3–14. Minneapolis: University of Minnesota Press.

Cohen, Cathy J. 1997. "Punks, Bulldaggers, and Welfare Queens: The Radical Potential of Queer Politics." *GLQ* 3 (4): 437–65. doi:10.1215/10642684-3-4-437.

Combahee River Collective. 1983. "A Black Feminist Statement." In *This Bridge Called My Back: Writings by Radical Women of Color,* edited by Cherríe Moraga and Gloria Anzaldúa, 210–18. 2nd ed. New York: Kitchen Table Press.

Crenshaw, Kimberle. 1991. "Mapping the Margins: Intersectionality, Identity Politics, and Violence against Women of Color." *Stanford Law Review* 43 (6), 1241–99. doi:10.2307/1229039.

Dames Making Games. n.d. "Dames Making Games: DMG Toronto: About." *Dames Making Games.* Accessed January 20, 2020. https://dmg.to/about.

DMG Toronto. 2019. "Member Manual." *Dames Making Games.* https://manual.dmg.to/.

Dunbar-Hester, Christina. 2008. "Geeks, Meta-Geeks, and Gender Trouble: Activism, Identity, and Low-Power FM Radio." *Social Studies of Science* 38 (2): 201–32. doi:10.1177/0306312707082954.

Dunbar-Hester, Christina. 2014. "Radical Inclusion? Locating Accountability in Technical DIY." In *DIY Citizenship: Critical Making and Social Media,* edited by Matt Ratto and Megan Boler, 75–88. Cambridge, MA: MIT Press.

Faulkner, Susan, and Anne McClard. 2014. "Making Change: Can Ethnographic Research about Women Makers Change the Future of Computing?" *Ethnographic Praxis in Industry Conference Proceedings,* 187–98. doi:10.1111/1559-8918.01026.

Ferguson, Roderick A. 2004. *Aberrations in Black: Toward a Queer of Color Critique.* Minneapolis: University of Minnesota Press.

Fisher, Stephanie, and Alison, Harvey. 2013. "Intervention for Inclusivity: Gender Politics and Indie Game Development." *Loading . . . The Journal of the Canadian Game Studies Association* 7 (11): 25–40. http://journals.sfu.ca/loading/index.php/loading/issue/view/11.

Halberstam, J. Jack. 2005. *In a Queer Time and Place: Transgender Bodies, Subcultural Lives.* New York: New York University Press.

Halberstam, J. Jack. 2011. *The Queer Art of Failure.* Durham, NC: Duke University Press.

Halperin, David M. 2003. "The Normalization of Queer Theory." *Journal of Homosexuality* 45 (2–4): 339–43. doi:10.1300/J082v45n02_17.

Harvey, Alison. 2014. "Twine's Revolution: Democratization, Depoliticization, and the Queering of Game Design." *GAME: The Italian Journal of Game Studies* 3: 95–107. https://www.gamejournal.it/3_harvey/.

Hawisher, Gail E., and Cynthia L. Selfe. 1991. "The Rhetoric of Technology and the Electronic Writing Class." *College Composition and Communication* 42 (1): 55–65.

Hendren, Sara. 2011. "Towards an Ethics of Estrangement." *Organs Everywhere* 3: 52–63. https://organseverywhere.com/pdf/OE_3_Cyborgs-and-Monsters.pdf.

Iadarola, Alexander. 2017. "Porpentine Charity Heartscape." *Mask Magazine,* May 2017. http://maskmagazine.com/the-carnal-issue/work/porpentine-charity-heartscape.

Johnson, E. Patrick. 2001. "'Quare' Studies, or (Almost) Everything I Know about Queer Studies I Learned from My Grandmother." *Text and Performance Quarterly* 21 (1): 1–25. doi:10.1080/10462930128119.

Kaszor, Daniel. 2014. "Jennie Faber Explains How DMG (Dames Making Games) Aims to Make Game Creation More Inclusive." *Financial Post*, October 24, 2014. https://business.financialpost.com/technology/gaming/jennie-faber-explains-how-dmg-dames-making-games-aims-to-make-game-creation-more-inclusive.

Keeling, Kara. 2014. "Queer OS." *Cinema Journal* 53 (2): 152–57.

Keogh, Brendan. 2013. "Just Making Things and Being Alive about It: The Queer Games Scene." *Polygon*, May 24, 2013. https://www.polygon.com/features/2013/5/24/4341042/the-queer-games-scene.

kopas, merritt, ed. 2015. *Videogames for Humans: Twine Authors in Conversation*. New York: Instar Books. http://www.instarbooks.com/books/videogames-for-humans.html.

Marcotte, Jess. 2017. "GGJ 2017: Some Thoughts on Rustle Your Leaves to Me Softly." *Adventures in Gaming, Critical Making, Game Jams, Research*, January 28, 2017. http://tag.hexagram.ca/jekagames/ggj-2017-some-thoughts-on-rustle-your-leaves-to-me-softly/.

Marcotte, Jess. 2018. "Queering Control(lers) Through Reflective Game Design Practices." *Game Studies: The International Journal of Computer Game Research* 18 (3). http://gamestudies.org/1803/articles/marcotte.

Muñoz, José Esteban. 1999. *Disidentifications: Queers of Color and the Performance of Politics*. Minneapolis: University of Minnesota Press.

Muñoz, José Esteban. 2009. *Cruising Utopia: The Then and There of Queer Futurity*. New York: New York University Press.

Nafus, Dawn. 2011. "'Patches Don't Have Gender': What Is Not Open in Open Source Software." *New Media & Society* 14 (4): 669–83. doi:10.1177/1461444811422887.

Nakamura, Lisa. 2016. "'Putting Our Hearts into It': Gaming's Many Social Justice Warriors and the Quest for Accessible Games." In *Diversifying Barbie and Mortal Kombat: Intersectional Perspectives and Inclusive Designs in Gaming*, edited by Yasmin B. Kafai, Gabriela T. Richard, and Brendesha M. Tynes, 35–47. Pittsburgh, PA: Carnegie Mellon ETC Press.

Orton-Johnson, Kate. 2014. "DIY Citizenship, Critical Making, and Community." In *DIY Citizenship: Critical Making and Social Media*, edited by Matt Ratto and Megan Boler, 141–55. Cambridge, MA: MIT Press.

Petit, Carolyn. 2013. "An Industry Divided." *GameSpot*, April 1, 2013. https://www.gamespot.com/articles/an-industry-divided/1100-6406257/.

Porpentine. 2013. "Parasite." *The New Inquiry*, June 25, 2013. https://thenewinquiry.com/parasite/.

Porter, James E. 2010. "Rhetoric in (as) a Digital Economy." In *Rhetorics and Technologies: New Directions in Writing and Communication*, edited by Stuart A. Selber, 173–97. Columbia: University of South Carolina Press.

Pozo, Teddy. 2018. "Queer Games After Empathy: Feminism and Haptic Game Design Aesthetics from Consent to Cuteness to the Radically Soft." *Game Studies: The International Journal of Computer Game Research* 18 (3). http://gamestudies.org/1803/articles/pozo.

Pritchard, Eric Darnell. 2017. *Fashioning Lives: Black Queers and the Politics of Literacy*. Carbondale: Southern Illinois University Press.

Ratto, Matt. 2011. "Critical Making: Conceptual and Material Studies in Technology and Social Life." *The Information Society: An International Journal* 27 (4): 252–60. doi:10.1080/01972243.2011.583819.

Ratto, Matt, and Garnet Hertz. 2015. "Defining Critical Making." In *Conversations in Critical Making*, edited by Garnet Hertz, 35–55. n.p.: CTheory Books. https://dspace.library.uvic.ca/handle/1828/7070.

Ratto, Matt, and Stephen Hockema. 2009. "Flwr Pwr—Tending the Walled Garden." In *Walled Garden*, edited by Annet Dekker and Annette Wolfsberger, 51–62. Amsterdam, NL: Virtueel Platform. http://aaaan.net/walled-garden/.

Rawson, K. J. 2014. "Transgender Worldmaking in Cyberspace: Historical Activism on the Internet." *QED: A Journal in GLBTQ Worldmaking* 1 (2): 38–60.

Rieder, David M. 2017. *Suasive Iterations: Rhetoric, Writing, and Physical Computing*. Anderson, SC: Parlor Press.

Rogers, Melissa. 2017. "Making Queer Feminisms Matter: A Transdisciplinary Makerspace for the Rest of Us." In *Making Things and Drawing Boundaries: Experiments in the Digital Humanities*, edited by Jentery Sayers, 234–48. Minneapolis: University of Minnesota Press. https://dhdebates.gc.cuny.edu/projects/making-things-and-drawing-boundaries.

Ruberg, Bonnie. 2015. "No Fun: The Queer Potential of Video Games that Annoy, Anger, Disappoint, Sadden, and Hurt." *QED: A Journal in GLBTQ Worldmaking* 2 (2): 108–24.

Ruberg, Bonnie. 2018. "Queer Indie Video Games as an Alternative Digital Humanities: Counterstrategies for Cultural Critique through Interactive Media." *American Quarterly* 70 (3): 417–38. doi:10.1353/aq.2018.0029.

Ruberg, Bonnie, and Amanda Phillips. 2018. "Not Gay as in Happy: Queer Resistance and Video Games (Introduction)." *Game Studies: The International Journal of Computer Game Research* 18 (3). http://gamestudies.org/1803/articles/phillips_ruberg.

Ruberg, Bonnie, and Adrienne Shaw, eds. 2017. *Queer Game Studies*. Minneapolis: University of Minnesota Press.

Sayers, Jentery, ed. 2017. *Making Things and Drawing Boundaries: Experiments in the Digital Humanities*. Minneapolis: University of Minnesota Press. https://dhdebates.gc.cuny.edu/projects/making-things-and-drawing-boundaries.

Sedgwick, Eve Kosofsky. 2003. *Touching Feeling: Affect, Pedagogy, Performativity*. Durham, NC: Duke University Press.

Sengers, Phoebe, and Garnet Hertz. 2015. "Critical Technical Practice and Critical Making." In *Conversations in Critical Making*, edited by Garnet Hertz, 9–21. n.p.: CTheory Books. https://dspace.library.uvic.ca/handle/1828/7070.

Shaw, Adrienne. 2015. "Circles, Charmed and Magic: Queering Game Studies." *QED: A Journal in GLBTQ Worldmaking* 2 (2): 64–97.

Sheridan, David M. 2016. "A Maker Mentality toward Writing." *Digital Rhetoric Collaborative*, March 28, 2016. http://www.digitalrhetoriccollaborative.org/2016/03/28/a-maker-mentality-toward-writing/.

Shivers-McNair, Ann. 2018. "Making and Mattering." In *Rhetorics Change/Rhetoric's Change*, edited by Jenny Rice, Chelsea Graham, and Eric Detweiler. Anderson, SC: Parlor Press & Intermezzo. http://intermezzo.enculturation.net/07-rsa-2016-proceedings.htm.

Sicart, Miguel. 2017. "Queering the Controller." *Analog Game Studies* 4 (4). http://analoggamestudies.org/2017/07/queering-the-controller/.

Squinkifer, Dietrich [Squinky]. 2017. "Rustle Your Leaves to Me Softly." YouTube video, 2:34, November 8, 2017. https://youtu.be/CFZkinLhDY4.

Stone, Kara. 2018. "Time and Reparative Game Design: Queerness, Disability, and Affect." *Game Studies: The International Journal of Computer Game Research* 18 (3). http://gamestudies.org/1803/articles/stone.

Walters, Shannon. 2015. "Graphic Disruptions: Comics, Disability and De-Canonizing Composition." *Composition Studies* 43 (1): 174–77.

Warner, Michael. 1993. "Introduction." In *Fear of a Queer Planet: Queer Politics and Social Theory*, edited by Michael Warner, vii–xxxi. Minneapolis: University of Minnesota Press.

Westecott, Emma. 2013. "Independent Game Development as Craft." *Loading . . . The Journal of the Canadian Game Studies Association* 7 (11): 78–91. http://journals.sfu.ca/loading/index.php/loading/issue/view/11.

11

DEVELOPING *A STRONG FIRE*
Bridging Critical Making, Participatory Design, and Game Design

Wendi Sierra
Texas Christian University

INTRODUCTION

While Native American representation has historically been low across all media, critical making as a method offers a great deal of promise in considering how to increase representation. Indeed, Matt Ratto (2011) posits that "using a shared process of making as a common space for experimentation encourages the development of a collective frame while allowing disciplinary and epistemic differences to be both highlighted and hopefully overcome," making it an engaging way to tackle wicked problems (253). Moreover, the proliferation of low/no-cost tools and systems for non-expert users to engage with has begun to shift how scholars, practitioners, and the public conceive of design itself, eroding the split between content creators and content users (Stappers, Visser, and Kistenmaker 2011). However, as Sarah Fox and Christopher Le Dantec (2014) note, "The reality of how [these tools] enable more diverse participation is far more complex" (786). Indeed, Michael J. Faris (this collection) explores more fully how utopian approaches to critical making ignore systemic inequalities in opportunity and support. This is particularly true in the case of game design, where issues of access, community support, and mentorship are often difficult to find in underrepresented communities.

This chapter presents the *A Strong Fire* project as a model of the hybrid, community-based approach possible when blending key practices from critical making, participatory design, and Indigenously determined game design. I refer to this as a project and not a finished object, as the design is meant to both develop an Indigenously determined video game *and* create opportunities to build capacity for the next generation of makers, addressing two key areas of underrepresentation.

https://doi.org/10.7330/9781646422586.c011

Indigenously determined game design, a practice informed by and in line with community-based critical making, can offer one avenue toward beginning to shift representation, both in the games themselves and in the community of designers.

SITUATING INDIGENOUS REPRESENTATION IN GAMES AND GAMING

Prior to discussing the particulars of this model, it will be useful to discuss the issues surrounding Indigenous representation and gaming that lead to the development of this model. Dmitri Williams et al. (2009), in their "Virtual Census" of video games, identify two important reasons to consider representation (or, in this case, the striking absence of representation) in all media, including gaming. First, "Imagery that is viewed or played repeatedly is more accessible when a person is attempting to recall information about that class of social objects" (819). In other words, media consumers draw on a knowledge store that includes both their own personal experiences *and* what they have observed in media when thinking about social objects. Stereotypical or problematic representations lead to these being more frequently recalled, while a relative absence of representation can lead to groups seeming invisible or no longer relevant.

For group members, viewing characters like oneself in media "serves as a marker for members [of a group] to know that they carry weight in society," while a lack of such characters can "lead to a feeling of relative unimportance and powerlessness" (Williams et al. 2009, 820). Thus, finding oneself in media artifacts can confirm for users their place both as a target audience and in a broader social context. Absence suggests the opposite, and systemic absence codifies these beliefs. Given the role representation plays both for group members and nonmembers, it is distressing to find that, in Williams et al.'s study of 133 titles released between March 2005 and February 2006, Native Americans were underrepresented by 90 percent when comparing the population of game characters to US census data, revealing a profound absence in the gaming landscape (Williams et al. 2009, 824). Even more striking, their study did not find a single *playable* Native American character in this time period. Perhaps unsurprisingly, these results align with representation in film and TV, where Native American characters represent only 0.3 percent of characters and are primarily supporting or background characters (Kopacz and Lawton 2011, 243).[1]

While Williams et al.'s census gives us a clear picture of Native American characters in the gaming landscape, it does not explore *how*

Native American characters are depicted when they do appear (as this is obviously outside the scope of their census). Nicholas Wisniewski's (2018) survey, which found a mere thirty games with any Native American characters (playable or not) from 2005 to 2015, explores in more detail how games depict these characters. It should be noted that his survey excluded nonhuman races and characters that could be construed as based on Native American cultures (like the Tauren race in *World of Warcraft*) and games based on real world sports representing actual athletes. In this study, Wisniewski found two concerning trends.

First, games set in contemporary (and even future) time periods overwhelmingly "lock the character to the reservation or tribal land . . . clothed in ceremonial costumes or clothing that imitated traditional styles, and carrying primitive weapons like bows and tomahawks" (Wisniewski 2018, 87). This echoes portrayal in traditional media, where Native Americans "are rarely (if ever) seen as contemporary figures in the media, which means they are absent from depictions of mainstream public spaces, such as schools and hospitals, and from many professional positions, such as teachers, professors, doctors, and lawyers" (Leavitt et al. 2015, 41). Indeed, Peter Leavitt et al. observe that while many groups experience underrepresentation (they note Latino Americans, LGBT+, and the working class as examples), Native Americans are unique in their absence *specifically* from modern time periods. Thus, in the rare instances when Native Americans do appear in contemporary media, their representation remains yoked to premodern professions, beliefs, identities. These representations further undercut their relevance and importance in modern society, both for Native Americans themselves and for media consumers.

In contrast to games portraying modern or future time periods, Wisniewski (2018) found that games set in the past often occupy "quasi-historical time periods that are essentialized and packed with exotic imagery," blending together various time periods and cultures to present players with a fantastical experience of 'Native-ness' that does not directly correspond to any actual place (30). In these portrayals, we see a "tourist" mode of representation, where features of a culture or environment that are in some way unusual or exotic to contemporary viewers are overemphasized (Hawkins 2005, 52). In blurring historical events, geographical features, and cultural elements, these games often both minimize historical atrocities and exoticize Native cultures and peoples. Wisniewski highlights *Assassin's Creed III* as an example of this minimizing. While the game does portray violence against the Mohawk tribe, it pointedly avoids depicting on screen any mention of *American* atrocities, instead depicting the British as villains in this scenario (Wisniewski 2018, 88). By avoiding the topics of land theft and displacement, or by attributing these actions to any entity but the American government, historical games invite players to act as tourists without fear of guilt or concern.

Of course, these trends are not unique to gaming or popular media. Many encyclopedias and textbooks, sources that many turn to for education, continue to represent Native Americans as a purely historical population with a relatively monolithic culture. Richard A. Grounds (2003), a Yuchi/Seminole scholar, recalls reading on two separate occasions about his tribe's extinction, first in an encyclopedia from 1881 and then again in a compendium of tribes published in 1979 (290). Tony Sanchez's (2007) survey of secondary education American history textbooks finds that many present "Native Americans as forgotten people whose disappearance from history after 1890 remains unexplained and without a foundation from the past that links to the present" (315). Jeffery Hawkins's (2005) survey of a similar set of textbooks found American education on Native Americans limited to five general (and highly stereotypical) topics: "Past nomadic lives of the plains, the significance of females in the northeast, the reservation settlements, the urbanization of Native Americans, and, now, casino dwelling Indians" (54).

Clearly there is a problem, and one that reaches far beyond game design and development. Of course, this is not to suggest that the games industry writ large is willfully racist. Indeed, some recent games, particularly *Never Alone* and *Assassin's Creed III*, demonstrate a significant effort on the part of major developers to do representation correctly. However, as Sam Srauy's (2019) survey of game designers found, many companies are under intense pressure to make games that are "easily relatable and quickly understandable," leaning toward "adopting narratives from past successful titles" (493). Srauy's study suggests that these twin tendencies have the side effect, intended or not, of leading many studios to shy away from including more diverse main characters in their games. Indeed, Srauy argues that "racism in North American video games may be at once invisible and *economically rational*" (494, emphasis added).

To summarize, then, there is a lack both in regard to texts that depict Indigenous peoples and in game designers. Critical making often does not result in a polished project (as the *making* is the primary intent and not the finished object). For this reason, I argue for a model incorporating both critical making and participatory design, informed by Indigenously determined game design, as one that addresses both sides of this issue, creating profound opportunity for making and developing a polished consumer product.

INDIGENOUS PERSPECTIVES ON CRITICAL MAKING

Indigenous scholar and game designer Elizabeth LaPensée, reflecting on her own history as a gamer and scholar, remarks, "I first started off critiquing these representations but recognized that if I was ever going to get to play a game that I wanted to play, I'd have to do it myself" (qtd in Martinson 2017, n.p.). This DIY ethos is a cornerstone of both the critical making movement and the Indigenous philosophy of survivance.[2] And, with the proliferation of free tools, from game engines to art and music assets, and the increasing ease of online distribution for games, it is easier than it has ever been for one to simply make the game one wants to see. However, there is a wide gulf between "easier than ever before" and easy—while new tools and distribution models have ameliorated some of the previous challenges to independent game development, game design and production remains a challenging endeavor. Indeed, in Erling Björgvinsson, Pelle Ehn, and Per-Anders Hillgred's (2012) review of contemporary design literature, which they characterize as enamored with terms like "co-creation" and "user-driven innovation," they criticize the market and industry-driven nature of these concepts, which "[tend] to be technical and expert oriented" rather than capacity-building (128). This section explores participatory and co-design practices as essential for supporting underrepresented communities.

While the games industry, following generally accepted practices in entertainment, has a long history of involving users in playtesting and evaluation, this usually comes fairly late in the design process (Khaled and Vasalou 2014, 93). A playable prototype is made, and user input is taken. Of course, at the point at which a prototype is completed, many crucial design decisions have already been made. Essentially, traditional game design practices are quite comfortable with user involvement to assist in creating a product *for* a user. Clay Spinuzzi (2005) articulates the difference between this and participatory design: "What distinguishes participatory design from related approaches such as user-centered design is that the latter supposes only that the research and design work is done on behalf of the users; in participatory design, this work must be done with the users" (165). Indeed, Ratto, Sara Ann Wylie, and Kirk Jalbert (2014) situate participatory design under the larger umbrella of critical making, in addition to critical technical practice, values in design, critical design, and tactile media. It's worth observing, however, that Ratto has previously defined critical making as explicitly not concerned with a final project. As both Faris and Kendall Gerdes demonstrate in this collection, the products of critical making can be incredibly important and powerful artifacts for communities.

One of the most widely praised games about an Indigenous community, *Never Alone*, demonstrates the opportunities and limitations of participatory design, particularly with regard to game design. *Never Alone*, based on Iñupiaq folklore, was a joint venture between the Cook Inlet Tribal Council (CITC) and E-Line Media, a developer experienced in serious games. Deborah L. Madsen (2017) praises the game's narrative and mechanics for "enact[ing] survivance," which is "a manner of living with Indigenous integrity while resisting by transcending the assimilative pressure applied by the dominant settler-colonial community" (92).

The Iñupiaq community was involved from project genesis to final completion. Even the selection of E-Line Media as a partner was based on tribal values. Gloria O'Neill, president and CEO of CITC, notes that throughout the development process, "it was really important that CITC's values of interdependence, resilience, accountability, and respect, were at the core of the [design] experience" (CITCAlaska 2016). Furthermore, the participatory design process called on a wide diversity of tribal members: "We're bringing elders, storytellers, young people, linguists from our community to be a part of the process" (CITCAlaska 2016). Amy Freedan, lead cultural advisor for E-Line Media reflected about the commitment to sustained engagement with the community, remarking, "It couldn't be a check in at the beginning and a check in at the end. Every step of the way we were calling community members, we were engaging them and asking them questions" (IGN 2014). And indeed, the tribe was given a great deal of control over the final product—at various points in the design process the community rejected concept art that felt "too Disney" and mechanics that violated traditional worldviews (CITCAlaska 2016). In addition to consulting with tribe members on the folklore that would lead to the story of the game, Alan Gershenfeld, president of E-Line Media, highlights that the game's genre (puzzle-platformer) was a community decision (his initial preference being an adventure game).

Nonetheless, it is worth noting where tribe members contributed to the design and where they did not. Tribe members were involved in the predevelopment design phase to help refine the concept, throughout the process as playtesters, and in production as voice actors and writers. They were not involved in the creation of the art or the actual coding of the game. Thus, while the CTIC and E-Line Media collaboration could be considered participatory design, they did not employ more classical elements of critical making. As previously noted, critical making often eschews the creation of polished consumer products: The "goal is not to create objects that in their apprehension open new visions and possibilities for observers. Instead . . . our main focus is on the act of shared construction itself as an activity" (Ratto 2011, 254). For *Never Alone* and the Iñupiaq community, the entire purpose of the project was to create a game that would "open new visions and possibilities for observers." For underrepresented groups seeking to make their stories known, final products are crucially important.

This rub, between process and product, reflects a serious challenge that Rilla Khaled and Asimina Vasalou (2014) find in incorporating more participatory aspects into game design: "Multiplicity of design needs serious games must fulfil ramps up the difficulty of designing them" (93). Serious games must be both informative and entertaining, meaning that requires participatory design experiences to leverage (1) knowledge about a particular subject, (2) best practices in game and mechanic design, and (3) technical proficiencies. Adding in the ability for community engaged exploration through critical making further complicates the design process. Thus, development like that seen in *Never Alone* offers a model for how to create engaging and authentic representations in media, but it does not build capacity within the community in ways that traditional critical making enables.

In contrast, LaPensée, Jason E. Lewis, and Skawennati Fragnito's "Skins 1.0" offers a model to "encourage First Nations youth to be more than consumers of digital media; rather, we wished to show them how they themselves could be creators with a cultural, critical approach to video games" (LaPensée, Lewis, and Fragnito 2010, 106). The Skins workshop, which ran for an entire academic year for Mohawk youth ages fifteen to seventeen, aimed to provide First Nations students with a pathway from consumers of media to critically, culturally engaged creators of media via a curriculum focused on "traditional stories and storytelling techniques, which [served] to both encourage youth to reflect on how stories are transmitted in their community and how they themselves can participate in the preservation, evolution, and future transmission of those stories" (LaPensée, Lewis, and Fragnito 2010, 107). Thus, the workshop blended traditional ways of knowing with game design and technical instruction and do so over an extended time period to allow a community of practitioners to form amongst the students. The workshop presents a solid model of "shared construction, joint conversation, and reflection" around Indigenous ways of knowing and representing knowledge, identity, and experience in digital media, both games and other. Moreover, as with the original intent of critical making, the purpose of the workshop was to find meaning in the sustained engagement and collaboration in making. While student projects were eventually submitted to various competitions, the development of a consumer product was not the primary intended outcome of the Skins workshop.

At this point I turn to the *A Strong Fire* project, and how it attempts to blend these two models, bringing in elements both of community feedback/participatory design *and* mentoring/capacity-building via critical making. In doing so, we hope to both produce a finished game that will share with the public Oneida's story of itself and create pathways for Oneida artists, musicians, and hopeful game designers to become makers themselves.

A STRONG FIRE: CONCEPT AND MENTORSHIP

The *A Strong Fire* project was intentionally designed with goals tied both to development and distribution. In the development phases, goals include engaging Oneida tribe members in both participatory design and critical making. Following the participatory model of *Never Alone*, tribal members and organizations have been extensively involved in the genesis and planning of the project, and will contribute throughout development, testing, and project completion. Taking inspiration from the critical making–focused *Skins 1.0* model, making opportunities are embedded at key points in the game's development. As a finished product, the game has two goals: For tribal members, the game introduces and reinforces vocabulary in the Oneida language, which is endangered, and for those not interested in learning the language, the game shares Oneida's cultural heritage via a focus on the Haudenosaunee[3] Thanksgiving Address. The address is a part of the culture intended to be shared and spread to all peoples. Indeed, Taiaike Alfred (1992) argues that this particular address is of crucial importance to all audiences today, explaining that as part of the address, "this knowledge [in the address] brings with it responsibility, and in that, we have a profound responsibility to ensure that we demonstrate respect and promote balance and harmony in all of our relationships" (8). The address itself, coupled with the interactive narratives throughout the game, shares this philosophy with a larger audience.

Developing Collaboration

The concept for *A Strong Fire* is the result of roughly six months of brainstorming and dialog amongst a variety of stakeholders. I began the process first with discussions among my own family members, exploring what need or want an Oneida game might satisfy. As these ideas began to take shape, I extended the conversations to include tribal organizations. I reviewed both Indigenously determined games and apps made for other tribes and Oneida's own apps and physical resources. After several visits and discussions with community members, one key idea emerged: the idea of a game that could reinforce language learning for children, introduce some elements of the language to adults, and generally share the Oneida concept of the Good Mind began to take hold.

Prior to addressing this particular project, it will be useful to briefly contextualize the state of the Oneida language. The Oneida Nation has been working tirelessly for many years to preserve their language, but it is an intense struggle. Like many Indigenous languages, Oneida is endangered. The Endangered Languages Project estimates of the nearly twenty thousand enrolled tribal members between three Oneida communities, there are only two hundred native speakers. The Oneida Nation has been working for almost a century now on language preservation. Several stories were audio recorded in the 1930s as part of a Works Progress Administration project to preserve narratives in Oneida. In the 1970s, a program was initiated to train language educators. Currently, Oneida of Wisconsin runs an immersion preschool program and language classes in the curriculum at the Oneida Nation Elementary and Middle School. More recently, the tribe has received a grant from the Administration for Native Americans Native Languages Preservation and Maintenance program to support the training of new Oneida language teachers.

With language support and expansion in mind as a possible focus for the game, communication began in earnest with various tribal organizations, including the Arts Center, the Community Education Center, the Oneida Nation Elementary School, and the Oneida Cultural Heritage Center, to name but a few. These meetings explored both other language and culture games and other games featuring Indigenous cultures. From this wealth of information, a rough sketch was developed for presentation and refinement.

At this stage, we ran into an interesting challenge. While the language educators and the Arts center were very eager to move forward with this idea, we met striking indifference from a number of other stakeholders. Interestingly, Fox and Le Dantec (2014) encountered a similar issue working with children to develop a serious game. They report that participatory design struggled in the early phases, when many of their participants struggled to fully conceptualize the idea or process, leading to shallow and unhelpful feedback. Similarly, we found that, while not opposed to the idea of an Oneida game, many community members (particularly those unfamiliar with digital educational games) lacked a conceptual framework to imagine the possibilities and were apathetic about engaging in the design process. To address this challenge, I rapidly developed two vocabulary minigames to act as proof of concept. While they did not include the narrative component essential to the cultural aspect of the proposal, they did model how a language game might engage adults and children alike. The short games, which teach the words for basic animals, feature a child's voice saying the names of the animals and display the words in the Oneida alphabet. They were posted online and shared with tribal members and departments. These prototypes helped stakeholders see the creative potential and imagine what a more extensive design might look like. The ideas generated from sessions following the development of these short games formed the basis for the game design document and solidified the choice of the Thanksgiving Address as the central organizing theme.

A Strong Fire Game Concept: The Thanksgiving Address

As previously mentioned, the Thanksgiving Address has both historical and contemporary value. Traditionally the address has been recited to open both formal events and social occasions. For the Oneida, this currently includes everything from General Tribal Council (the governing body) meetings to high school assemblies. There are several versions of the Thanksgiving Address in Oneida, from the simple preschool version posted on the tribe's own website[4] to extended poetic versions. The address is a cornerstone of both the tribe's attempt at language revitalization and its celebration of Oneida tradition and culture. A Mohawk elder (Porter 2008, 5) explains:[5]

> The Thanksgiving Address is the most important prayer—if you want to call it a prayer—that we, the Iroquois, have. It includes the essence of all our existence. It includes all the relationships to the earth that we are in. That's what it does.
> I often refer to it as the skeleton key. You know, in the older days when they used to refer to the key that could open every door? Well, that Thanksgiving address is for our six nations, our human people, it's our skeleton key to the world that we live in. And it'll open up all things for us.

The Thanksgiving Address encapsulates the grounding philosophy of the Haudenosaunee mindset. The various ways the name of the address has been translated help to contextualize the complicated feelings the address is meant to invoke. In Mohawk, the address is called *Ohen:ton Karihwatehkwen*, which is directly translated as "what we say before we do anything important" (Porter 2008, 8). As this translation demonstrates, the sentiment of thankfulness is not an accurate parallel for what the address conveys. Indeed, translators have also used the words "blessing" or "greeting" instead of address. There are certainly aspects of both intents, but again, neither is accurate. Drawing on a translation from the Seneca, the address is meant to evoke a "more general feeling of happiness over the existence of something or someone" (Chafe 1961, 1). As a piece of oral literature, it is difficult to date the Thanksgiving Address. A printed copy produced by the Native Self Sufficiency Center in 1993 claims the address is over one thousand years old, a claim supported by linguistic and archeological evidence that identifies the Northern Iroquoian language group splitting from the larger Proto-Iroquoian language family around this time (Snow 1996). A truncated version of the address is included below. It should be noted that there are many versions of the address, several of which are substantially longer and include more specific acknowledgments for each element.

> *Teyethinuhwela·túkwehokúha*
> We extend our greetings, thankfulness and love to them All Of The People
> *Teyethinuhwela·tú· Yukhinulhá ohwátsya²*
> We extend our greetings, thankfulness and love to her Our Mother The Earth
> *Teyethinuhwela·tú· Ohnekanushokúha*
> We extend our greetings, thankfulness and love to them All Of The Waters
> *Teyethinuhwela·tú Nya²tekátsyake*
> We extend our greetings, thankfulness and love to them The Different Kinds Of Fish
> *Tetwanuhwela·tú· Ka²niyohutés awáhihte²*
> We extend our greetings, thankfulness and love to it The Short Plant Strawberry
> *Tetwanuhwela·tú· Oyu²kwa²u·wé*
> We extend our greetings, thankfulness and love to it The Native Tobacco
> *Teyethinuhwela·tú· Áshʌ na²tekutʌhnu·téhle²*
> We extend our greetings, thankfulness and love to them The Three Sisters
> *Teyethinuhwela·tú· Onekli²shúha*
> We extend our greetings, thankfulness and love to them The Grasses
> *Teyethinuhwela·tú· Onuhkwa²thokúha*
> We extend our greetings, thankfulness and love to them All Of The Medicines
> *Teyethinuhwela·tú· Nya²tekalu·táke*

We extend our greetings, thankfulness and love to them The Different
   Kinds Of Trees
Teyethinuhwela·tú· Nya?tekatsi?nu·wáke
We extend our greetings, thankfulness and love to them The Different
   Kinds Of Insects
Teyethinuhwela·tú· Kutilyo?okúha
We extend our greetings, thankfulness and love to them All Of The
   Indigenous Animals
Teyethinuhwela·tú· Otsi?tʌhashúha
We extend our greetings, thankfulness and love to them The Birds
Teyethinuhwela·tú· Owela?shúha
We extend our greetings, thankfulness and love to them The Winds
Teyethinuhwela·tú· Latihsakayu·téhse?
We extend our greetings, thankfulness and love to them The Thunderers
Teyethinuhwela·tú· Yukhihsótha? wehní·tahle?
We extend our greetings, thankfulness and love to her Our Grandmother
   The Moon
Tehethwanuhwela·tú· Shukwahtsíha wehní·tahle?
We extend our greetings, thankfulness and love to him Our Elder Brother
   The Sun
Teyethinuhwela·tú· Yotsistohkwa·lú·
We extend our greetings, thankfulness and love to them The Fires Scattered
   About
Teyethinuhwela·tú· Kayé nihʌnukwé·take
We extend our greetings, thankfulness and love to them the Four Beings
Tehethwanuhwela·tú· Skanyatali·yó
We extend our greetings, thankfulness and love to him Handsome Lake
Tehethwanuhwela·tú· Shukwaya?tísu
We extend our greetings, thankfulness and love to him The Creator
Ta· tho niyohtúhak yukwa?nikúhla?. Táne? tho.
This is how it shall be our minds. That's it.

Paul Williams (1992) argues the address has contemporary relevance for all peoples: "In its basic call to humility, honor, and respect, we have been given the key to understanding the way to the survival of ourselves and the natural environment" (14). Fuller versions of the address conclude with the speaker asking the listener's forgiveness for any mistakes made in the recitation, requesting listeners fill in any omissions: "We have now arrived at the place where we end our words. Of all the things we have named, it was not our intention to leave anything out. If something was forgotten, we leave it to each individual to send such greetings and thanks in their own way." The humility shown in this closing is important, as it models the humility all people should show: "We must always remember that we are not perfect, that we can and will make mistakes" (Lickers 1992, 160).

From Concept to Development

With a practical goal (language learning) and a guiding philosophy (the Thanksgiving Address), the game architecture began to take shape. Given its importance in Oneida/Haudenosaunee culture, the Thanksgiving Address acts as an organizing mechanism for *A Strong Fire*. Players log into a home page resembling a map with characters/animals representing each piece of the address. These buttons lead players to interactive narratives based on stories and folklore related to each verse. The story of Sky Woman, accessed from the section honoring the Earth Mother, demonstrates both how Good Mind will be shared through stories and how targeted vocabulary will be introduced to players. Each indent below marks a short interactive scene. During game play, the English words in parenthesis are not included, as players see the images and understand the words from context and repetition.

> We encourage each other to live with Ka'nikuhli:yo (a good mind). This story explains how we came to live on Turtle Island.
>
> Long ago, before there was any land here, there were creatures that lived in the Ohne:kanus (water) and birds that flew above it. Above, there was a land called Kaluhya:ke thʌ:tlu (Sky World) where people with supernatural powers lived. In the middle of the land, a great tree with many different fruits that gave them light existed.
>
> The rule was that if anyone harmed the tree, they would be punished greatly. A young woman and Lu:kwe (man) were expecting a baby. She craved the roots and bark from the tree and asked Lu:kwe (man) to get these for her but he was afraid of being punished.
>
> Lu:kwe (man) waited for all the people to leave, and started digging but the ground suddenly caved in leaving a big hole in the ground by the tree. The young woman found out he did not get what she wanted because he got scared, so she got mad and said she would get it herself. She went to the hole and when looking in, she saw water below. She fell in, grabbed ground and tree roots on her way down.
>
> Birds and water animals saw light through the hole, and saw Yotsitsya'ahs (Sky Woman) falling from Kaluhya:ke thʌ:tlu (Sky World) toward Ohne:kanus (water). One bird flew up to asheya'takenha (help) her. They sent a bird to see if a water animal could support her. The water animals discussed it and A'no:wal (turtle) agreed so all the birds safely brought her down to A'no:wal (turtle) back. The frightened Yotsitsya'ahs (Sky Woman), who fainted. She woke up on A'no:wal's (turtle) back. Seeing only Ohne:kanus (water), water animals and birds, she asked where she was and they pointed to the hole and said they saw her fall through it.

She asked if there was any mud or dirt to mix with what she grabbed, but the animals weren't sure. Each of the water animals tried to asheya'takenha (help her).

First otter dove into ohne:kanus (water) but he did not get any mud. Yotsitsya'ahs (Sky Woman) wathonuhwela:tu (gave thanks) to otter. Then loon dove into ohne:kanus (water). He didn't bring back any mud, but Yotsitsya'ahs (Sky Woman) wathonuhwela:tu (gave thanks) to him for wahoya'takenha (helping her). Next beaver dove into ohne:kanus (water). He wanted to asheya'takenha (help her) and was sad that he did not find any. Yotsitsya'ahs (Sky Woman) wathonuhwela:tu (gave thanks) and said, "You tried your best." Muskrat dove into ohne:kanus (water), and was gone a long time, worrying the others. Then he floated to the top with a little dirt between his claws, which Yotsitsya'ahs (Sky Woman) mixed with her dirt.

She placed it on the back of A'no:wal (turtle). The dirt grew and grew, and different things began to grow. Now that Yotsitsya'ahs (Sky Woman) was safe, she began to live with Ka'nikuhli:yo (a good mind) in our home, turtle island.

Now you have heard the story of Yotsitsya'ahs (Sky Woman), and how we came to live on Turtle Island. This is why we Tehetwanuhwela:tu (give thanks to him, the creator), and live with Ka'nikuhli:yo (a good mind).

Conceptually, Oneida's Sky Woman story demonstrates the Good Mind, modeling how each creature helps to the best of its ability. The geese cannot find earth, but they can help Sky Woman to the water. Otter and loon are unsuccessful, but nonetheless help come up with the idea and attempt to find earth. Turtle volunteers his sturdy shell as the basis for land. Each animal contributes in their own way to help Sky Woman and, by extension, all of humanity. No single animal is responsible for their success; all are necessary.

As the player moves through the interactive narrative, the stories are read aloud by an Oneida elder. The Sky Woman story, as written above, follows best practices in language acquisition pedagogy. The story is limited to eight words in Oneida, words that are easily decipherable given the context and visual imagery apparent as they are introduced. Further, each word is repeated at least three times throughout the short story. Importantly, by tying narratives and game events to the vocabulary-learning occurring in each short story, children will have an experience to tie their word recognition to, driving increased comprehension.

Following each interactive narrative, players unlock a series of mini-games emphasizing vocabulary and providing practice opportunities. These vary in genre and style, including memory-style card games and match-three games. We have made a conscious choice to minimize lose conditions and fail states, both in these mini-games and in the interactive narratives. In talks with various community members and organizations it became clear that these game elements were incongruous with our goal of emphasizing the principles of the Good Mind. Similarly, embedding challenge and winning scenarios in the postnarrative mini-games balances reinforcing the vocabulary and preserving the emotional resonance of the stories.

Community Mentorship

The previous section defines why this particular design emerged for this particular community, and how the principles of participatory design were used to generate and refine the concept. At this point, we successfully applied for a National Endowment for the Humanities Digital Projects for the Public grant that would enable us to fully develop and distribute the project. While critical making suggests that the end product is less important than the creation process, all of our constituent organizations are eager to see a polished game shared widely. The striking lack of representation in gaming leads us to desire something that can find a public audience beyond the Oneida community. Participatory design offers us a framework for substantially involving community members in the design process while working with technical professionals at an educational games company to finish the work. Despite these virtues, participatory design does not offer us a pathway toward greater participation in game design by community members. Thus, we incorporate critical making at key moments, using the design process as an opportunity to introduce making during targeted design phases. These practices can be split into two groupings: those that will result in direct contributions and those with more abstract contributions.

For concept art and game music, tribal members will be directly designing assets to be included in the final game. We have used the development process as an opportunity to establish apprentice-style mentorships to design concept art and game music. One established musician and one established artist will each work with an up-and-coming young tribe member. Our established artists are both Native, though not both Oneida, and while they are recognized in their respective styles, they are unfamiliar with games and game design. The apprentices, in contrast, are both very familiar with games but new to their respective fields. Thus, the project is an opportunity to both connect young people with expert practitioners in their field and bring both experts and novices into a new compositional medium. In the design process our apprentices are learning professionalization, technique, and exploring issues akin to those discussed in the Skins workshop—the representation of Indigenous identities and ways of knowing (in our particular case, Oneida's concept of the Good Mind) via art, music, and game design. While these opportunities are limited to only one young artist and one young musician, we felt the extended ability to collaborate with a mentor and to contribute directly to the project was worth the limitation.

In addition to establishing direct one-on-one mentoring opportunities related to both music and art, we are developing a series of short game design workshops for Oneida youth as part of the Community Center summer activities program. These workshops, inspired by but substantially shorter than the Skins program, will not contribute directly to the game production. Instead, they will use the game prototype as a model to explore and expand on. Like the Skins workshop, our workshops will focus on grounding game design principles and technology instruction in traditional storytelling techniques and content. We do not yet have the capacity to offer a full-year curriculum, but beginning with these shorter summer workshops will, hopefully, plant a seed of interest both in the Oneida youth and in Oneida organizations. Given the shorter nature of these workshops, it seems unlikely that any substantial production will take place; however, for these particular camps the final product is not the goal. Instead, we intend to use the resources we have available, both human and technological, to expose students to a variety of tools and technologies and provide a supportive community for initial forays development.

Ultimately, lack of representation is a wicked problem with no easy solution. The games industry, made up of fiscally conservative entities, is slowly bending toward change, as the involvement of cultural advisors in *Assassin's Creed III* demonstrates. However, such examples are few and far between. *Never Alone*, a highly praised game drawing on an Indigenous community extensively throughout the design process, offers a great deal of promise for participatory design, but did not, as part of the development model, involve substantial elements of critical making. Conversely, the Skins 1.0 workshop was a triumph of critical making and capacity building but did not produce anything for broader consumption (nor was it meant to). Thus, a hybrid approach combining participatory design (to produce more and better games) and critical making (to engage Native youth in the design process and offer pathways to development) potentially offers a great deal of promise for tackling this issue.

PRACTICAL REFLECTION: ON (NOT) MAKING PRACTICES AND INDIGENOUSLY DETERMINED GAME DESIGN

Throughout this chapter I have situated *A Strong Fire*'s design process as informed by a hybrid of critical making and participatory design. This method is a small step toward addressing representation both on screen and in the design community. One game obviously will not solve either problem; nonetheless, this method is one of many such efforts taken by the larger community of Indigenous scholars and game designers down a long road of envisioning another path. Importantly, this is a particular method, developed for a particular project, with the strengths and needs of a particular community at the forefront of all decisions. If, as a reader, you find yourself wishing to explore similar challenges in your own community, the model of the *A Strong Fire* development process may offer an example of how these ideas might be situated within particular goals and contexts.

However, as my emphasis above hopefully suggests, the problem of Indigenous representation in games is a complex one, and not something that is perhaps appropriate to be tackled in shorter critical making assignments outside a well-defined context. Indeed, approaching Indigenous representation requires Indigenous peoples and communities to be foregrounded and must proceed with respect and reciprocity as guiding values. For assignments and models of critical making and game design pedagogy, particularly in a rhetoric or first year composition context, several collections already exist. While not all of these selections use the phrase critical making in their framework, the principles of learning through creation and design as rhetorical and scholarly practice are extensively explored. These suggestions include everything from game design documents, which ask students to explore and outline a game without making one (Colby 2014), to board game assignments (Martin et al. 2019), to digital game assignments in Twine or Scratch (Brown and Alexander 2016). Indeed, elsewhere I have written a manifesto-esque call for making games in composition classes and provided a tool overview (Sierra 2018). All of these would be excellent next steps for an instructor looking to find some hands-on guidance to game design assignments in a classroom context, divorced from the Indigenous context discussed throughout the piece.

Outside the engaged frame of a community project, I might humbly suggest that issues related to Indigenous representation and making are better addressed via listening. I would heartily encourage instructors to consider how students might analyze and interpret issues related to representation in gaming. *Seven Myths of Native American History* is an excellent starting point that addresses many of the misconceptions that arise from how American history and cultural artifacts depict Native Americans. In addition to *Never Alone*, Indigenously determined games including *When Rivers Were Trails* and *Māori Pā Wars* offer examples of Indigenous creators defining their own representations and ways of knowing/being. Finally, research questions might include more specific investigations into the games industry. Who makes our games? Whose values get coded into them?

NOTES

1. It is worth noting that, while a substantial amount of scholarly work has been done analyzing representations of Native Americans in film and TV, Native American representations in games is an emerging field. The Williams et al. census, while dated, remains the definitive large-scale study looking broadly at representation.
2. Survivance, a term popularized by Vizenor (1999) in *Manifest Manors*, refers to Indigenous ways of existing and reclaiming community and identity under colonialism. LaPensée more fully explores the concept of survivance as it relates to game design in "Survivance as an Indigenously Determined Game" (2014b) and "Survivance Among Social Impact Games" (2014a).
3. The Haudenosaunee are also known as the Iroquois, a derogatory name that seems to refer to them as either snakes or killers (though the exact origins of this epitaph are debated). Throughout this chapter I will use the confederation's own name for itself, which translates into "People of the Longhouse." This name does not just refer to the dwellings the Haudenosaunee lived in, but is also reference to the mindset, traditions, and values that the people should embody. Longhouses were traditionally multigenerational, and at times multifamily, dwellings. Thus, the name is a call to remember the interconnected nature of all people.
4. https://oneida-nsn.gov/our-ways/our-story/ceremonies/.
5. Porter, a leader in the Mohawk community, had a dream in which his grandma told him, "Whatever we told you over the years since you were born—our beliefs and our way of looking at the world . . . write it down . . . our young people need grandmas. Our younger generation, they need to know things, and nobody's telling them" (2008, 3). Porter's book *And Grandma Said: Iroquois Teachings As Passed Down Through The Oral Tradition* is just that, Porter's attempt to preserve for future generations the cultural knowledge that is rapidly fading after generations of colonialization.

REFERENCES

Alfred, Taiaike, 1992. "The People." In *Words That Come Before All Else: Environmental Philosophies of the Haudenosaunee*, edited by the Haudenosaunee Environmental Task Force. Ontario, Canada: Native North American Travelling College.

Björgvinsson, Erling, Pelle Ehn, and Per-Anders Hillgren. 2012. "Agonistic Participatory Design: Working with Marginalized Social Movements." *CoDesign* 8 (2–3): 127–44. doi: 10.1080/15710882.2012.672577.

Brown, James J., and Eric Alexander. 2016. "Procedural Rhetoric, Proairesis, Game Design, and the Revaluing of Invention." In *Play/Write: Digital Rhetoric, Writing, Games*, edited by Douglas Eyman and Andrea D. Davis, 270–87. Anderson, SC: Parlor Press.

Chafe, Wallace L. 1961. *Smithsonian Institution Bureau of American Ethnology Bulletin 183: Seneca Thanksgiving Rituals*. Washington, DC: US Government Printing Office. https://repository.si.edu/handle/10088/15475.

CITCAlaska. 2016. "Never Alone: The Making Of—2016." YouTube video, 12:00, Nov. 16, 2016. https://youtu.be/dgndBVFrc2U.

Colby, Richard. 2014. "Writing and Assessing Procedural Rhetoric in Student-Produced Video Games." *Computers and Composition* 31: 43–52. doi:10.1016/j.compcom.2013.12.003.

Fox, Sarah, and Christopher Le Dantec. 2014. "Community Historians: Scaffolding Community Engagement through Culture and Heritage." In *Proceedings of the 2014 Conference on Designing Interactive Systems*, 785–94. ACM. doi:10.1145/2598510.2598563.

Grounds, Richard A. 2003. "Yuchi Travels Up and Down the Academic Road to Disappearance." In *Native Voices: American Indian Identity & Resistance*, edited by Richard A. Gounds, George E. Tinkger, and David E. Wilkins, 290–317. Lawrence: University of Kansas Press.

Hawkins, Jeffrey. 2005. "Smoke Signals, Sitting Bulls, and Slot Machines: A New Stereotype of Native Americans?" *Multicultural Perspectives* 7 (3): 51–54. doi:10.1207/s15327892mcp0703_9.

IGN. 2014. "Never Alone: Implementing Culture in Video Games." YouTube video, 3:04, December 9, 2014. https://youtu.be/zU138DiQtKw.

Khaled, Rilla, and Asimina Vasalou. 2014. "Bridging Serious Games and Participatory Design." *International Journal of Child-Computer Interaction* 2 (2): 93–100. doi:10.1016/j.ijcci.2014.03.001.

Kopacz, Maria A., and Bessie Lee Lawton. 2011. "Rating the YouTube Indian: Viewer Ratings of Native American Portrayals on a Viral Video Site." *American Indian Quarterly* 35 (2): 241–57. doi:10.5250/amerindiquar.35.2.0241.

LaPensée, Elizabeth. 2014a. "Survivance Among Social Impact Games." *Loading* . . . 8 (13): 43–60. https://journals.sfu.ca/loading/index.php/loading/article/view/141.

LaPensée, Elizabeth. 2014b. "Survivance as an Indigenously Determined Game." *AlterNative: An International Journal of Indigenous Peoples* 10 (3): 263–75. doi:10.1177/117718011401000305.

LaPensée, Elizabeth, Jason E. Lewis, and Skawennati Fragnito. 2010. "Skins 1.0: A Curriculum for Designing Games with First Nations Youth." In *Proceedings of the International Academic Conference on the Future of Game Design and Technology*, 105–112. ACM. doi:10.1145/1920778.1920793.

Leavitt, Peter A., Rebecca Covarrubias, Yvonne A. Perez, and Stephanie A. Fryberg. 2015. " 'Frozen in Time': The Impact of Native American Media Representations on Identity and Self-understanding." *Journal of Social Issues* 71 (1): 39–53. doi:10.1111/josi.12095.

Lickers, F. Henry. 1992. "The Creator." In *Words That Come Before All Else: Environmental Philosophies of the Haudenosaunee*, edited by Haudenosaunee Environmental Task Force. Ontario, Canada: Native North American Travelling College.

Madsen, Deborah Lea. 2017. "The Mechanics of Survivance in Indigenously-Determined Video-Games: Invaders and Never Alone." *Transmotion* 3 (2): 79–110. https://archive-ouverte.unige.ch/unige:100383.

Martin, Cathlena, with Stephen Gilbert, Jesus "Chuy" Guizar, Will Kirkpatrick, and Sara Perry. 2019. "Roanoke: A Post-Mortem on Undergraduate Research Game Design." *OneShot: A Journal of Critical Games & Play* 1. http://oneshotjournal.com/roanoke-a-post-mortem-on-undergraduate-research-game-design/.

Martinson, Patti. 2017. "Expressing Herself in Design and Art: Elizabeth LaPensée." *Sequential Tart*, January 16, 2017. http://www.sequentialtart.com/article.php?id=3021.

Porter, Tom. 2008. *And Grandma Said: Iroquois Teachings, as Passed Down through the Oral Tradition.* Bloomington, IN: Xlibris Corporation.
Ratto, Matt. 2011. "Critical Making: Conceptual and Material Studies in Technology and Social Life." *The Information Society* 27 (4): 252–60. doi:10.1080/01972243.2011.583819.
Ratto, Matt, Sara Ann Wylie, and Kirk Jalbert. 2014. "Introduction to the Special Forum on Critical Making as Research Program." *The Information Society* 30 (2): 85–95. doi:10.1080/01972243.2014.875767.
Sanchez, Tony R. 2007. "The Depiction of Native Americans in Recent (1991–2004) Secondary American History Textbooks: How Far Have We Come?" *Equity & Excellence in Education* 40 (4): 311–20. doi:10.1080/10665680701493565.
Sierra, Wendi. 2018. "Creating Space: Building Digital Games." In *Proceedings of the Annual Computers and Writing Conference, Volume 1, 2016–2017*, edited by Cheryl E. Ball, Chen Chen, Kristopher Purzycki, and Lydia Wilkes, 66–73. Fort Collins, CO: The WAC Clearinghouse. https://wac.colostate.edu/resources/wac/proceedings/cw2016-2017/.
Snow, Dean. 1996. *The Iroquois.* Hoboken, NJ: Wiley-Blackwell.
Spinuzzi, Clay. 2005. "The Methodology of Participatory Design." *Technical Communication* 52 (2): 163–74.
Srauy, Sam. 2019. "Professional Norms and Race in the North American Video Game Industry." *Games and Culture* 14 (5): 478–97. doi:10.1177/1555412017708936.
Stappers, Pieter Jan, Froukje Sleeswijk Visser, and Sandra Kistemaker. 2011. "Creation and Co: User Participation in Design." In *Open Design Now: Why Design Cannot Remain Exclusive*, edited by Bas Van Abel, Lucas Evers, Roel Klaassen, and Peter Troxler, 140–48. Amsterdam: BIS Publishers.
Vizenor, Gerald Robert. 1999. *Manifest Manners: Narratives on Postindian Survivance.* Lincoln: University of Nebraska Press.
Williams, Dmitri, Nicole Martins, Mia Consalvo, and James D. Ivory. 2009. "The Virtual Census: Representations of Gender, Race and Age in Video Games." *New Media & Society* 11 (5): 815–34. doi:10.1177/1461444809105354.
Williams, Paul. 1992. "The Stars." In *Words That Come Before All Else: Environmental Philosophies of the Haudenosaunee*, edited by Haudenosaunee Environmental Task Force. Ontario, Canada: Native North American Travelling College.
Wisniewski, Nicholas. 2018. "Playing with Culture: The Representation of Native Americans in Video Games." Master's thesis, Northern Arizona University. https://search.proquest.com/docview/2054024791.

12

TWISTED TOGETHER
Twine Games as Solidarity Machines

Kendall Gerdes
University of Utah

INTRODUCTION

In late 2014, video games became the contemporary site of yet another culture war, a contentious battle over shared understanding of and identification with video games culture. GamerGate erupted as a coordinated campaign of harassment against queer and feminist game developers over who would be permitted to claim an identity as a gamer.[1] Ostensibly, GamerGate was touched off by the success of a free, hypertext-based video game made with an open-source software tool called Twine. Twine games often challenge the commonplace criteria for a good video game: The player is not always a hero, killing enemies and solving puzzles, and the games don't always resolve in a clear final win. In the case of Zoë Quinn's Twine game *Depression Quest*, at least some GamerGators viewed these differences as a critique of gamer cultural values. Their purported concern was "ethics in videogames journalism," a flimsy pretext derived from the embittered ranting of the ex-boyfriend of *Depression Quest* developer Zoë Quinn. For months, Quinn and other high-profile women, feminists, and cultural critics in video games were targeted by GamerGators with threats and sustained harassment, including sexually violent messages. Their personal contact information (and that of loved ones) was posted publicly online. GamerGators viewed the success of *Depression Quest* as illegitimate and unearned. Perhaps unsurprisingly, GamerGate demonstrated that video games can be the site of violent identification and intensely policed cultural belonging. But in this chapter, I contend that video games—Twine games in particular—can also present an alternative to this violence of identification.

GamerGate reflected the intensification of preexisting and pervasive antifeminist and misogynist affects that were already circulating in gamer culture (see Evans and Janish 2015). *Depression Quest* touched a nerve

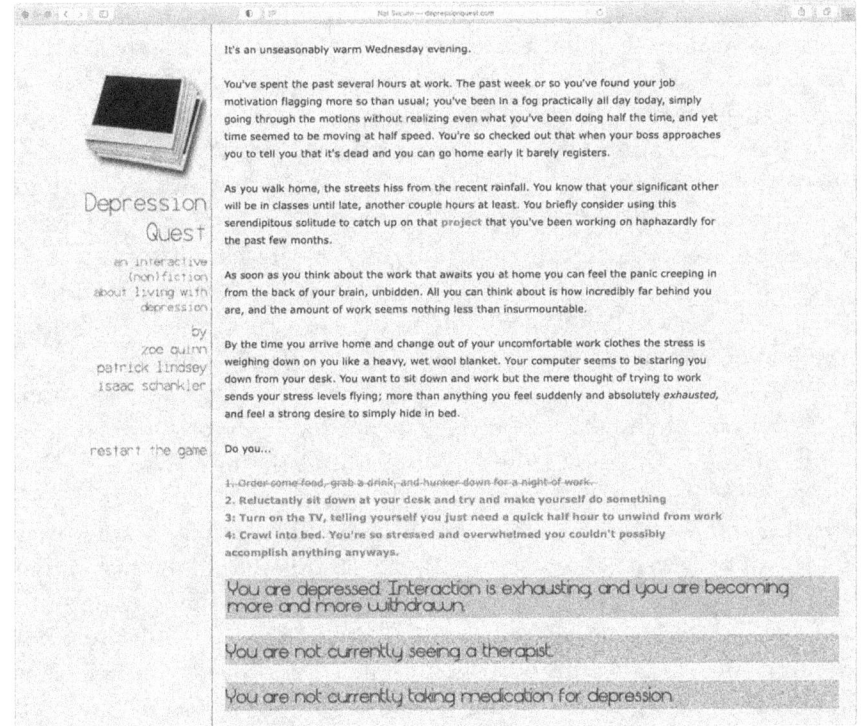

Figure 12.1. Your limited and diminishing choices in Depression Quest. *Screenshot by author.*

simply by eschewing the violent, corporate, and masculinist values that many gamers hold passionately. I want to introduce the issues this chapter will explore by saying a bit more about how *Depression Quest* frames some key rhetorical issues in video games and gamer culture. Indeed, *Depression Quest* challenges the highly policed definition of what counts as a video game. Quinn's game doesn't make you feel like the powerful, daring action hero in traditional video games. You can't kill anyone (including yourself). There is no clear win condition. There aren't even really puzzles to solve. When you launch *Depression Quest* on a computer, which you can do through the popular digital game distribution software called Steam, you are greeted by a description of depression penned by the late David Foster Wallace overlaid on a background that looks like a staticky TV screen. A lengthy description of the game contextualizes the goals of the game: to simulate depression, so that "other sufferers will come to know that they aren't alone," and to make people without depression aware of what living with depression actually feels like.

When a player begins the game, the player is addressed in the second person—"You are a mid-twenties human being"—in a way that invites the player to identify with the protagonist of the game. Characteristic of a game made with Twine, the text that appears onscreen is laden with hyperlinks that connect to other nodes in the game. At crucial junctures, *Depression Quest* presents you with choices to make about what to do and how to care for yourself, but in order to simulate the limitations of living with depression, some of these options expire—instead of appearing as a bold purple hyperlink, the text appears in red with a strike through. The game assesses your choices, and tracks the state of your depression, which eliminates more possible choices as the depression deepens.

When rhetoricians make arguments about video games, we often draw on Ian Bogost's framework of procedural rhetoric (see Bogost 2007; King 2010; Thevenin 2017). Procedural rhetoric describes how the rules of game both constrain and enable the players' agency. In *Depression Quest*, the rules also purposefully diminish it, taking away choices as your well-being worsens. It may be apparent to the player that an action you can no longer take is preferable to the options available to you, but depression may deprive you of better choices—that's the point of the game. Anna Anthropy (2012) argues that procedural rhetoric is what makes video games a unique medium: "Since games are composed of rules, they're uniquely suited to exploring systems and dynamics" and "especially good at communicating relationships" between "actions or choices and their consequences" (20).

But recently, some scholars have criticized procedural rhetoric for occluding the affective dimension of video games. In his book on video games' procedurality and the production of embodied habits, Steve Holmes (2017) argues that affective design is "an important site through which players' bodies are habituated in relationship to broader social and political contexts" (101). Because Holmes examines corporate games, his analysis focuses on how players' affective bonds keep them immersed in games, building feelings of mastery (121). Twine's accessibility and the culture of making among Twine players might alter this imagining of player as consumer. In a special issue of *QED: A Journal in GLBTQ Worldmaking* devoted to queer games, Bo Ruberg (2015) draws our attention to the frustrating, annoying, maddening, disappointing, heartbreaking, and hurtful affects in games as a way of critiquing the normative belief that games should only be fun. They contend, "Looking at games that go beyond fun creates new spaces for players, games, and queer worlds at the margins" (11). Making games is a way of making worlds (see Faris on reparative making in this collection).

Looking at Twine games through the lens of procedural rhetoric can produce some critical insights, and in this chapter, I want to show how a studying game's procedures can also help us understand its affective rhetoric. In the context of this collection on critical making, I want to suggest that the affective force of the Twine games I examine here is what makes them theoretically significant for rhetoricians. Matt Ratto (2014) argues that critical making involves "materially productive, hands-on work" that aims to "uncover and explore conceptual uncertainties, parse the world in ways that language cannot, and disseminate the results of these explorations through embodied, material forms" (227). Hypertext games might test the limits of this view of materiality and embodiment, but I'm especially struck by Ratto's claim that critical making can "parse"—that is, analyze into units of meaning—a world that exceeds language's meaning-making powers. Rhetoricians know that language has other powers, too. David Sheridan (2016), in "A Maker Mentality Toward Writing," argues that rhetorical force is not always or exclusively communicated through semantic meaning.

Rhetoricians have long understood identification to be a (if not *the*) central dynamic governing rhetorical interaction. For Kenneth Burke (1969), of course, identification is figured as a way of crossing a chasm that contains and separates individuals from one another. Identification bridges the gulf of our differences with something shared. But some scholars have argued that empathy and imagination are not enough to create identifications between those with power and those with less of it. As Avital Ronell (2002) contends, "The ability to put oneself in the place of the other does not preclude violence" (206). Diane Davis (2010) criticizes the Burkean view of identification as "the search for confirmation, in the other, of one's own structures of understanding" (73). Identification effaces the difference of others, since whatever you can't grasp escapes you. As Ronell has illustrated, it's "not a matter of ability or aptitude" (Ronell 2002, 206), not as if you can simply be good enough or smart enough to avoid committing the violence of overwriting the other with your own understanding.[2] The violation is definitional, central to the act of identification: Putting yourself in the other's place is always an "intrusion" (206)—even if the substitution is figured as sacred. Identification alone can't preclude violence.

Instead, the Twine games I study here demonstrate how the structure of identification can be twisted toward solidarity with others. In *Inessential Solidarity*, Davis (2010) argues that the ethical problem of identification is not how to make a connection with an/other, "but '*how to make a disconnection*'" (25)—how to secure a distance between self and other in which the other's difference from me is not effaced.³ Failing to identify, Davis argues, can open one "to the other as other" (35)—although this relation doesn't exactly promise harmony. Dissociation "installs a bewildering and (temporarily) ineffaceable distance" (34) between self and other, but it is in this distance that an ethical relation to alterity can form. It is this ethical relation that Davis names "solidarity." As my title suggests, I hazard that Twine games are uniquely suited to function as *solidarity machines*. James J. Brown Jr. (2014b) argues that machines are simply generative "whatsits" (496). If solidarity is not captured, contained, or exhausted by the act of identification, then Twine games that amplify minoritarian logics, *resist* understanding, and *undermine* identification might qualify as solidarity machines. While we might easily understand video game design as a form of critical making, I also want to draw your attention to the rhetorical force of Twine games in excess of their meaning: A solidarity machine can also be understood as an agent of critical making.

I want to comment briefly on the style of this chapter, namely the second-person address. Though the informality of direct address is not usually preferred in scholarly writing, addressing you, my readers, as "you" serves an important rhetorical function here. It places my authorial voice somewhere in the same neighborhood as the developers of many hypertext games (for more on "you" in hypertext fiction, see Bell and Ensslin 2011). It also aligns me with an expansive view of video games in which language too can be a text-based game: governed by interpretive rules, temporally precarious should you be interrupted (or decide to quit), and most importantly for my purposes here, relational. In a 2012 blog post that's part technical documentation for Twine and part gamemaking manifesto, Porpentine Charity Heartscape (cited by only her first name at this point in time) asserts: "Our global network is composed of human minds uploaded into word form. On this plane the word is the most potent unit of force. It costs a couple of keystrokes to control someone else's brain for a second, and longer if you do it right" (Porpentine 2012). This is an electrifying claim about the power of language to direct and connect people, and about language as a rhetorical force that opens us to one another. If you will accept even provisionally that addressed language—as Burke says, rhetoric—can be a game, then one of its rules must be that players can be affected by and in language (see Sierra in this collection for more on how games can help preserve and sustain language and the possibility of this relation). As you read this chapter, I invite you to play at least a few of the games I discuss, some of which are available online for free.

Throughout this chapter, I use video games themselves as a form of evidence, a source for premises in my arguments, since as Mary Flanagan (2013) has argued, play too can be critical. By drawing on my own experiences as a player, I want to make legible the scenes of the failed identification, nonunderstanding, and yet irremissible inclination that Twine games can construct. As Patrick Jagoda (2017) has argued, critical making can be a way of building on critical theory without privileging acts of interpretation (357). My argument proceeds by way of video games: games that pose problems for the field's traditional thinking about identification; games that institute a space of ethical distance between self and other in which a relation of solidarity can take hold. By looking at several games designed by queer and trans women game developers, I explore the scenes of intimacy, violence, trauma, healing, ephemerality, and apocalypse that these games open up—sometimes all at once. I begin by examining Twine itself as a procedural program with constraints and affordances made possible by its design. I then examine two sets of Twine games in order to show how their affective forces interact with procedural rhetoric and rhetorical theories of identification to allow them to function as solidarity machines.

TWINE'S PROCEDURAL RHETORICS

Like video games, Twine itself is a rule-governed digital program, and it too has a procedural rhetoric delimited by the operations it can perform. Developed in 2009 by Chris Klimas, Twine enables its users to design games without requiring much knowledge of programming languages or computer code. Users of Twine can create a game by making hyperlinks that connect passages of text. Twine can be complexified using protocols like HTML, CSS, and JavaScript, but its simplicity means that virtually anyone with computer access can use it to make a game, even those who don't have access to the resources or coding knowledge possessed by industry insiders. The simplest way of making Twine games is to join passages of text together using hyperlinks. Nodes that are connected by hyperlinks are called passages. Although nothing in Twine's design restricts it to proceeding mainly by text rather than, say, images, or even sounds, Jane Friedhoff points out that the visual, spatial layout of Twine's passages resembles writers' planning and organizing technologies, from notecards to Scrivener (Friedhoff 2014; for more on using Twine in the writing classroom, see Gerdes, Beal, and Cain, 2020).[4] But the connections between passages in Twine also resemble an organic structure: Porpentine compares Twine games to "creatures under a microscope or root networks carrying information" (qtd. in Friedhoff 2014, 4). Porpentine even created a Tumblr account that posts pictures of Twine "node maps," fittingly called "Twine Garden."

Anna Anthropy (2015) argues, "All you need to know to make a game in Twine is how to write. And the same way that people who don't think of themselves as artists know how to doodle on napkins, folks who don't think of themselves as writers know how to scrawl in a journal" (36). Using Twine can make video game design more personal, ephemeral, and accessible than ever before. Twine has been called "the video game technology for all" because it has opened the rhetorical space of video game design to voices that have previously been marginalized or altogether silenced (Hudson 2014; see also Kareem 2015). In addition to Porpentine's (2012) Twine instructional manifesto ("Creation Under Capitalism"), there are online forums, wikis, and even print essays, manuals, and anthologies aimed at getting people to make things with Twine (see Anthropy 2019; kopas 2015b). Friedhoff (2014) observes that such reference materials focus first on "answering '*why* would you make a game at all?'" (3). Twine's documentation is oriented toward what Friedhoff calls "the expressive potential" of hypertext. As Friedhoff contends, "These tutorials act as a call-to-arms for potential developers, validating and encouraging the use of individual experience as the subject of a game" (3).

The validity of writing about your own experience is a key part of Twine's accessibility: Friedhoff argues that Twine lends itself well to writing in the genre of the vignette, which she defines, citing Bogost, as a "brief, indefinite, evocative description or account of a person or situation" (Friedhoff 2014, 6). Sometimes called "personal games," Friedhoff argues that Twine is a particularly fruitful tool for queer game developers, for people without much experience coding, and for creating games that deal with "taboo" topics and traumas that would never see the light of day in a commercial game store. When you make a game in Twine, the software exports the game as an HTML file, so many Twine games are published as websites, but they can also be made available for download and played offline. Twine games can be uploaded on personal websites or to a free Twine hosting site like http://philome.la/, sold on indie games and art sites like http://itch.io/ or http://gumroad.com/, or even emailed directly to players.[5] Friedhoff argues that "Twine's distribution model is key to its support of non-mainstream games" (7). While developing a game for iPhone or Xbox may require designers to pay for access to a development kit, obtain permission to release work in the format and/or to sell work released in the format, and to submit their games to a review process that explicitly bars certain content, Twine

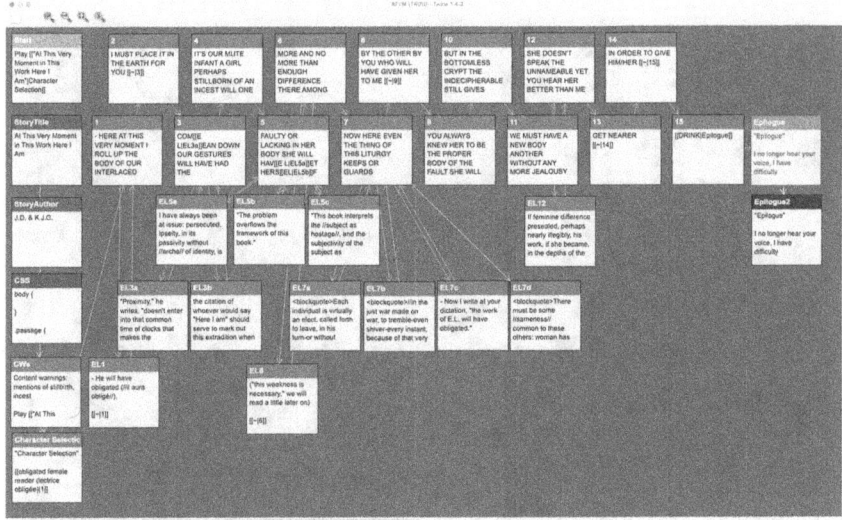

Figure 12.2. Node map of a game in Twine 1.4.2. Screenshot by author.

has none of these barriers. It can be used for free in any web browser. There's no customs authority reviewing your game's content. That kind of freedom from policing, built into the program's procedures, has made Twine the site of rhetorically inventive work in the medium of video games, reinventing what video games can be and do and how they can make you feel and respond. Twine games have leveraged the procedures of the platform to engage players affectively, engaging topics and emotions that mainstream games can't touch, producing new channels for the force of rhetorical affection to produce solidarity.

SEX AND VIOLENCE

I turn now to a set of Twine games that illustrate specific possibilities opened up by their departures from the norms of more mainstream video games. I want to show how these games can challenge the centrality of identification to rhetoric by focusing on the way their procedures create and exert affective force—and to do this, I have to get specific about the affects in question. Instead of cultivating empathy or establishing commonality between players and the characters these games position us to identify with, these Twine games generate an inessential solidarity through their production of an addressed affective force—addressed to you, the reader/player. In this section, I look at three Twine games that reinvent what sex and/or violence can do in video games.

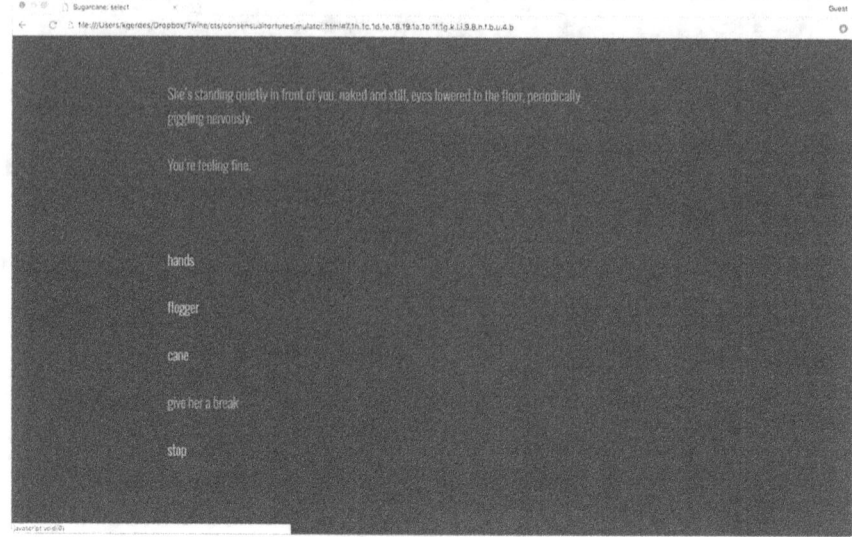

Figure 12.3. A decision point in Consensual Torture Simulator. *Screenshot by author.*

In merritt kopas's provocatively titled *Consensual Torture Simulator*, you have agreed to do a sadomasochist scene with your girlfriend in which you will use impact play (slapping, spanking, flogging, hitting) to make your partner cry. Getting her to cry is the goal of the scene, but it is not necessarily the goal of the game: you can stop the scene at any time, and your decisions can escalate or de-escalate the scene's intensity. Notice the highlighted link in figure 12.3, which says "give her a break."

When I played *Consensual Torture Simulator* for the first time, I had read a lot about it, so I thought I knew what would happen in the game, and I thought I had a sense of the argument it could make. I suppose I thought I was emotionally prepared to play it, but I was surprised by my own resistance to escalating the scene. I found that I had to exceed the level of sustained impact I was comfortable with in order to make my partner cry. Moreover, her tears come in varying intensities—"crying" isn't a simple or obvious win condition—so I felt it was up to me to decide when and if we had met our goal. In the denouement of the game, you administer aftercare to your partner, praising and comforting her and getting her a glass of water or fixing her a cup of tea. It's remarkable how much responsibility you are made to feel, how you must confront your own normative beliefs about sexuality, violence, partnership, negotiation, consent, and care in just the few minutes it takes to play the game. The game's affective force can be overwhelming; it can open players up to their responsibility for affecting others.

What you do in *Consensual Torture Simulator* that you can't do in a traditional video game is negotiate sex and consent as physical and emotional activities instead of as a conquest. In the blockbuster game of 2015, *Fallout 4*, for example, sex is essentially the result of stat check. During dialogue with a potential partner, the option to press a button to "flirt" appears. There's no description of what strategies you might use to flirt; no multiple ways to approach flirting as a task. You can't even see the specific dialogue your character will use. You just press the button for "flirt," and the game checks your character's charisma stat. If it's high enough, your partner is receptive to your advances. This is basically a pickup artist's idea of the interplay between sex and procedural rhetoric: seduction is a procedure performed on a target object, and sex is a reward for correct performance. *Consensual Torture Simulator* disperses this transactional view of sex with a scene of trust, risk, and decision. Rather than procedures that orient your partner as a target or objective, *Consensual Torture Simulator*'s procedures exert their force and affect on *you*, making you *feel* the stakes of the scene: your responsibility, your desire for intimacy, your own personal relationships to violence, sex, and care.

In an essay that discusses the game, kopas (2015a) points out that video games have tended since their inception to be about conflict, competition, overcoming resistance and obstacles, and solving problems or puzzles. These tendencies make it difficult, she argues, to develop games that incorporate sexuality without turning sex into another conflict or challenge. Answering the claim that it's just technologically easier to develop ballistic mechanics rather than responsive dialogue or to show bodies ripped apart and dying rather than embracing, kopas reminds us that even if this claim were true, "development technologies aren't neutral; they're informed by the interests of their creators, often in subtle and totally unpredictable ways" (223). In other words, the procedural rhetoric of mainstream studio games has been shaped over time to privilege and normalize violence—and it could, in theory, be reshaped in other directions. Echoing kopas, Brown (2014b) reminds us that "rules are not only followed. They are also *authored*" (497). *Consensual Torture Simulator* rewrites the rules about what kind of violence is commonplace, normative, and acceptable in video games.

The link between intimacy and violence is also explored in Christine Love's *Even Cowgirls Bleed*. In the game, you are a cocksure young gunslinger on her way to San Francisco. The game changes your cursor into a gunsight, and there's no need to click on hyperlinks since your gun fires whenever you roll the sights over a link. Whenever you have the option to holster your gun instead, it appears on the left or right side of the screen, alternating sides, as if you showily toss your weapon from hand to hand just to put it away. Eventually, line breaks in the game's text form a ring of links around the option to surrender the gun. Rather than present you with the moral or ethical decision, *Even Cowgirls Bleed* strips you of choices, forcing you to engage in violence to advance the game. Such choke points in Twine games are another way of illuminating the normalcy of using violence as a game mechanic. They can also insert distance between you and a player-character you can't fully control.

The cartoonish violence of traditional video games, the structural violence of capitalism, and the intimate violence of domestic abuse converge in Porpentine's *Ultra Business Tycoon III*. The Twine game opens with a simple and familiar interface that mimics a 1990s shareware game: you may start a new game, load a game, view options, read the guide, view the credits, or quit. You can get an early hint that *Ultra Business Tycoon III* is not what it purports to be if you try to load a game—clicking the link for the only saved file results in an italicized message: "*Your big sister's save file. Best not to fuck with it.*" As you wend your way through the parodically violent business world, amassing tens of thousands of dollars and eluding cops, more italicized reflections create incursions into your opportunity to identify with the ultra business tycoon, and the "you" of tycoon-game separates from the "you" of Porpentine's game, and you learn that the tycoon power fantasy makes a kind of dissociative escape for the player, a young trans girl growing up in a violent home.

In a 2014 conference presentation, James J. Brown Jr. reads *Ultra Business Tycoon III* as an example of what he calls "obfuscated mapping," a turn on the work of cognitive mapping done by graphical user interfaces. "The vast majority of games teach us that what we do makes sense," Brown contends, "that we can control an environment, and that there is a clear relationship between actions and results" (Brown 2014a, 3). *Ultra Business Tycoon III* is not like this. This game is home to a "realistic Bee AI," subterranean trash zones, animals that ooze, and unmonetizable vomit. Brown argues that obfuscation in the game's interface can "productively defamiliarize" your interaction with the game, and with its subjects—trauma, trash, video games, and capitalism—and you can learn to look *at* the screens that organize meaning that you're usually taught to look through. Of course, Twine is one of those screens. As a tycoon, you are powerless against a degrading wastescape cordoned off from cleanliness, order, and safety. You face non sequitur consequences

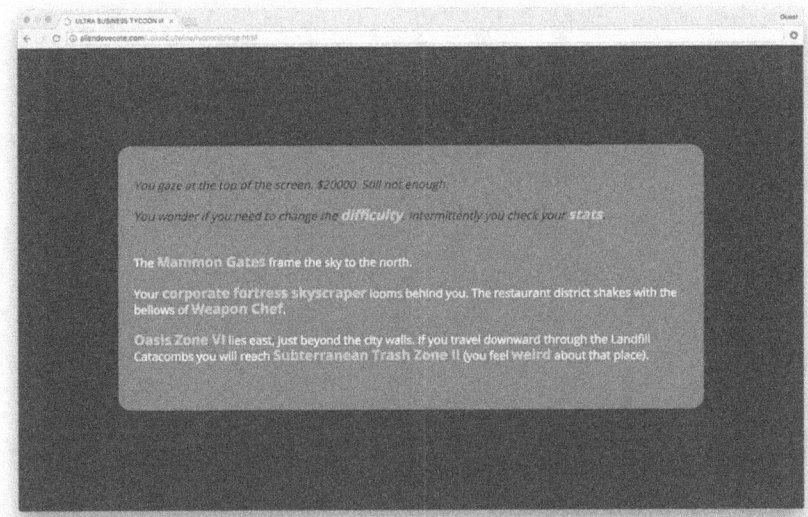

Figure 12.4. Your italicized thoughts on playing Ultra Business Tycoon III, *inside Porpentine's* Ultra Business Tycoon III. *Screenshot by author.*

for both logical and illogical actions. You might die as often as you find a reward. But the player-character of Porpentine's *Ultra Business Tycoon III* is not a tycoon. You're a child, and your powerlessness to secure yourself and your sibling from a chaotic and violent parent is reflected in the game within the game.

Brown closes his argument by positioning obfuscated mapping as not simply a neutral rhetorical practice of sense-making (or unmaking) but as inflected by the "array of practices and strategies deployed by LGBTQ writers" who "have used Twine to describe lives in which concepts like agency and control are especially fraught" (Brown 2014a, 10). The games I discuss here tell highly specific, even minoritarian stories, yet they remain open for audiences that their makers could not have anticipated. To the extent that these games reflect the experiences of their queer women makers, these experiences become available to audiences who differ from them not if you simply put aside your differences and focus on what you have in common, but only if *you* the reader will also become "you," the game's addressee. You aren't just receiving meaning; you are receiving an address. Solidarity depends on hearing and responding to a call that you can never really be sure is meant for you. Nothing is assured: there is no ethical purity in what I am describing; amplifying minoritarian logics shares a border with appropriating

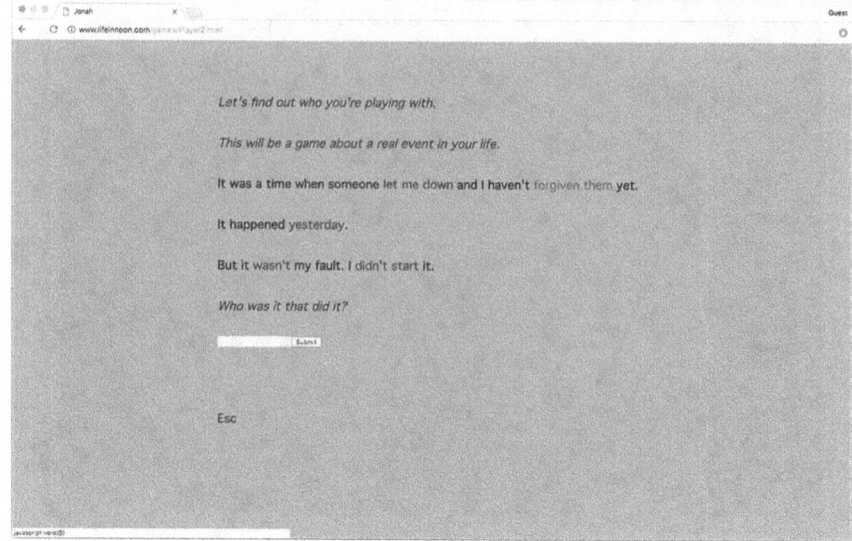

Figure 12.5. You control the situation in Player 2. *Screenshot by author.*

alterity to one's own understanding. Solidarity is only an invitation, a condition of possibility for an ethical relation to open.

EPHEMERALITY, FORGIVENESS, AND STARTING OVER

I want to extend this analysis in one further direction, to look at another set of games focused on ephemerality. As mainstream studio games get exponentially bigger, featuring fifty-hour revenge quests with dozens of repeatable side quests to make you stronger and richer, brief games that emphasize forgiveness, healing, and starting over stand out in sharp contrast. These games make internal difference—or we might say growth, or more simply, change—a central procedural mechanic. Like the games discussed in the previous section, they may push identification and even understanding to its limits, but they nevertheless implicate their players through their force of rhetorical affection. I focus on the worldmaking of the games in this section.

Lydia Neon's *Player 2* is a reconciliation game. The game begins by announcing that your playthrough will be kept safely anonymous and explains how you can download the game and play offline if you prefer. "Player 2" isn't a party to the game, not directly; instead, you choose someone to fill the role, someone who has hurt or wronged you. The game puts you in control of how you represent and respond to them.

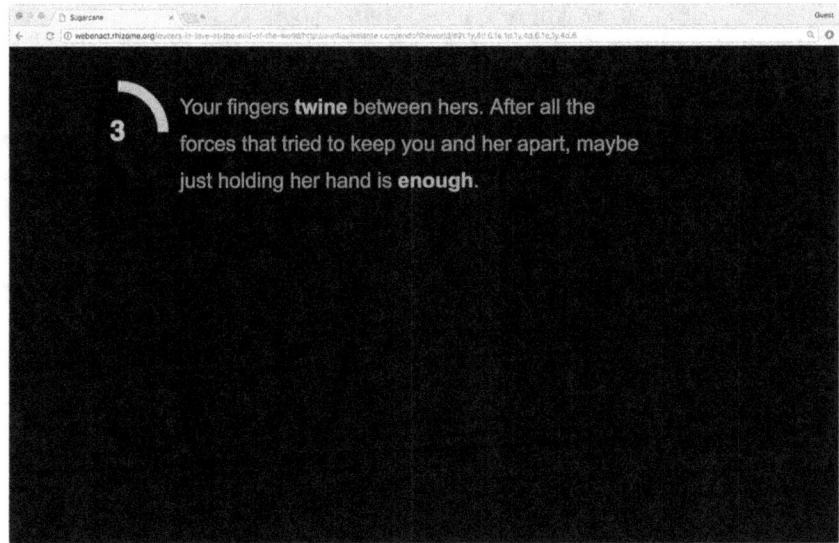

Figure 12.6. Time is running out in Queers in Love at the End of the World. *Screenshot by author.*

Most of the hyperlinks don't direct you to entirely new passages but substitute new words or phrases into the sentences when you click them. For example, the first sentence reads: "It was a time when someone let me down and I haven't dealt with it yet." Clicking "let me down" changes the hypertext to "insulted me," then, "hurt me," then "belittled me," and so on. Clicking "dealt with it" changes it to "confronted them about it," then "fixed it," then "resolved it," and so on. Clicking causes most of the hyperlinked phrases to cycle through a list of choices—a game mechanic that limits how you can describe and deal with painful circumstances while suggesting possibilities that might put language to your emotions.

The game takes your input and walks you through a process of forgiveness, or at least of letting go of negative feelings and moving on from an interpersonal conflict that's weighing you down. When the game is over, none of your grievances are saved. Your slate is wiped clean. The game's ephemerality is itself an argument for healing and moving on. Instead of obliterating your enemies with a barrage of bullets, *Player 2* makes the relationality of a wound into a game mechanic—one you can, for a moment at least, safely control. The game depends on cultivating attentiveness to your own exposedness to others.

In Anna Anthropy's *Queers in Love at the End of the World*, as in *Player 2*, nothing is saved: a timer counts down ten seconds before the apocalypse wipes everything away. There are 180 unique passages in the game—far more than even the speediest reader could navigate in the allotted time. When the game ends, an option to restart appears—I find myself replaying this game five or six times in a row every time I play it. Each playthrough offers you a tiny slice of a vast and varied world. A brief moment of possibility opens right before, and because, "everything is wiped away." *Queers in Love at the End of the World* lets you explore that possibility. The sense of magnitude is humbling, especially against the apocalyptic backdrop: you'll only see as much of this world as you can in the limited time you have left. Even though a path can end before the timer runs out, you can never exhaust all the possibilities in one run.

If you use the last ten seconds of history to hold hands with your girlfriend, you can arrive at the passage in figure 12.6, which says: "Your fingers twine between hers. After all the forces that tried to keep you and her apart, maybe just holding her hand is enough." If you click the hyperlink "twine," the passage reads: "What a powerful mode of expression." There are no hyperlinks out of that next passage; if you arrive with any time left, you'd just watch the world end. And yet, even in the face of total annihilation, you're always offered the chance to restart.

Queers in Love at the End of the World is a testimony to the force of even ephemeral love to outlast—to outlast both the norms and laws that have prohibited it, and the shared world that made it possible. The procedurality of the ticking clock could have the rhetorical effect of engendering panic, but after a few playthroughs, the consistency with which the world is destroyed can also allay your sense of urgency. It can be kind of nice to know it's all about to be over, and your piece of it was partial and incomplete, but you still get to sail into oblivion with someone you deeply love. And there's no one right way to finally do that. Your experience of togetherness is what matters, even as the (game)world that made it possible is obliterated by time.

The queer worlds in which forgiveness and starting over become possible are not always brief, however. Porpentine's brilliantly written *With Those We Love Alive* crafts a time-bound world in which you play as a smith or "artificer" who makes weapons, crowns, and prostheses for a monstrous queen. You're something of a captive at the empress's castle, though you can visit the city, enter the temple, and drink in the barlike dream distillery. You can walk through the royal gardens or go down to the lake and meditate. When you go to your workshop, you can only work if you are rested, and you sometimes have to go back to your room to sleep. In my playthrough, I bounced erratically, and a bit bitterly, between drinking in the distillery, cursing the silence of the gods in the

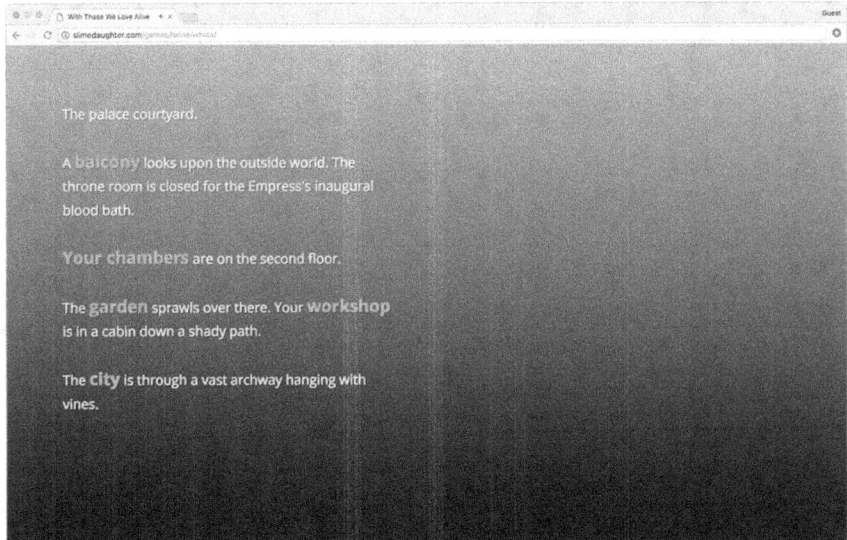

Figure 12.7. The places you can explore in With Those We Love Alive. *Screenshot by author.*

temple, and sulking when I was unable to work. Sleep governs the passage of time in the game.

With Those We Love Alive is home to many rituals, some optional and some required. Every so often, before you can sleep, you have to take your hormones. Refusing to do so will keep you up—no matter for how long you spend visiting other places—and when you exhaust the temple, bar, and workshop, you finally have to attend to your body. You administer a gendering regimen that's almost incidental to the game's plot, but it is integral to your health and well-being and advancement through the game. Hormones don't alter your character in any apparent way; it's more like they keep you self-similar. Perhaps this ritual may activate your own visceral knowledge of navigating illness, or age, or rigid gender binarism. Being unable to sleep (and thus to continue gameplay) without completing this act of self-care spotlights the structuring influence of gender, age, and health on your body. *With Those We Love Alive* asks you to attend not only to the body of your character within the game, but also the own body that you are using to play the game.

At various junctures, you are asked to draw a sigil on your arm representative of some moment or decision point in the game. The stack of sigils you inscribe on your arm is echoed by the inscribed arms that are housed in jars in the game's temple. Through the physical act

258   KENDALL GERDES

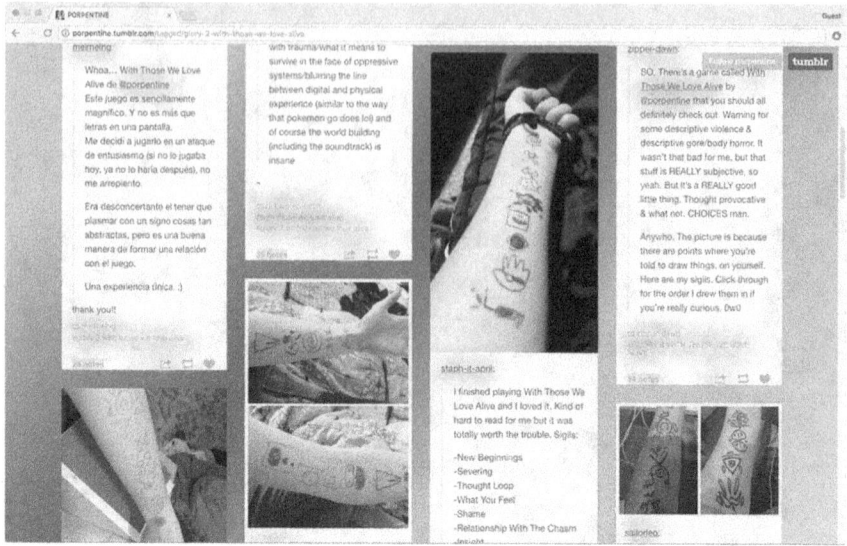

Figure 12.8. Photos of sigils drawn on the arms of With Those We Love Alive *players, published on Tumblr. Screenshot by author.*

of inscription, you are invited on the one hand to identify with your character—to imagine that your sigils are *her* sigils, too—but on the other hand, your attention is drawn to how you are not the player character: how your inscriptions remain outside of the game, on your own flesh, and how it maybe hurts if you chose too sharp a pen to draw with, or how long your drawings will last (probably days after you're done playing the game) if you chose a softer nibbed marker. When you complete the game, a link can take you to Porpentine's blog, where she posts photos that players have submitted of their sigils. (It's worth noting the photos are overwhelmingly, though not exclusively, white.) None of the sigils will look quite the same as yours, and they may represent emotions or choices that differ from your own. In fact, most will likely be unrecognizable to you: Even though you have been through the same game—even played as the same person—your experiences, and the symbols you created to navigate them, are totally alien to others. *With Those We Love Alive* puts you into relation through your differences from her, from yourself, and from each other. The difference between you remains inassimilable, undigestible, unabsorbable; your reckoning with the trauma, violence, and gender in the game institutes a space of ethical distance by calling your attention to your *inability* to put yourself fully in the place of the other.

BECOMING TWISTED TOGETHER

Failing to identify can open up a different sort of relation to difference than identification allows. As Davis (2010) argues, the distance between self and other required to institute a space for ethical relation only opens "in the failure of identification, each time," since in that failure " 'I' am opened to the other *as* other and get the chance to experience something like responsibility for the other that exceeds (and conflicts with) 'my' narcissistic passions" (35). That is, when you fail to fully identify with an other, your failure opens a chasm across which some relation to difference—other than its total assimilation—becomes possible. Rather than bridging the gap with something essentially shared, solidarity can take hold across this distance, not in spite of but because of the differences. The very force of being affected rhetorically—that is, touched or moved in language by someone who is *other than* me—is what makes solidarity possible. Twine games like those I have discussed here can teach us that solidarity is not based on identification or even shared meaning; a Twine can twist us together through the address of gameplay, through engendering address in the act of gamemaking, and through the rhetorical relation that obtains between existents, inviting alteration in what's possible in our worlds, virtual and otherwise.

Thinking of Twine games as solidarity machines—as games that are able to generate solidarity through rhetorical address, through affective force, and through worldmaking—brings into relief how making games can be a critical intervention both in scholarly conversations around procedural rhetoric and rhetorical identification and in the ordinary experience of Twine game players. Failing to identify with the games' protagonists doesn't stop a rhetorical relationship from forming, one based on the solidarity of rhetorical affection rather than the achievement of identification. Video games can be a resource for the rhetorical invention of theory, not only by analyzing their procedures, but by examining, experiencing, and authoring their affective force. Critical making should not be limited to work performed by scholars or by those with specialized technical knowledge of their media, and it should not exclude theory. Sharon Crowley (2006) argues that the Sophist Gorgias could be understood as a theorist of rhetoric because his rhetorical performances were instructive models for students of rhetoric. She writes, "Theories are rhetorical inventions: depictions or assessments produced by and within specific times and locations as a means of opening other ways of believing and acting" (28). By treating Twine games as a site for theory-making, I contend that these solidarity machines are more ordinary, accessible, intimate, and powerful than their usual treatment in either rhetoric or games scholarship has previously suggested.

NOTES

1. For more on GamerGate's specious claims, antifeminism and misogyny, and the patterns of harassment and abuse within gamer culture, see Mortensen 2016.
2. I cite Ronell's incisive scholarship on the ethics of identification here, even as I am cognizant of an NYU Title IX investigation finding her responsible for hostile environment harassment (made public in 2018).
3. Bryan Crable (2009) makes a parallel argument in a more Burkean register, contending that pure persuasion is the ultimate rhetorical motive, and that the divisions bridged by identification must first be rhetorically produced.
4. *Black Mirror: Bandersnatch* (2018), Netflix's choose-your-own-adventure-style "interactive film," was drafted using both Scrivener and Twine (Rubin 2018).
5. Philome.la closed to new submissions in September 2019.

REFERENCES

Anthropy, Anna. 2012. *Rise of the Videogame Zinesters: How Freaks, Normals, Amateurs, Artists, Dreamers, Dropouts, Queers, and People Like You Are Taking Back an Art Form*. New York: Seven Stories Press.

Anthropy, Anna. 2015. "Love, Twine, and the End of the World." In *The State of Play: Creators and Critics on Videogame Culture*, edited by Daniel Goldberg and Linus Larsson, 31–51. New York: Seven Stories Press.

Anthropy, Anna. 2019. *Make Your Own Twine Games!* San Francisco, CA: No Starch Press.

Bell, Alice, and Astrid Ensslin. 2011. "'I Know What It Was. You Know What It Was': Second-Person Narration in Hypertext Fiction." *Narrative* 19 (3): 311–29. doi:10.1353/nar.2011.0020.

Bogost, Ian. 2007. *Persuasive Games: The Persuasive Power of Videogames*. Cambridge, MA: MIT Press.

Brown, James J., Jr. 2014a. "Obfuscated Mapping." Paper presented at the Society for Literature, Science, and the Arts Conference, October 10, 2014, Dallas, TX. https://web.archive.org/web/20160319092534/http://clinamen.jamesjbrownjr.net/2014/10/15/obfuscated-mapping/.

Brown, James J., Jr. 2014b. "The Machine That Therefore I Am." *Philosophy & Rhetoric* 47 (4): 494–514.

Burke, Kenneth. 1969. *A Rhetoric of Motives*. Berkeley: University of California Press.

Crable, Bryan. 2009. "Distance as Ultimate Motive: A Dialectical Interpretation of *A Rhetoric of Motives*." *Rhetorical Society Quarterly* 39 (3): 213–39. doi:10.1080/02773940902991445.

Crowley, Sharon. 2006. *Toward a Civil Discourse: Rhetoric and Fundamentalism*. Pittsburgh, PA: University of Pittsburgh Press.

Davis, Diane. 2010. *Inessential Solidarity: Rhetoric and Foreigner Relations*. Pittsburgh, PA: University of Pittsburgh Press.

Evans, Sarah Beth, and Elyse Janish. 2015. "#INeedDiverseGames: How the Queer Backlash to GamerGate Enables Nonbinary Coalition." *QED: A Journal in GLBTQ Worldmaking* 2 (2): 125–50.

Flanagan, Mary. 2013. *Critical Play: Radical Game Design*. Cambridge, MA: MIT Press.

Friedhoff, Jane. 2014. "Untangling Twine: A Platform Study." In *Proceedings of DiGRA 2013: DeFragging Game Studies, Atlanta, GA*, edited by Celia Pearce, John Sharp, and Helen

Kennedy, n.p.: Digital Games Research Association DiGRA. http://www.digra.org/digital-library/publications/untangling-twine-a-platform-study/.

Gerdes, Kendall, Melissa Beal, and Sean Cain. 2020. "Writing a Videogame: Rhetoric, Revision, and Reflection." *Prompt: A Journal of Academic Writing Assignments* 4 (2): 3–12. doi:10.31719/pjaw.v4i2.64.

Holmes, Steve. 2017. *The Rhetoric of Videogames as Embodied Practice: Procedural Habits*. London: Routledge.

Hudson, Laura. 2014. "Twine, the Videogame Technology for All." *The New York Times Magazine*, November 19, 2014. www.nytimes.com/2014/11/23/magazine/twine-the-video-game-technology-for-all.html.

Jagoda, Patrick. 2017. "Critique and Critical Making." *PMLA* 132 (2): 356–63. doi:10.1632/pmla.2017.132.2.356.

Kareem, Soha. 2015. "Tying in Diversity with Twine Games." *Broken Pencil* 66 (March): 27. brokenpencil.com/news/tying-in-diversity-with-twine-games/.

King, Matt. 2010. "Procedural Rhetoric—Rhetoric's Procedures: Rhetorical Peaks and What It Means to Win the Game." *Currents in Electronic Literacy*. https://currents.dwrl.utexas.edu/2010/king_procedural_rhetorics_rhetorics_procedures.html.

kopas, merritt. 2015a. "Ludus Interruptus." In *The State of Play: Creators and Critics on Videogame Culture*, edited by Daniel Goldberg and Linus Larsson. New York: Seven Stories Press.

kopas, merritt, ed. 2015b. *Videogames for Humans: Twine Authors in Conversation*. New York: Instar Books.

Mortensen, Torill Elvira. 2016. "Anger, Fear, and Games: The Long Event of #GamerGate." *Games and Culture* 13 (8): 787–806. doi:10.1177/1555412016640408.

Porpentine. 2012. "Creation Under Capitalism and the Twine Revolution." https://web.archive.org/web/20131114013954/http://nightmaremode.net/2012/11/creation-under-capitalism-23422/.

Ratto, Matt. 2014. "Textual Doppelgangers: Critical Issues in the Study of Technology." In *DIY Citizenship: Critical Making and Social Media*, edited by Matt Ratto and Megan Boler, 227–36. Cambridge, MA: MIT Press.

Ronell, Avital. 2002. *Stupidity*. Champaign: University of Illinois Press.

Ruberg, Bonnie. 2015. "No Fun: The Queer Potential of Video Games that Annoy, Anger, Disappoint, Sadden, and Hurt." *QED: A Journal in GLBTQ Worldmaking* 2 (2): 108–24.

Rubin, Peter. 2018. "How the Surprise New Interactive Black Mirror Came Together." *Wired*, December 28, 2018. https://www.wired.com/story/black-mirror-bandersnatch-interactive-episode/.

Sheridan, David M. 2016. "A Maker Mentality Toward Writing." Gayle Morris Sweetland Digital Rhetoric Collaborative. March 28, 2016. https://www.digitalrhetoriccollaborative.org/2016/03/28/a-maker-mentality-toward-writing/.

Thevenin, Benjamin. 2017. "Twine as Alternative Media: Video Games, the Culture Industry and Social Change." *Teaching Media Quarterly* 5 (2): 1–13. https://pubs.lib.umn.edu/index.php/tmq/article/view/652.

SECTION 5

Critical Making as Instructional Design

13
CULTIVATING CRITICAL MAKERS
Crafting with Paper-Electronic Circuits in an Online First Year Composition Course

Bree McGregor
American University

In the fall of 2017, I set out to reimagine an accelerated, online, first year composition course as an experiment in critical making pedagogy. Composition has a rich history of multimodality (Arola and Wysocki 2012; Bowen and Whithaus 2013; Braun 2013; Davis and Shadle 2007; Grouling and Grutsch McKinney 2016; Handa 2004; Hobbs 2006; Kress 2010; Lutkewitte 2013; Palmeri 2012; Shipka 2011; Shipka 2016; van Leeuwen and Kress 2011; Wysocki 2004; Yancey 2004), and critical making and first year writing courses have the characteristics of tinkering and exploring in common (Bowen and Whithaus 2013; Ellis 2013; Grabill and Hicks 2005; Sirc 2004). Critical making assignments offered another opportunity to emphasize how the writing process is both recursive and multimodal, and to help foster a sense of community within an online class through collective learning experiences.

In this chapter, I share the background context for the student population with whom I put critical making pedagogy into practice, a rationale for the course in critical making scholars David Sheridan (2015) and Mark Hatch (2013), and a detailed description of how I designed a critical making–themed course and implemented maker assignments in an accelerated, online first year composition course. I will also share samples of student work and students' end-of-semester reflections on critical making. In doing so, I will demonstrate how critical making isn't just for traditional eighteen- to twenty-two-year-old college students or restricted to campuses with makerspaces. Critical making has a role to play in any composition classroom. Similar to the concepts that John Jones explores in his chapter in this collection, "The Syllabus as Carpentry," this chapter also explores how the creation of objects for both inventive and analytical purposes supports a writing and rhetoric pedagogy.

BACKGROUND DESCRIPTION OF STUDENTS

From 2010 to 2017, I taught undergraduate writing courses for Arizona Western College in a variety of formats, but I predominantly taught online. Arizona Western College is a Hispanic-serving institution located in the desert southwest farming community of Yuma, Arizona. Sharing borders with California and Mexico, Yuma is also home to an Army proving ground and a Marine Corps air station. This amalgamation of factors results in a richly diverse student population that includes migrant farming and active-duty military communities. More than six thousand students attend Arizona Western College each semester, and most are classified as nontraditional. In the fall of 2018, Arizona Western College reported that 71 percent of its students were enrolled part-time, and 64 percent were first-generation college students (Stofft et al. 2018, 1). Yet Arizona Western College students are no longer unique in these ways. According to the US Department of Education (2015), nontraditional students have surpassed traditional students as the majority of undergraduates in the United States. These students typically have more time constraints than traditional students, and online classes often provide the flexibility they need. However, through informal, end-of-semester feedback, I found that many of my students' expectations for online learning did not align with their experiences. These students did not always feel prepared for the technology-related tasks associated with online learning. In particular, my students had described feeling apprehensive about learning to use technical tools or platforms outside of Blackboard. They had also expressed concerns about participating in collaborative assignments in an asynchronous, online course. Because of this feedback I had received over the years, I gave special consideration to the design of critical making assignments, which would be collaborative and technical in nature.

CRITICAL MAKING FOR COLLABORATION AND COLLECTIVE LEARNING

With students' prior online learning experiences in mind, in the fall of 2017 I designed an accelerated, eight-week online first year composition course as a pedagogical experiment in critical making. This decision was motivated in part by recently adopted syllabus changes, which emphasize the use of a variety of composing technologies and engagement in flexible and recursive strategies (Arizona Western College 2017). I designed maker assignments to meet the syllabus's updated learning outcomes, and in the process, I hoped to create learning communities successfully engaged in low-stakes, collaborative, critical-making experiences.

Critical making provided another means of scaffolding *process* into student writing. Creating space for the writing process was particularly important in my accelerated courses, where the abbreviated academic schedule often creates a frenetic pace for both students and the instructor. Rather than moving directly from the researching and note-taking stage into the drafting stage, I envisioned maker assignments as an in-between step to help students pause, step back from *writing* as a mode of production, and to re-see their work from a different perspective, through the tactile affordances of *crafting*. Students would handcraft paper-electronic circuits using conductive copper tape, coin cell batteries, and LED stickers to visually represent connections between their research questions and the key points that emerged from research findings as an integrated step in the writing process.

Because critical making assignments served as part of the process—rather than the final product—they were designated as low-stakes assignments. Students' maker assignments were evaluated on students' willingness to explore ideas and tinker with a different mode of composing, not on their technical success in creating light-up paper-electronic circuits. While a low-stakes designation helped reduce students' resistance to critical making assignments, it was also important for me to create decentered, collective learning opportunities by making space for students and instructor to ask questions, share ideas, and receive formative feedback throughout the semester. Hatch (2013) asserts that sharing maker projects with other makers, as not just the final products but works in various stages of progress, is fundamental to critical making. He describes a "sharing philosophy" that emerges when a group of learners participate alongside one another in a makerspace: "People show off their creations knowing criticism was left at the front door, and everyone feels comfortable asking for help, guidance, and input into projects as they go through the build process. Sharing makes a makerspace a community" (17–18). Echoing the critical-making-and-learning philosophy of Hatch, I hoped, through the affordances of a collaborative learning environment, and the low-stakes designation of maker assignments, to build a strong sense of community within the class, something I had found challenging to foster in asynchronous online courses. Sheridan (2015) suggests that if learning is to be collaborative, a space needs to be designed to accommodate interactive, social activities. However, because our online class did not meet in person at any point during the semester, and because Arizona Western College did not have a dedicated makerspace on campus, I needed to rethink traditional notions of space as something other than traditional, on-campus classrooms filled with tidy rows of desks. Sheridan emphasizes the importance of informal learning spaces, or *nonclassrooms*, where "space participates in and helps nurture larger networks that are supportive of learning" (Sheridan 2015). To meet the needs of my students, I needed to create a dedicated space hosted online, one that supported learning and collaboration.

Using WordPress, a free blogging and website content management system, I created the public website ENG 101 Freshman Composition (http://eng101maker.wordpress.com), which hosts the course description, maker assignments, required materials, video tutorials, and a collaborative student-created blog. I aimed to empower students by providing them with an opportunity to compose in a public, digital space for audiences beyond the instructor (Anson and Miller-Cochran 2009; Bowen and Whithaus 2013; Clark 2010; Yancey 2004). Within the blog, students built an authorial identity (Clark 2010) by publishing photos and videos of their paper-electronic circuits, sharing links to maker resources and inspiration, and discussing ideas that emerged in relation to their research projects. Our class website provided a centralized, digital repository of information as well as a "means to highlight new avenues of communication introduced by digital culture and technologies" (Jones, this volume). Within the site, I incorporated small, weekly maker assignments into semester-long, inquiry-based research projects as brainstorming and reflection activities (see McGregor 2017 to access these assignments).

PRIORITIZING RESOURCES AND SUPPORT

As I considered how to design a critical making–themed course, two major concerns were anchored in the forefront of my mind: students' technical abilities and their access to resources and support. I anticipated that most—if not all—students would be unfamiliar with paper-electronic circuits. The maker assignments needed to be simple and straightforward, while at the same time providing opportunities for students to stretch themselves creatively. Because the course met online and the college lacked an on-campus makerspace, students did not have the experience of face-to-face collaboration and support or the option of utilizing on-campus resources. It was imperative that the WordPress website was simple in design and easy to navigate. It also needed to contain a powerhouse of beginner-level tutorials and resources. In addition to the tutorials I initially provided, I created opportunities for students to collaboratively build a repository of maker resources and inspiration within our website and to engage in reflective conversations about their maker experiences.

A final and equally important but less tangible consideration loomed: striking a tone that would empower students in a fledgling critical making–themed course. Providing students with a roadmap to the course and making it clear that maker assignments were low-stakes and experimental (and perhaps even fun!) was key to diffusing anxiety students might have as they entered the course and key to shaping positive attitudes about maker assignments as the semester progressed. Two weeks before the semester started, I invited students into the Blackboard course and WordPress website, where they found detailed information about course policies, grade weights, assignments and due dates, the nontraditional materials they would need to purchase for critical making assignments, and introductory paper-electronic circuits tutorial videos. In my welcome email to students, I emphasized the low-stakes nature of critical making assignments and ensured students they would earn full credit for these weekly arts-and-crafts-style activities even if their paper-electronic circuits didn't function properly. By designating maker assignments low-stakes, a strategy proven successful when integrating new technical tools into a writing course (Anderson 2008; Bowen and Whithaus 2013), I hoped to enhance student participation and encourage students to take risks with their maker assignments.

To pique students' interest in critical making and help them connect with a much larger community of makers throughout the semester, I first introduced them to other stories about critical making outside the context of post-secondary education, including the *KQED* article "How A Makerspace in Juvenile Hall Helps Young People See Their Value" (Schwartz 2016) and the *Time* article "Why the Maker Movement Is Important to America's Future" (Bajarin 2014). These articles positioned critical making outside of academic context and highlighted its value in local communities and professional organizations. I also linked short, engaging, beginner-level informational videos like "What is a MakerSpace?" (Explee 2014) and "Circuit Stickers" (Qi 2014) that offered quick, easy paper-electronic circuit-crafting tutorials. My goal was to help students realize that when they began creating and publishing their own paper-electronic circuit-making experiences to our website, they were joining a vast digital community of makers.

Finally, I emphasized the role of collective learning, describing the importance of building a supportive learning community within our own course by helping one another and actively engaging in the maker assignments each week. To that end, I introduced a Classroom Participation and Citizenship Policy (Getto 2013) and used language like *colleagues* and *teammates* to describe students' relationships with one another and to help foster mutual respect and trust, and I emphasized the role of social learning and peer support in a distance-learning course as one that would create a richer and more meaningful learning experience for all students (Wenger 1999). Because this was a newly designed course and experimental in nature, I aimed to decenter myself as the course *expert* (Fordham and Oakes 2013; Frost 2011; Graban, Charlton, and Charlton 2013) and create an environment where I would learn alongside my students. Our collective participation in the WordPress website, the content contributed through thinking, writing, making, sharing, and reflecting, would shape the design and success of the course.

DESIGNING SIMPLE PAPER-ELECTRONIC CIRCUIT ASSIGNMENTS

I introduced students to the paper-electronic circuit assignments as a way of reenvisioning what it means to compose with an emphasis on the mediums we use to communicate, explaining that a *medium* is a go-between: a way for information to be conveyed from one person to another. Marshall McLuhan (1967), in *The Medium Is the Message* has suggested that the medium itself—even more than the content it carries—shapes how we interact and see the world. By working with different mediums and modes of communication, students will reconsider how to reach audiences and achieve rhetorical goals. These critical, self-reflective lessons can be transferred back to print-based writing, which has always been multimodal to begin with.

The course was designed around a semester-long research project, which seeks to answer the question: *How do the people in a social group see themselves in their world?* Situated within the project were weekly research and writing tasks that culminate in maker assignments. The maker assignments were designed to help students reflect on the research and writing process through the tactile processes of creating simple paper-electronic circuits. A paper-electronic circuit is "a low-voltage electronic circuit that is created on paper or cardboard using conductive copper tape, LEDs and a power source such as a coin-cell battery" (Makerspaces n.d.). In addition to learning the basics of how circuits function, paper-electronic circuits can be a useful way to help students slow down and reflect on their research findings through the tactile process of making. Through the research and maker assignments, students engaged in different types of research and critical thinking and reflection practices, which in turn helped them foster dialectical, recursive writing habits. These are valuable skills and habits that support students learning success in writing courses as well as writing and communication activities in all areas of their lives.

At the conclusion of the semester, after students' final grades were submitted to Arizona Western College and after receiving approval from both George Mason University's and Arizona Western College's institutional review boards, I sent a recruitment email to students asking for volunteers to complete a brief digital survey and to allow me to use excerpts from their blog posts as findings in this chapter. This chapter only shares survey responses and examples of student paper-electronic circuit assignments from those students who agreed to share their work as research participants. Although the blog is a public site and some students elected to share their first and last names in the content they published to the site, I have nonetheless obscured identifying information from screenshots shared in this chapter.

Over the course of the eight-week semester, students published five posts to the student blog in response to assignment prompts. Students' posts contained reflective writing responses, images, videos, and hyperlinks. In the first blog assignment, students were asked to introduce themselves and, as a trial run, create a simple, paper-electronic circuit through the following steps:

1. Tell us a little about yourself! What are a few things we should know to better understand who you are as both a writer and a college student? What's one interesting or unique thing (a hobby, a skill, an interest, or experience) that sets you apart from the crowd?
2. Do an Internet search for cool maker projects that use copper tape and LED lights. Share the link to one of the resources you find. Tell us about it briefly. What kind of resource is this, and why do you think it will be useful as we experiment with our own maker projects this semester?
3. At home, create a simple circuit, and share a photo or video of your work. Option: print and use this handout. Additional help: here is a video that shows you how to create a simple circuit, step by step.

To support them in this task, they were directed to tutorials published in the course website that would help them with the technical requirements for both writing and publishing a blog post in WordPress and for creating a simple paper-electronic circuit. Finally, as a first step toward engaging students in a collaborative learning community, students were asked to welcome their colleagues to the course through the comments

Figure 13.1. Eduardo's paper-electronic lightsaber.

feature available at the bottom of each student's blog post. In examples presented in figures 13.1 and 13.2 below, the eighteen-year-old student from Chihuahua, Mexico, Eduardo C., created a lightsaber to delight fellow Star Wars fans and offer encouragement to his course colleagues. He described his hope that maker projects would help him feel connected with his classmates: "I believe these maker projects will give us a better understanding of [each other], since most of [us] are posting stuff that relates to [ourselves] in some sort of way." In figure 13.1, Eduardo presented his final product for his audience, and in figure 13.2, he documented the process he used to create the paper-electronic circuit.

Eduardo's introduction blog post is one example of students' ability to successfully create a simple paper-electronic circuit in the first week of the semester, and it also underscores the way that creating for public audiences helped the student think about his rhetorical goals, which resulted in Eduardo's creation of a lightsaber.

Following the introduction assignment, three critical making assignments were used as reflections on the weekly research and writing tasks. These tasks included primary and secondary research activities

Cultivating Critical Makers   273

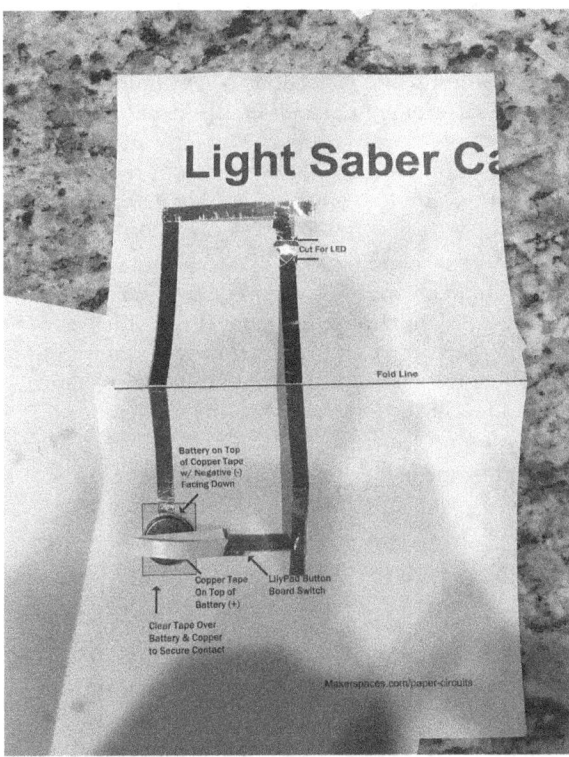

Figure 13.2. Eduardo's technical process used to create the paper-electronic lightsaber.

for their semester-long research project, and the maker assignments helped students reflect on the research they conducted each week. Students selected one or two "ah-ha!" research moments, essentially helping them narrow their research to a couple of key findings that in turn helped them begin to answer their research question. Using conductive copper tape, LED stickers, and coin cell batteries, they literally illuminated their key findings of the week. Similar to the introductions assignment, once they crafted their paper-electronic circuits, students took photos or short boomerang videos of their findings and published them to the WordPress blog.

For their first research assignment, students conducted observations, and their initial observation findings led students to their research proposal. Students were given step-by-step instructions:

Step 1: Conduct your 1-hour observation and record your notes using the dual-entry method from Week 1 readings in *TCW*. Once you complete your observation, review your notes and write a brief reflection on the experience. What was most interesting? What confused you or surprised you? What seems really important to you?

Step 2: After you conduct your 1-hour observation, set your notes aside for a day and give yourself a little time and space to process the experience.

Step 3: Review your observation notes and reflection. What findings stand out to you? What was most memorable about this experience? Any new thoughts to add or connections to make? Identify and narrow down to two or three most interesting discoveries from your observation experience. These should be discoveries that complicate your own understanding of your topic and lead to more questions.

Step 4: Using your maker materials, illuminate these discoveries for us. What are the connections between your observation findings and your research topic? Show us. You might choose to create a traditional *mind map* or *web* on paper, using the copper tape and LED lights along with text and drawings. Or you might choose a different method for sharing your findings, or a unique surface to work on—one that represents, or is linked to, your topic.

Step 5: Take photos or video of your maker project and publish it here on our WordPress blog, along with brief responses to the following questions:
- What is your research topic and focus?
- Where and when did you conduct your observation, and why?
- What made these discoveries so fascinating to you?
- What important research questions do they raise?

Change the "category" of this post from "uncategorized" to "observation maker assignment" and publish.

Step 6: Read and respond to at least three colleagues' posts. Let them know what you like about their maker project and why. Ask a question. Share an additional idea or make a suggestion. *Add something new to the conversation.*

As depicted in figure 13.3 below, most students elected to create their maker assignments on paper, using pencils, pens, and colored markers. They also included words, hand-drawn images, and arrows that indicated the directional flow of ideas. The final products were simple but effective means of helping students reflect on their research and connect findings back to their overarching question about the group they were studying. In figure 13.3, Sara presented a paper-electronic circuit that illuminated her key finding in response to her research questions on the topic of college students and mental health. In her post, she explained that she observed students on campus at lunch because it enabled her to watch social interactions between students. She reflected that "on this day a few fights occurred, as well as the daily gossip," which helped her better focus her inquiry project's framing questions as a result.

When creating weekly maker assignment instructions, I relied on simplicity and repetition, and subsequent maker assignment followed a set of instructions nearly identical to their first observation maker

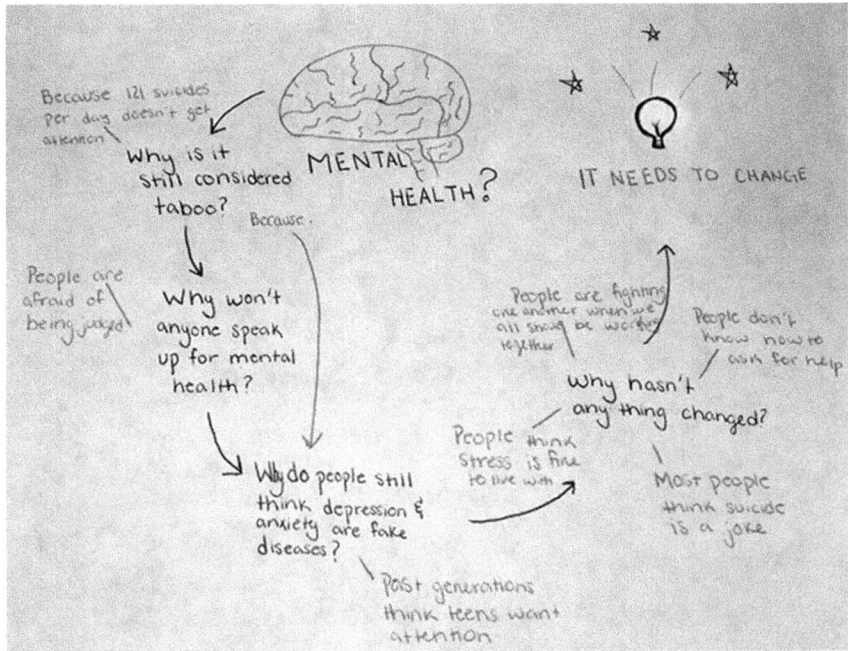

Figure 13.3. Student's paper-electronic circuit illuminated a key finding from an observation she conducted for her students' mental health study. (https://eng101maker.wordpress.com/2017/10/30/observations-2/)

assignment. The final maker assignment blog post was different, however. For the final blog post, I asked students to share resources, inspiration, and creative or helpful ideas for the final maker assignment. The final maker assignment would by submitted with the completed research project as part of their digital writing portfolio. Students responded to the resources and inspiration blogging assignment with a variety of inspiring images, videos, and links to paper-electronic circuit designs. Two examples are depicted in figures 13.4 and 13.5. In figure 13.4, Alejandra successfully created a paper-electronic circuit with three glowing lights. Alejandra explained that she had struggled to successfully create paper-electronic assignments, but nevertheless, for her final maker assignment, she wanted "to go above my simple creations." The resource she shared appealed to her because it offered tutorials of increasing difficulty that she could work her way up to as her confidence grew.

In figure 13.5, Kenya shared a resource offering alternative models for creating simple circuits. This assignment empowered her to undertake a new simple-circuit design for her final project, and she said, "I loved it so much that I showed my kids how to make them."

Figure 13.4. In the final maker assignment, Alejandra shared a resource that helped her create more complex paper-electronic circuits.

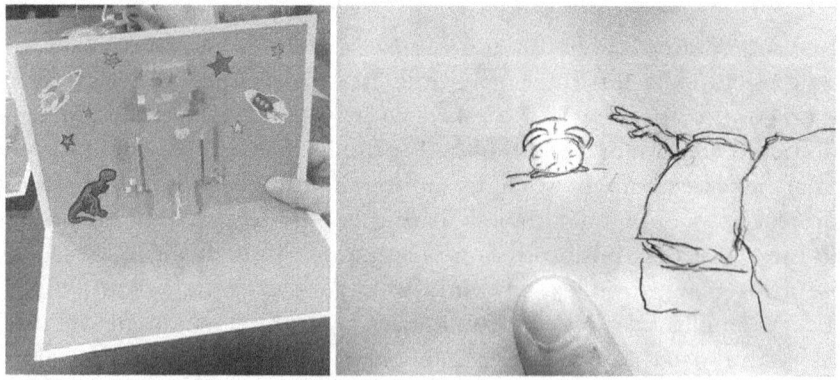

Figure 13.5. Kenya demonstrated how to create three-dimensional "pop-up" paper-electronic circuits in preparation for her final project.

Students, through resource sharing and experimenting, demonstrated the wide variety of resources and inspiration available online and the vast network of people who are involved in critical making. Their written responses also underscore their own experiences with maker assignments (trial and error as well as successes) over the course of the semester and their burgeoning enthusiasm for critical making. Students' collaborative learning experiences had created a sense of community membership among students who were all experiencing the joys and frustrations of maker assignments for the first time together.

STUDENTS' THOUGHTS ON A CRITICAL MAKING-THEMED FIRST YEAR COMPOSITION COURSE

After the semester concluded and final grades submitted to the college, I sent an email survey to students, requesting their feedback on the course (see appendix). The survey was sent only to the fifteen students who completed the course, and eleven students completed the survey. While the study sample is small, it does offer useful insights into students' firsthand experiences and their preferences for critical making in online learning environments.

Students' survey responses were overwhelmingly positive and indicated a genuine interest in maker assignments and a critical making–themed course. Their candid feedback was both insightful and encouraging for anyone thinking about using maker assignments in a composition course: all of the students surveyed said they were interested in the maker assignments, and 100 percent of students surveyed said they would recommend a critical making–themed first year composition course to other students. Students also found maker assignments beneficial to their research projects. Ten out of eleven students reported that maker assignments helped them reflect on their research, and 100 percent of students surveyed reported that maker assignments helped them connect their findings to their research question and helped them feel more prepared to write about those research findings.

I posed several questions to students about the usefulness of maker resources, the technical difficulty of maker assignments, and their comfort with the assignments. Students unanimously reported that course materials and videos in the course website helped them learn how to create the maker assignments. Likewise, 100 percent said the materials were suitable for nonexperts or individuals not familiar with the technology, and 100 percent said they became increasingly comfortable with maker assignments as the semester progressed.

The survey contained three open-ended questions designed to gauge what students liked most about maker assignments, what they enjoyed least, and what impact maker assignments had on rethinking how to reach audiences and achieve their rhetorical goals. In response to the question "What did you enjoy most about maker assignments?," three themes emerged: being creative, reflecting, and crafting.

Being Creative. Students described maker assignments as creatively stimulating. One student wrote, "The maker assignments really stretched me creatively." Another stated, "I loved how creative I got in my maker assignment." A third student explained, "I liked how I was able to reflect [on] my research in a new creative way."

Reflecting. One student wrote, "The creative process created a gateway for my writing to expand and connect with the information I've gathered for my research." Another connected the experience to specific learning preferences: "It tied my thoughts together from the week in a visual way. I am a visual learner, the more I see it the more I understand it." A third student described being creative as a means to reflect on findings "in a new and innovative way."

Crafting. Finally, seven students said they enjoyed the act of crafting paper-electronic circuits. One student wrote, "They gave me something fun to do at the end of the week that wrapped up all the work I had done." Another student explained, "What I enjoyed about the maker assignments was that they were more of a relaxing way to finish your work for the week and reflect on what you have done." Five students described the acts of "drawing" and "using electronic circuits" as the most enjoyable part of the maker assignments. One student explained, "I really enjoyed being able to draw and light up different things using circuits as a way to depict the findings I discovered." Another student added, "The best part of any maker assignment is seeing your thoughts illuminated by the LED lights."

In response to the question *"What did you enjoy least about maker assignments?,"* two themes emerged: *learning a new technology* and *ongoing technical difficulties*. Three students wrote about the challenge of learning to create paper-electronic circuits. One student explained, "Getting started was a little hard. It was a lot of trial and error, but I wouldn't say I didn't like that. All new things need work." Another student wrote, "The thing I enjoyed least was that I wasn't able to understand how to make the [paper-electronic circuits] work the first time." A third student echoed a similar sentiment: "The least enjoyable thing was probably getting the material to work for me, though every beginning is hard!"

Technical difficulties occurred regularly for students during the semester. Five students described struggling to properly applying the conductive copper tape and successfully illuminate the LED stickers. One student explained, "the copper tape sometimes ripped & I would need to start again." Another simply described the task of "trying to make the lights work and keep them on" as the greatest challenge. A third student provided further detail: "Trying to make the LED lights stay on after I drew out my thoughts was the most annoying thing I've encountered. I believe it is due to my own error but it was a constant battle for me just to get my maker assignment to illuminate." Four out of the five students who described technical difficulties went on to explain how they resolved the issues. One student wrote, "At first getting the maker assignments to light up was tricky it took me too long to make the maker assignment. Then, after a few tutorials I got the hang of it and had no trouble at all making them." Another student explained, "Sometimes, it was very difficult for me to get my circuits to work because either the tape wasn't connecting right or the LED wasn't positioned correctly. But in the end, I got it done and presented my ideas beautifully." Yet another student described the process as "a lot of trial and error." A testament to the nature of critical making and collective learning, these comments indicate how students were consistently willing to troubleshoot, tinker, and experiment until the issues were resolved.

Three students wrote that they enjoyed all parts of the maker assignments and there were no "least enjoyable" aspects to the work. Only one student pointed to the need to be creative as the least enjoyable aspect, stating, "I am not creative, so I had to think outside of the box, which was hard."

When asked, "In what ways did the maker assignments help you rethink how to reach your audience and achieve your rhetorical goals using visual/interactive design?," students' responses fell into three general categories: rethinking how students learn; rethinking what it means to compose; and rethinking rhetorical goals.

Rethinking Learning. Three students said maker assignments stretched their capacity for learning. One student wrote, "Students might benefit from this approach because it will definitely allow them to look at learning through a different lens." A second student suggested, "This would give the student a sort of edge over their peers and can help them come out on top within their educational career." A third student explained, "The maker assignment allowed me to use my mind and think in different ways."

Rethinking Composition. Four students wrote about rethinking what it means to compose. One student described the course as "inventive" because "it pushes the student to step out of the 'normal' way of learning in an English course." Another student wrote, "I believe most students have always written . . . and this gives a different perspective of putting your own thoughts on paper." A third student explained, "It gives students a more different approach than just plain old type-base writing." Finally, a fourth student summed it up this way: "A 'maker mentality' can help a student gain confidence as a reader and writer."

Rethinking Rhetorical Goals. Four students described factors that improved their ability to think about how to reach their audience and achieve their rhetorical goals. One student summed it up as an opportunity for students "to be creative, think outside the box, try new things, new methods of expressing themselves."

Perhaps most surprising of all was students' responses to the final question, which asked if they see themselves as a *maker*. One hundred percent of students surveyed said that, by the end of the course, they'd come to see themselves as *makers*, as someone who tinkers, experiments, and creates.

FINAL THOUGHTS: RE-SEEING THE CRITICAL MAKING–THEMED FIRST YEAR COMPOSITION COURSE

At the conclusion of the semester, as I reviewed students' final writing portfolios, and I reflected on their blog posts and survey responses, I felt that the critical making–themed course was a success because it had privileged experimentation as a legitimate form of learning and facilitated an informal and social learning dynamic shared between students and instructor. Rather than creating weblogs that "sit on the periphery of the physical classroom" (Brooke 2005), the student blog became a routine communication space where students collaborated to support one another for an outward-facing, public audience. They blogged to share their experiences—their triumphs and challenges and resources—with one another and with outside audiences like future students who might happen upon the course website.

In the process of critical making, collaborating with course colleagues, and reflecting on the composing process, students demonstrated an increased sense of metacognition—of their writing process and what it means to compose. If I were to teach the course again, I would continue to refine the assignments in ways that further encourage collaboration and community-building among students. Blogs are social media, yet constant prompting is required to help students move beyond one-way information dissemination and create genuine conversation among students. In future iterations of the course, I would find ways to increase use of the comments feature in the blog posts. I would also connect with maker communities outside of our course. At the time of this writing, the course website has over 3,300 hits, yet few visitors have left comments on the website, and no visitors have commented in response to student blog posts. To increase blog traffic during the semester in the hope that students might connect with critical makers outside the of our course, I would encourage students to conclude blog posts with questions that might stimulate discussion and to tag their posts with relevant keywords. These tactics would further encourage students to think about the rhetorical goals of each blog post as means of connecting with specific audiences beyond our course.

As Jones has also argued in his chapter of *Reprogrammable Rhetoric*, the collective move in higher education away from paper to digital technologies requires us to give increased consideration to the principles of information architecture when we design our courses. The design of my online critical maker–themed course offers a road map for instructors who wish to design a similar course. The final question of the end-of-semester survey asked students if a maker themed–course should be offered in future semesters. Students were unified in their response: Yes. Students described the maker assignments as a creative outlet that made research less stressful, enabled them to reflect on their writing and research in a new way, and helped them reconsider how to engage in a recursive writing process.

APPENDIX 13.A

ENG 101 SURVEY: YOUR MAKER EXPERIENCE

1. RESEARCH PROCEDURES: This research is being conducted to better understand student experiences with critical making assignments in an online undergraduate composition course. If you agree to participate in the survey, your responses will be used to illustrate student experiences in a maker-themed composition course. This survey takes approximately 10 minutes to complete. This survey will close on May 31, 2018.

    RISKS: There are no foreseeable risks for participating in this research.

    BENEFITS: There are no benefits to you as a participant other than to further research in critical making studies and composition pedagogy.

    CONFIDENTIALITY: If you choose to participate, your survey responses will be used anonymously. Your consent is optional and has no impact on your course grade. While it is understood that no computer transmission can be perfectly secure, reasonable efforts will be made to protect the confidentiality of your transmission.

    PARTICIPATION: Your participation is voluntary, and you may withdraw from the study at any time and for any reason. If you decide not to participate or if you withdraw from the study, there is no penalty or loss of benefits to which you are otherwise entitled. There are no costs to you or any other party. Your decision to participate will not impact the grade you receive for the course.

CONTACT: This research is being conducted Bree McGregor, student in the Writing & Rhetoric PhD program at George Mason University, IRB #9740981. She may be reached at 9285020249 or bmcgrego@gmu.edu, and principal investigator Dr. Steve Holmes may be reached at 7029931180 for questions or to report a research-related problem. You may contact the George Mason University Office of Research Integrity & Assurance at 7039934121 if you have questions or comments regarding your rights as a participant in the research. This research has been reviewed according to George Mason University procedures governing your participation in this research.

CONSENT: I have read this form, all of my questions have been answered by the research staff, and I agree to participate in this study.
　__ Yes, I acknowledge my role as a research subject and give my consent.

2. Rate your interest in the maker assignments:
　　No interest whatsoever __1 __2 __3 __4 __5 Highly interested

3. Did the online course materials, videos, etc. provided by instructor help you learn the maker assignments?
　　__ Yes
　　__ No

4. Were the materials suitable for nonexperts or individuals not familiar with the technology?
　　__ Yes
　　__ No

5. As the semester progressed, did you become more comfortable with the maker assignments?
　　__ Yes
　　__ No

6. Did maker assignments help you reflect on your research findings?
　　__ Yes
　　__ No

7. Did maker assignments help you connect your findings to your research question?
　　__ Yes
　　__ No

8. Did maker assignments help you feel more prepared to write about your research findings?
　　__ Yes
　　__ No

9. What did you enjoy most about the maker assignments?
_____

10. What did you enjoy least about the maker assignments?
_____

11. Did maker assignments help you rethink how to reach your audience and achieve your rhetorical goals through the expanded rhetorical affordances of visual and interactive design?
    __ Yes
    __ No

12. In what way(s) did maker assignments help you rethink how to reach your audience and achieve your rhetorical goals using visual/interactive design?
_____

13. How did the maker assignments impact the way you think about print based writing?
_____

14. At the conclusion of this course, do you see yourself as a "maker" as someone who tinkers, experiments, and creates?
    __ Yes
    __ No

15. Why might students benefit from cultivating a "maker mentality" in a freshman composition course?
_____

16. Is a maker themed course one you would recommend to other students?
    __ Yes
    __ No

REFERENCES

Anderson, Daniel. 2008. "The Low Bridge to High Benefits: Entry-Level Multimedia, Literacies and Motivation." *Computers and Composition* 25 (1): 40–60. doi:10.1016/j.compcom.2007.09.006.

Anson, Chris M., and Susan K. Miller-Cochran. 2009. "Contrails of Learning: Using New Technologies for Vertical Knowledge-Building." *Computers and Composition*, 26 (1): 38–48. doi:10.1016/j.compcom.2008.11.002.

Arizona Western College. 2017. "Arizona Western College Syllabus: ENG 101 Freshman Composition I /GE." Arizona Western College. https://web.archive.org/web/20210415224630/https://www.azwestern.edu/sites/default/files/syllabi/ENG-101%206101.pdf.

Arola, Kristin L., and Anne Frances Wysocki, eds. 2012. *Composing (Media) = Composing (Embodiment): Bodies, Technologies, Writing, the Teaching of Writing*. Logan: Utah State University Press.

Bajarin, Tim. 2014. "Maker Faire: Why the Maker Movement Is Important to America's Future." *Time*, May 19, 2014. http://time.com/104210/maker-faire-maker-movement/.

Bowen, Tracey, and Carl Whithaus, eds. 2013. *Multimodal Literacies and Emerging Genres*. Pittsburgh, PA: University of Pittsburgh Press.

Braun, Catherine C. 2013. *Cultivating Ecologies for Digital Media Work: The Case of English Studies*. Carbondale: Southern Illinois University Press.
Brooke, Collin. 2005. "Weblogs as Deictic Systems: Centripetal, Centrifugal, and Small-World Blogging." *Computers and Composition Online*. http://cconlinejournal.org/brooke/brooke.htm.
Clark, J. Elizabeth. 2010. "The Digital Imperative: Making the Case for a 21st-Century Pedagogy." *Computers and Composition* 27 (1): 27–35. doi:10.1016/j.compcom.2009.12.004.
Davis, Robert L., and Mark F. Shadle. 2007. *Teaching Multiwriting: Researching and Composing with Multiple Genres, Media, Disciplines, and Cultures*. Carbondale: Southern Illinois University Press.
Ellis, Erik. 2013. "Back to the Future? The Pedagogical Promise of the (Multimodal) Essay." In *Multimodal Literacies and Emerging genres*, edited by Tracey Bowen and Carl Whithaus, 37–72. Pittsburgh, PA: University of Pittsburgh Press.
Explee. 2014. "What Is a MakerSpace?" *YouTube video*, July 13, 2014. https://youtu.be/NLEJLOB6fDw.
Fordham, Traci, and Hillory Oakes. 2013. "Rhetoric across Modes, Rhetoric across Campus: Faculty and Students Building a Multimodal Curriculum." In *Multimodal Literacies and Emerging Genres*, edited by Tracey Bowen and Carl Whithaus, 313–35. Pittsburgh, PA: University of Pittsburgh Press.
Frost, Erin. A. 2011. "Why Teachers Must Learn: Student Innovation as a Driving Factor in the Future of the Web." *Computers and Composition* 28 (4): 269–75. doi:10.1016/j.compcom.2011.10.002.
Getto, Guiseppe. 2013. "Rights/Powers." *Guiseppe Getto*. https://web.archive.org/web/20170617051733/http://guiseppegetto.com/pwr393/syllabus-2/rightspowers/.
Graban, Tarez Samara, Colin Charlton, and Jonikka Charlton. 2013. "Multivalent Composition and the Reinvention of Expertise." In *Multimodal Literacies and Emerging Genres*, edited by Tracey Bowen and Carl Whithaus, 248–81. Pittsburgh, PA: University of Pittsburgh Press.
Grabill, Jeffrey T., and Troy Hicks. 2005. "Multiliteracies Meet Methods: The Case for Digital Writing in English Education." *English Education*, 37 (4): 301–11.
Grouling, Jennifer, and Jackie Grutsch McKinney. 2016. "Taking Stock: Multimodality in Writing Center Users' Texts." *Computers and Composition* 41: 56–67. doi:10.1016/j.compcom.2016.04.003.
Handa, Carolyn. 2004. "Introduction: Placing the Visual in the Writing Classroom." In *Visual Rhetoric in a Digital World: A Critical Sourcebook*, edited by Carolyn Handa, 1–5. New York: Bedford/St. Martin's.
Hatch, Mark. 2013. *The Maker Movement Manifesto: Rules for Innovation in the New World of Crafters, Hackers, and Tinkerers*. New York: McGraw Hill Professional.
Hobbs, Renee. 2006. "Multiple Visions of Multimedia Literacy: Emerging Areas of Synthesis." In *International Handbook of Literacy and Technology*, vol. 2, edited by Michael C. McKenna, Linda D. Labbo, Ronald D. Kieffer, and David Reinking, 15–28. New York: Routledge.
Kress, Gunther R. 2010. *Multimodality: A Social Semiotic Approach to Contemporary Communication*. New York: Routledge.
Lutkewitte, Claire. 2013. *Web 2.0 Applications for Composition Classrooms*. Southlake, TX: Fountainhead Press.
Makerspaces. n.d. "*Paper Circuits for Makerspaces*." Makerspaces. Accessed July 20, 2017. https://www.makerspaces.com/paper-circuits/.
McGregor, Bree. 2017. "Research Projects, Assignments, Portfolio." *English 101 Freshman Composition*. https://eng101maker.wordpress.com/project-overview-assignments/.
McLuhan, Marshall. 1967. *The Medium is the Massage: An Inventory of Effects*. London: Penguin Books.

Palmeri, Jason. 2012. *Remixing Composition: A History of Multimodal Writing Pedagogy*. Carbondale: Southern Illinois University Press.

Qi, Jie. 2014. "Circuit Stickers." *Jie Qi*. http://technolojie.com/circuit-stickers/.

Schwartz, Katrina. 2016. "How a Makerspace in Juvenile Hall Helps Young People See Their Value." *KQED News*, August 25, 2016. https://www.kqed.org/mindshift/46177/how-a-makerspace-in-juvenile-hall-helps-young-people-see-their-value.

Sheridan. David. M. 2015. "Digital Composing as Distributed, Emergent Process: Technology-Rich Spaces and Learning Ecologies." In *Making Space: Writing Instruction, Infrastructure, and Multiliteracies*, edited by James P. Purdy and Dànielle Nicole DeVoss. Ann Arbor: Sweetland Digital Rhetoric Collaborative and University of Michigan Press. https://www.digitalrhetoriccollaborative.org/makingspace/ch5.html.

Shipka, Jody. 2011. *Toward a Composition Made Whole*. Pittsburgh, PA: University of Pittsburgh Press.

Shipka, Jody. 2016. "Transmodality in/and Processes of Making: Changing Dispositions and Practice." *College English* 78 (3): 250–57.

Sirc, Geoffrey. 2004. "Box-logic." In *Writing New Media: Theory and Applications for Expanding the Teaching of Composition*, by Anne Frances Wysocki, Johndan Johnson-Eilola, and Geoffrey Sirc, 111–46. Logan: Utah State University Press.

Stofft, Lorraine, Betty Lopez, Robyn Torres, and Marisela Dickman. 2018. *Fall 2018 Standard Reports*, December 19, 2018. Arizona Western College. https://www.azwestern.edu/sites/default/files/awc/institutional-research/Fall%202018%20Standard%20Reports%20%28as%20of%20census%20day%29.pdf.

US Department of Education. 2015. *Demographic and Enrollment Characteristics of Nontraditional Undergraduates: 2011–12*, September 2015. https://nces.ed.gov/pubs2015/2015025.pdf.

van Leeuwen, Theo, and Gunther Kress. 2011. "Discourse Semiotics." In *Discourse Studies: A Multidisciplinary Introduction*, 2nd ed., edited by Teun A. Van Dijk, 107–25. Los Angeles: SAGE. doi:10.4135/9781446289068.n6.

Wenger, Etienne. 1999. *Communities of Practice: Learning, Meaning, and Identity*. Cambridge: Cambridge University Press.

Wysocki, Anne Frances. 2004. "Opening New Media to Writing: Openings and Justifications." In *Writing New Media: Theory and Applications for Expanding the Teaching of Composition*, by Anne Frances Wysocki, Johndan Johnson-Eilola, and Geoffrey Sirc, 1–41. Logan: Utah State University Press.

Yancey, Kathleen Blake. 2004. "Made Not Only in Words: Composition in a New Key." *College Composition and Communication* 56 (2): 297–328.

14

CRAFTING IN THE CLASSROOM
Carpentry and Pedagogy in Rhetoric and Composition

John Jones
The Ohio State University

This chapter explores how practices of making, such as interactive prototyping and other forms of nontextual rhetoric, can be a means of connecting practices of textual and nontextual rhetoric.[1] I will describe three approaches to the concept of carpentry—here understood as making things for inventional and analytical purposes—and discuss how these approaches can be a bridge between textual and nontextual rhetorics using the example of syllabus design. I then conclude by introducing the Syllabot, an interactive project built on the Arduino platform that invites students to engage with the syllabus in new ways.

In this chapter I approach nontextual rhetorics not as a corrective but a continuation of practices with a rich history that allows us to question existing assumptions about the practice and teaching of rhetoric and writing—such as assumptions about what counts as writing and rhetorical performance—and embrace multiple forms of rhetorical engagement and production. In her critique of feminist new materialism, Sara Ahmed (2008) argues that the movement ignores "historical materialism" (32), an omission that ignores or erases the work of scholars who are largely women, Indigenous, or people of color. Or, as Alison Ravenscroft (2018) puts it in their argument for the centrality of Indigenous materialism, "Not all humans require [new materialism's] correctives" (357). Similarly, pulling the thread of extralinguistic rhetoric reveals a rich tapestry of rhetorical practice that has long been engaged with the material effects of communication (cf. Goggin and Tobin 2009; Haas 2013; Powell et al. 2014), and as rhetoric and writing studies engage with prototyping and other forms of nontextual expression, this history is important for its insights and examples of rhetoric's material nature.

CARPENTRY IN THE WRITING CLASSROOM

In his description of carpentry, Ian Bogost (2012) makes the case for academics to "consider that writing is not the only method of engendering interest" (111) in our ideas, presenting carpentry as a widely applicable tool for the humanities and the sciences, domains where, according to him, "writing casts a pallid shadow over experimentation and construction" (88). This and other warnings against the stultifying nature of writing are certainly hyperbole on Bogost's part; he is nothing if not a savvy craftsperson of academic writing. Bogost's advocacy for expanded modes of academic communication will not seem new to rhetoric and writing scholars, as there have long been discussions on the value of multimodal communication in rhetoric and writing studies (cf. Alexander and Rhodes 2014; Ball 2004). Yet, the concept of carpentry provides an interesting avenue for exploring the connections between textual and nontextual rhetorics. Following Ravenscroft's (2018) reminder that the posthuman should not simply perpetuate the masculine "human" it replaces (354), we might also ask how practices of carpentry, or any theoretical or pedagogical tool, can be employed so as not to replicate systems of inequality in academia or the "white as default" designs of digital spaces tools (Arola 2017, 211). A small part of this work can be to explicitly foreground a broad range of practices for making meaning. The standard genres of academic writing are not the only means of being in the world, and thus not the only means of making meaning or making rhetoric (Brown and Rivers 2013).

Following Bogost (2012), James J. Brown Jr. and Nathaniel Rivers (2013), and Shannon Mattern (2012), I define carpentry as the practice of making things that do inventional and analytic work. Within this broad definition I identify three practices of carpentry that can serve as aids to rhetorical invention: connecting form with content, discovering new avenues of communication, and exploring material relations. These practices are not exclusive of each other; they can be explored on their own or combined in projects, either simple or complex. Rather, this taxonomy is intended to serve as an inventional aid for thinking about carpentry's relationship to multimodal authorship and exploring the rhetorical potential of things.

Here I purposefully expand the notion of carpentry to include text, albeit text considered not from the perspective of meaning but as a thing. This may seem odd, given that Bogost's goal for carpentry was to challenge academics to move beyond textual genres. As Jonathan Alexander and Jacqueline Rhodes (2014) argue, different modes have different rhetorical affordances, yet there are few modes that can be easily divorced from linguistic ways of knowing. If examples of carpentry cannot be disentangled from text—indeed, if the critical work of carpentry is a means of stimulating and inspiring texts like Bogost's—then perhaps its insights can be brought back to text (and other modes of making) as purposeful tools of invention. In this way, carpentry can be a theoretical bridge between textual and nontextual communication, scaffolding new forms of rhetorical production. Carpentry can aid students in their understandings of existing and emergent rhetorical encounters in a range of media, including multimodal communication, print and oral genres, and prototyping (Faris et al. 2018). As the now well-established fields of multimodal writing and visual rhetoric came to prominence in response to the graphics-centric internet, so the continued dispersion of computing into the environment underscores the need for rhetoric and writing pedagogy to explore the impacts of these tools on our communication. As Bree McGregor demonstrates in their chapter in this volume, when students become makers, they are also more proficient in exploring the boundaries and affordances of multimodal and other communication, and this invariably improves their ability to become critical authors and users of texts created by others. I would add that by incorporating interactive computing assignments into rhetoric and writing classrooms, students can become more adept at analyzing and using interactive technologies, through a better understanding of their affordances and by exploring their rhetorical potential. Grounding such explorations in understandings of the material world that are not narrowly focused on standard academic modes and genres can validate ways of knowing and being that students bring to the classroom.

THREE PRACTICES OF CARPENTRY

Connecting form with content is perhaps the most fundamental mode of carpentry. As Mattern (2012) puts it, carpentry that connects form with content involves utilizing the "modes of production" such that "the form" of the thing " 'enacts' the argument." From foundational works on new media writing (cf. Wysocki et al. 2004) to electronic journals like *Kairos*, rhetoric and writing studies has long explored and advocated for the inventional and expressive possibilities of new media forms. Scholars have also explored the rhetorical affordances of pre-digital media. One example of form connecting with content for a particular rhetorical end is Elizabeth Parker's needlework sampler.

Needlework samplers are generally a small piece of cloth on which individuals, mostly women, would "learn, practice, and record the available means of creation" with needle and thread "via choices of stitches, threads, materials, colors, motifs, and patterns" (Goggin 2009, 33). According to Maureen Daly Goggin, samplers were records of their creators' inventional explorations, making possible new rhetorical engagements. Goggin describes at length the sampler of Elizabeth Parker, who used the form to describe her traumatic experiences as a housemaid in stitched text (Goggin 2009, 37 ff.). In choosing the sampler to tell her story, rather than a traditional written genre, Parker engages in what Heather Pristash, Inez Schaechterle, and Sue Carter Wood (2009) describe as "coding"—where rhetoricians "make the statements that they needed to make to the people they wanted to talk to while hiding their intent from other, potentially hostile, audiences" (15). By coding her message in a sampler, it could be overlooked as a decoration or practice in stitching, thus lending Parker the privacy and freedom to share incredibly personal content. A similar example of blending form and content is the Willard Dress. Named after Frances Willard and designed for members of the women's temperance and suffrage organization she founded, the dress was designed to be less bulky and constrictive than other dresses of the day. Significantly, it also included a divided skirt—trousers—in a form that would appear to be a traditional dress. Pristash, Schaechterle, and Wood read this garment as a subversive one that, presented at a time when women's trousers were highly controversial, "appeared to conform to contemporary ideologies while at the same time pursuing radical, even subversive reforms" (2009, 25). The dress was a uniform and a disguise, a means of liberation in the guise of conformity. Both needlepoint samplers and the Willard Dress made possible the communication of subversive messages that would have been difficult or impossible for many women to share in traditional genres at the time they were created.

Carpentry invites creators to not simply enact arguments through form but to discover new ways of communicating by exploring the physical properties of the material world. That is, not simply to match form and content, but to use the properties of a medium to drive invention. Brown and Rivers (2013) argue that the inventional possibilities of such carpentry derive from play, from exploring the possibilities of a medium, device, or platform in order to discover what it might offer communicators. The discovery of new forms of communication requires students to search for the boundaries within which a tool or medium operates. David Rieder (2016) has outlined a heuristic process for revealing the communication potential of environmental sensors and actuators, digital devices that convert analog inputs into digital data or digital data into analog outputs, respectively (47–57). Rieder's seven-step process moves through exploring the history of the sensor or actuator, discovering its range of inputs or outputs, and visualizing the data it produces. Rieder focuses on digital tools, but a process of discovery focused on inputs and outputs and the range of interaction they allow for could be used with any sort of tool or material object. For example, small studies of office workers have shown that the availability of natural light can impact their energy levels and quality of sleep (Boubekri et al. 2014). If students wanted to draw attention to the design of a classroom, for example, they could place removable tape on the floor during class meetings on different days to track where light fell within the room. A collection of these markings over the course of a semester would show the availability of light in that space at that time of year and how the layout and other features of the room accommodated it, serving as a record of how the classroom space allows for its inhabitants to experience natural light.

Exploring material relations, the third carpentry practice, both builds on the other two and exhibits unique properties of its own. Material relations refers to the interactions or encounters between things, with an emphasis on exploring the aspects of those relations that are independent of human attention. Using the example of a paper screen dividing two rooms, Graham Harman (2002) notes that a mote of dust, photons of light, or a slicing blade would all encounter the screen differently, in ways that are unique to their own thinghood—as a barrier, a medium, or something to be sliced, respectively (30). Thinking about material relations speculatively provides us with a model for understanding tools in generative ways that can be useful for exploring extralinguistic rhetorics. Nathaniel Rivers and Lars Söderlund (2016) argue that these relations can be the basis for better understanding the usability of things, while Brown and Rivers (2013) call for rhetoricians to explore the "strange, alien conversations happening around us" (29) by building new tools that place these things in conversation.

Consider Robin Wall Kimmerer's (2013) essay "The Gift of Strawberries," in which she connects her childhood experience of discovering wild strawberries to the fruit's connection to Potawatomi culture. "Strawberries," Kimmerer recalls, "first shaped my view of a world full of gifts simply scattered at your feet" (23). From this starting point, Kimmerer works outward, exploring how an item's status as a gift (in contrast with something bought) changes relationships between people and things—for example, we would tend to not appreciate socks bought at a store as much as socks knitted by a family member (26). Combining memoir and analysis, Kimmerer traces the relationships between strawberries, cultural tradition, and the impact of capitalism on our relationship with things.

In contrast to Kimmerer's use of an essay to explore material relations, Brown and Rivers's (2013) water sphere is a physical tool for exploration. The water sphere, a 3D-printed container with adjustable interior locks like those on a dam, is given to students who are tasked with rearranging the dividers to channel the water in specific ways (for example, to fill a particular section). As part of this process, they experiment with both the mechanics of the sphere and its lock system as well as with the physical nature of water, as it constantly seeks its level. Although Kimmerer and Brown and Rivers embrace media beyond the digital, exploring the relationships between material tools has a direct application in the study of interactive digital devices. Given the ubiquity of networked computing and the Internet of Things, the relations between our things have become less abstract and speculative ("How does a photon encounter paper?") but programmed and managed, often in ways that are deliberately obscured for users. Seeking to clarify this obscurity through situating ourselves within our relationships with interactive tools and exploring their impact on our thinking, as Kimmerer does, and identifying the affordances of these tools, as do Brown and Rivers, will be a key feature of twenty-first-century digital literacy. Building interactive objects gives students a base for beginning this research.

USING CARPENTRY TO EXPLORE TEXTUAL RHETORICS

Here I apply the three types of carpentry to the course syllabus to explore how these practices can be useful to both textual and nontextual rhetorical practices. The syllabus is a unique tool: it is both central to the classroom experience in American universities and seen by many instructors as being largely ignored by students. The subject of memes ("It's in the syllabus!") and counter-memes ("Your students just want to talk to you!"), it serves purposes in the classroom that transcend its textual content. Charles Fornaciari and Kathy Lund Dean (2014) note that the most common metaphors for the syllabus describe it as a contract or as a means of reinforcing instructor power over students (705–8), frames that are reinforced by college administrations as a means of inoculating teachers and institutions from lawsuits (705). Such assumptions are no doubt widespread, even among instructors who see themselves as student advocates. Such assumptions cannot help but place a wedge between students and teachers, a problem exacerbated in the case of students with disabilities, first-generation students, or any student not habituated to the academic norms on which the syllabus is based.

To address these problems, Lund Dean and Fornaciari (2014) provide seven guidelines to inform syllabi construction for adult learners—signaling mutual respect through inclusive language, adapting syllabi to students' reading habits, making the schedule central, designing the syllabus for accessibility and engagement, providing paper and digital copies to accommodate multiple reading methods, establishing the centrality of the syllabus to the course by referring to it throughout the semester, and being willing to tolerate ambiguity (as opposed to contractual rigidity) in the syllabus as an avenue for responding to student needs. Together, these changes signal Lund Dean and Fornaciari's attempts to adapt syllabi to changing modes of engagement with text, refocusing the syllabus as a road map that will help orient students to the learning goals of the course and their progress in it. Drawing from these tips and other research, in this section I apply the three approaches to carpentry to rethink the syllabus and end with a rhetorically focused project that allows students to explore the syllabus in new modes via interactive prototyping.

Connecting Form with Content

In their web text "Accessible Syllabus," Anne-Marie Womack and coauthors (2016) provide examples that illustrate (and justify) the move from text-heavy syllabi—those that may rely on dense blocks of type that give little thought to their presentation—to syllabi that include images and design elements to help students navigate the document, access their information, and absorb it more effectively. Lund Dean and Fornaciari (Fornaciari and Lund Dean 2014; Lund Dean and Fornaciari 2014) argue that since students have multiple accessibility needs that cannot be addressed through a single mode or medium, instructors should work to meet those needs by providing syllabus content in multiple modes. Like Lund Dean and Fornaciari, Womack et al. urge instructors to provide students with syllabi in both print and digital forms, ensuring that print text is designed for easy reading and digital copy can be manipulated (for example, made larger for legibility) by readers. Whether through a learning management system, mobile devices like phones or tablets, pdfs, or paper, different methods of delivery can be used to address the accessibility needs of students and model practices that illustrate the carpentry principle of the thoughtful coupling of form and content. Insofar as the content of a syllabus should serve a central role in a course, it can model connecting form with content, and instructors can engage thoughtful design and multiple modes in crafting the syllabus. Womack et al. (2016) argue that images can be used to make syllabi more legible and easier to skim, providing alternative text or descriptions of images in the body of syllabi and helps to make these images accessible to a wider range of users. Syllabus materials that are delivered online, such as through a learning management system, can make use of online affordances, like linking and alt text to streamline content and meet students' needs. Document design should avoid large chunks of text and long line lengths, replacing them with shorter paragraphs, narrower columns, and thoughtful use of listing to aid in skimming and reading for students with visual or attentional needs. Such design choices mirror the evolution of web-based content and build on its familiarity, rejecting the idea that text is something readers look through for content and embracing the importance of design for understanding. While these forms are not ideal for all information, their emphasis on concision, modularity, and skimability (cf. Lynch 2008, 231 ff.) are a good formal fit with the needs of the syllabus.

Providing syllabus content using a combination of these approaches—print copies in class along with dynamic digital copies online that follow best practices for accessibility—can help to address the access needs of multiple students as well as serve as an entry point for instructors to discuss the connection of content and form beyond written text. To Fornaciari and Lund Dean's and Womack et al.'s arguments for access I would add that expanding the modes in which syllabus information is presented provides instructors with a model for expanding the acceptable forms of communication within a course. Like the examples of Parker's needlepoint sampler, the Willard Dress, or the water sphere, a multimodal syllabus can become a means for signaling that a course embraces modes of communication that may be meaningful to students outside of the classroom. For example, when instructors introduce the syllabus for the first time, they might draw attention to the multiple modes in which it is offered—print, digital copies, learning management system (LMS). By embracing multiple modes from the outset of the course, instructors can demonstrate both how they are able to serve different informational needs and show that rhetorical concepts and ideas are applicable to a range of communication practices. Using bullets to highlight a list or images to assist students in identifying course textbooks are choices in form that focus on the needs of the content and readers. Emphasizing this for students will give them examples of how to do the same as they move to projects that may rely less on text. More broadly, such moves illustrate the connections between form, mode, and use and the role each plays in effective communication.

Exploring Material Relations

As I have argued, one of the primary benefits of exploring relations is inventional. As Goggin notes with needlepoint samplers, these explorations can be both a means to their own end (Goggin 2009, 33) and also support explorations can serve human-centric goals, such as by providing a means for discovering new functionality, in tools and textual media (Rivers and Söderlund 2016). In this vein, instructors can follow Lund Dean and Fornaciari's (2014) suggestions to both streamline the syllabus and give it a more central role in the classroom. One primary avenue for this could be breaking the syllabus into components—the course description, assignment descriptions, policy statements, and schedule—and separating these components from a central document and relocating to where they will be most relevant. For example, inclusivity policies and statements about access to campus counseling services might be relocated to the home page of the course's learning management system so students have easy access to them and frequent reminders of their existence. For courses that rely heavily on online discussion boards, a statement on online decorum might reside as a widget next to the discussion window. Such moves would liberate this information from the basement of the course policies statement and signal its ongoing importance to the work of the class. Rather than being buried in a section that students might never read, or only read once and then forget, they would be central to the working of the course. A student who might quickly glance over information about campus counseling service in a two-page policy statement at the beginning of the semester and then forget about it may be more likely to use this service if it is ready-to-hand. While this relationship might be primarily textual, it provides a ground for instructors to illustrate the importance of identifying and making use of these relations in with nontextual media, including multimedia and interactive prototyping.

A similar method could be used for other parts of the syllabus. For example, detailed assignment sheets could connect back to learning objectives in the course description, making these objectives not seem rote, but rather connected to the students' day-to-day learning. The course objectives might be moved from their most common location—a bulleted list somewhere near the course description—to become a timeline that runs parallel with the schedule, demonstrating when objectives should be met and the activities and processes that lead to their achievement. This method directly relates policies to those situations, environments, or activities where they apply, rather than see them as the terms of a contract. Of course, these changes are not all unique. I am arguing here that by specifically employing these changes and discussing them with students, they can serve broader course goals in modeling the rhetorical deployment of nontextual affordances of communication.

New Avenues of Communication

The syllabus can also be a means to highlight new avenues of communication introduced by digital culture and technologies. There is little doubt that the new genres introduced by digital technologies—email, workplace chat, text messaging—have had a profound impact on communication practices. We can acknowledge the different habits of readers who are skilled in processing text online—for example, via social media—without succumbing to fearmongering about the death of reading or any other such techno-hysteria. Instructors can pursue new methods of content delivery or engagement with the content of the course. One example of this is the Syllabot, an Arduino project that mediates students' engagement with the syllabus via digital interaction, providing a new means of encountering syllabus content.

SYLLABOT

When teaching interactive prototyping in the writing classroom, it is important for students to engage with multiple levels of carpentry—not simply understanding a prototyping platform but engaging the affordances of that platform to make arguments. I have briefly sketched some means by which the carpentry concepts described here can be used in rethinking the role and form of the syllabus. In this section, I provide an example exploring new avenues of communication by translating information from the syllabus into an interactive prototyping project on the Arduino platform—the Syllabot.[2] Syllabot is an experiment in using the affordances of the Arduino platform to do the work of engaging students in course material.

Syllabot (see figure 14.1) is an Arduino project that delivers syllabus questions on an LCD screen in response to the user pressing a button. The function of Syllabot is adapted from an activity by Maryellen Weimer (2002; cited in Fornaciari and Lund Dean 2014), who suggests that rather than instructors reading or telling students about the syllabus, students should be given time in class to explore it on their own, following this with a low-stakes or ungraded quiz that "fosters a more in-depth examination" of the document (Fornaciari and Lund Dean 2014, 712). The design and coding of Syllabot is adapted from Scott Fitzgerald, Michael Shiloh, and Tom Igoe's "Crystal Ball" (2012, 114 ff.). The code to facilitate scrolling text for the LCD is adapted from build2master's (2015) "arduino-lcd-scroll-long-text."

Syllabot could be used in multiple ways. Before the start of class, the instructor preloads Syllabot with questions. Students are placed in small groups that correspond to the number of questions, and each group sends a representative to push the button to receive their question. This activity could be administered to the group, having individual students handle flipping the device to reveal the next question, or it could be treated as a kind of scavenger hunt, where students work in small groups to find answers to the question they are assigned when one of their members engages with the Syallabot. Alternatively, a group of students could generate questions from the syllabus materials and then load them into Syllabot for other students to answer. More ambitiously, the instructor might use the first class meeting of a digital prototyping course to build Syllabot with the students, walking them through the process of wiring the Arduino, sensors, and actuators, explaining the basics of the code, then having students engage with the Syllabot in order to kickstart a discussion of the syllabus.

This project can serve several purposes. Most generally, it follows Weimer's guidelines in providing space for students to "'discover' what the course is about" (Fornaciari and Lund Dean 2014, 712) by searching for answers to important questions raised by and about the syllabus, rather than the instructor reading the syllabus or describing it to them. In a rhetorically themed course on interactive prototyping, it can also serve as an introduction to the Arduino platform, illustrating some of the possible inputs and outputs available on the platform and the ways in which those inputs and outputs can be used to engage with the environment. The example here is both simple in concept but advanced in execution—the LCD has sixteen inputs, making the wiring more complex than many Arduino builds, and the code for the scrolling text makes use of multiple variables and coding concepts like loops that will require some explaining for beginners. Instructions and code for Syllabot are available on this collection's companion website at https://upcolorado.com/component/k2/item/6219-reprogrammable-rhetoric-supplemental-design-materials-and-programming-scripts. However,

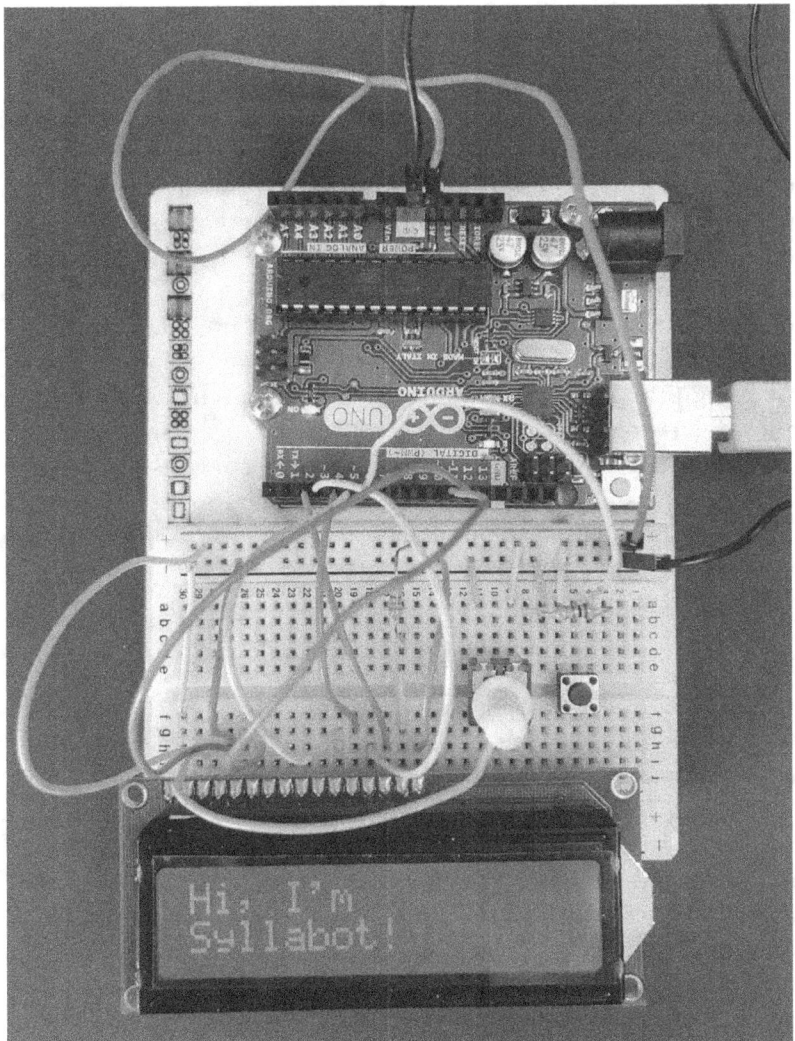

Figure 14.1. A photo of an Arduino and breadboard wired for the Syllabot project. The text on the LCD screen reads, "Hi, I'm Syllabot!" (photo by the author).

readers with access to the hardware can simply copy and modify the design and code in the appendices to suit their own needs.

Building and using Syllabot, or any Arduino project, will not be accessible to all students—assembling the project requires the manipulation of small wires and components and operating Syllabot requires being able to push a small button located in a tangle of wires and other components. Like any other media that isn't accessible to all users—audio that requires a transcript, video that requires captions and audio description, text that requires screen-readable digital versions—Arduino projects like the Syllabot require instructors to think about and provide accessible alternatives. Students could brainstorm ways of replacing the button with sensors that can be more easily triggered by persons with mobility impairments, such as a photoresistor that can identify when a hand or other object passes over it or a tilt sensor that responds to slight movements or taps.[3] In combining physical interaction with the course syllabus, however, this may be a project that can be built upon and modified so that students (and instructors) discover new connections between the course material and the physical environment, directly associating the course design with its content—in this case, interactive prototyping—and remediating the syllabus into an interactive tool. More significantly, it can be an invitation to students to bring new skills and new kinds of media into the writing classroom and providing a means for students to demonstrate their rhetorical skills in venues beyond the traditional paper.

CONCLUSION

This chapter engages carpentry as a bridging concept for connecting things and text and as a means of scaffolding students' interactions with nontextual rhetorics. Inviting the design of physical objects into the rhetoric and writing classroom, along with the code that powers digital interaction, is both a logical progression of the concerns of computers and writing and a tool for bridging textual and nontextual rhetorics in tangible ways. Things—strawberries, needlepoint, water—speak. Text has form (Lanham 2006) and can enter relations with other things and texts. As McGregor notes in this collection, students can be apprehensive about encountering new technologies in a writing classroom, from online course platforms to printable circuits. For a pedagogy of extra-linguistic rhetorics, we can draw on the long history of multimodal writing and visual rhetoric as well as a wealth of nontextual rhetorics in other media as a basis for exploring new forms of rhetorical engagement. Making is a natural counterpart to other pedagogies of nontextual rhetoric. Multimodal communication can be a logical bridge to the study of the rhetorical impacts of programming and the design of interactive objects, and as McGregor shows, this design can inform and support writing instruction. As computation continues its expansion into the environment, students of rhetoric will increasingly need expertise in critically engaging with a range of devices and the information that they produce. Exploring the interaction made possible by these devices will be a crucial part of this rhetorical training. The three approaches to carpentry I have provided here can serve as inventional aids, making them part of a larger effort to bridge the gap between their expectation for more traditional rhetoric and digital media fare and the compelling reasons for rhetoric and writing scholars to engage in this work.

NOTES

1. Much of my thinking in this chapter is deeply indebted to conversations with my colleagues in the Rhetoric, Writing, and Literacy program at The Ohio State University as well as the training provided by the Teaching Support Program at OSU's Michael V. Drake Institute for Teaching and Learning. I would also like to thank Lois Bennett, Emily Coble, Eric Detweiler, Madonna Fajardo Kemp, Alyson M. Lynn, and Elizabeth Williams for inviting me to visit their course and for their thoughtful feedback on this chapter.
2. I wish to thank Amy Sardone for naming Syllabot and for her feedback and suggestions on its function.
3. Fitzgerald, Shiloh, and Igoe's "Crystal Ball" (2012, 114 ff.), on which this project is based, uses a tilt sensor to change the text display.

REFERENCES

Ahmed, Sara. 2008. "Some Preliminary Remarks on the Founding Gestures of the 'New Materialism.'" *European Journal of Women's Studies* 15 (1): 23–39. doi:10.1177/1350506807084854.

Alexander, Jonathan, and Jacqueline Rhodes. 2014. *On Multimodality: New Media in Composition Studies*. Urbana, IL: Conference on College Composition and Communication/National Council of Teachers of English.

Arola, Kristin L. 2017. "Indigenous Interfaces." In *Social Writing/Social Media: Pedagogy, Presentations, and Publics*, edited by Douglas M. Walls and Stephanie Vie, 211–26. Fort Collins: The WAC Clearinghouse, University Press of Colorado. doi:10.37514/PER-B.2017.0063.2.11.

Ball, Cheryl E. 2004. "Show, Not Tell: The Value of New Media Scholarship." *Computers and Composition* 21 (4): 403–25. doi:10.1016/j.compcom.2004.08.001.

Bogost, Ian. 2012. *Alien Phenomenology, or, What It's Like to Be a Thing*. Minneapolis: University of Minnesota Press.

Boubekri, Mohamed, Ivy N. Cheung, Kathryn J. Reid, Chia-Hui Wang, and Phyllis C. Zee. 2014. "Impact of Windows and Daylight Exposure on Overall Health and Sleep Quality of Office Workers: A Case-Control Pilot Study." *Journal of Clinical Sleep Medicine* 10 (6): 603–11. doi: 10.5664/jcsm.3780.

Brown, James J., Jr., and Nathaniel A. Rivers. 2013. "Composing the Carpenter's Workshop." *O-Zone: A Journal of Object-Oriented Studies* 1: 27–36. https://www.academia.edu/18103756/Composing_the_Carpenter_s_Workshop.

build2master. 2015. "arduino-lcd-scroll-long-text." GitHub. https://github.com/build2master/arduino-lcd-scroll-long-text/blob/master/arduino-lcd-scroll-long-text.ino.

Faris, Michael J., Andrew M. Blick, Jack T. Labriola, Leslie Hankey, Jamie May, and Richard T. Mangum. 2018. "Building Rhetoric One Bit at a Time: A Case of Maker Rhetoric with littleBits." *Kairos: A Journal of Rhetoric, Technology, and Pedagogy* 22 (2). https://kairos.technorhetoric.net/22.2/praxis/faris-et-al/index.html.

Fitzgerald, Scott, Michael Shiloh, and Tom Igoe, eds. 2012. *Arduino Projects Book.* Torino, Italy: Arduino LLC.

Fornaciari, Charles J., and Kathy Lund Dean. 2014. "The 21st-Century Syllabus: From Pedagogy to Andragogy." *Journal of Management Education* 38 (5): 701–23. doi: 10.1177/1052562913504763.

Goggin, Maureen Daly. 2009. "Stitching a Life in 'Pen of Steele and Silken Inke': Elizabeth Parker's Circa 1830 Sampler." In *Women and the Material Culture of Needlework and Textiles, 1750–1950,* edited by Maureen Daly Goggin and Beth Fowkes Tobin, 31–49. Burlington, VT: Ashgate.

Goggin, Maureen Daly, and Beth Fowkes Tobin, eds. 2009. *Women and the Material Culture of Needlework and Textiles, 1750–1950.* Burlington, VT: Ashgate.

Haas, Christina. 2013. *Writing Technology: Studies on the Materiality of Literacy.* New York: Routledge.

Harman, Graham. 2002. *Tool-Being: Heidegger and the Metaphysics of Objects.* Chicago, IL: Open Court.

Kimmerer, Robin Wall. 2013. *Braiding Sweetgrass: Indigenous Wisdom, Scientific Knowledge and the Teachings of Plants.* Minneapolis, MN: Milkweed Editions.

Lanham, Richard A. 2006. *The Economics of Attention: Style and Substance in the Age of Information.* Chicago, IL: University of Chicago Press.

Lund Dean, Kathy, and Charles J. Fornaciari. 2014. "The 21st-Century Syllabus: Tips for Putting Andragogy into Practice." *Journal of Management Education* 38 (5): 724–32. doi:10.1177/1052562913504764.

Lynch, Patrick J. 2008. *Web Style Guide.* New Haven, CT: Yale University Press.

Mattern, Shannon. 2012. "Aliens to Armoires: Philosophical Carpentry." *Words in Space,* May 29, 2012. http://wordsinspace.net/shannon/2012/05/29/aliens-to-armoires-philosophical-carpentry/.

Powell, Malea, Daisy Levy, Andrea Riley-Mukavetz, Marilee Brooks-Gillies, Maria Novotny, and Jennifer Fisch-Ferguson. 2014. "Our Story Begins Here: Constellating Cultural Rhetorics." *Enculturation: A Journal of Rhetoric, Writing, and Culture* 25. http://enculturation.net/our-story-begins-here.

Pristash, Heather, Inez Schaechterle, and Sue Carter Wood. 2009. "The Needle as the Pen: Intentionality, Needlework, and the Production of Alternate Discourses of Power." In *Women and the Material Culture of Needlework and Textiles, 1750–1950,* edited by Maureen Daly Goggin and Beth Fowkes Tobin, 13–30. Burlington, VT: Ashgate.

Ravenscroft, Alison. 2018. "Strange Weather: Indigenous Materialisms, New Materialism, and Colonialism." *Cambridge Journal of Postcolonial Literary Inquiry* 5 (3): 353–70. doi:10.1017/pli.2018.9.

Rieder, David M. 2016. *Suasive Iterations: Rhetoric, Writing, and Physical Computing.* Anderson, SC: Parlor Press.

Rivers, Nathaniel, and Lars Söderlund. 2016. "Speculative Usability." *Journal of Technical Writing and Communication* 46 (1): 125–46. doi: 10.1177/0047281615600635.

Weimer, Maryellen. 2002. *Learner-Centered Teaching: Five Key Changes to Practice.* San Francisco, CA: Jossey-Bass.

Womack, Anne-Marie, Annelise Blanchard, Cassie Wang, and Mary Catherine Jessee. 2016. "Accessible Syllabus." https://www.accessiblesyllabus.com/.

Wysocki, Anne Frances, Johndan Johnson-Eilola, Cynthia L. Selfe, and Geoffrey Sirc. 2004. *Writing New Media: Theory and Applications for Expanding the Teaching of Composition.* Logan: Utah State University Press.

INDEX

Page numbers followed by f indicate figures.

able-bodiedness. *See* ability; compulsory able-bodiedness
ability, 9, 35, 201, 202, 203, 204, 214. *See also* ability/disability system; disability
ability/disability system, 35. *See also* compulsory able-bodiedness
access, 12, 17, 16, 130, 133, 134, 138, 139, 201, 202, 204–5, 208–9, 214–25, 216, 222, 244, 247–48, 260; critical, 204, 215; experiential, 204, 215; financial, 129; functional, 204, 215; material, 201, 204, 214; meaningful, 201; transformative, 204–5, 214–15
accessibility, 15, 30, 32, 56, 212–13, 292–3, 297–98. *See also* adaptive technologies
actionable media, 184–85, 194, 195
actionable monuments, 180, 184–85, 194, 195. *See also* Roadkill Tollbooth Actionable Monument
#actionclimate, 131
activism: augmented reality, 19, 180; civic, 52; coalitions and, 85; critical making and, 8, 76, 86, 95, 204, 214; critical text mining and, 17, 93, 94; data curation and, 9; equality and, 97, 99; hashtags and, 98, 99, 101, 103; students and, 68, 100; yarn-bombing, 4
Adafruit Industries, 146, 148f, 154, 161f
adaptive technologies, 32, 41
Administration for Native Americans Native Languages Preservation and Maintenance, 230
adversarial design, 7
aesthetic, 55, 56, 57, 59, 97, 168
affect, 14, 22, 109, 111, 182, 186, 213, 218, 244–45, 247, 249, 251, 254, 259
affect theory, 37
affective rhetoric, 245
agency, 175, 253; authorial, 77; networks of, 95; nonhuman, 9, 37–38, 39, 86; user, 18; video game players', 244
Ahmed, Sara, 202, 286
AIDS Memorial Quilt, 50
Alexander, Jonathan, 217, 287

Alfred, Taiaiake, 229
algorithm, 81, 95, 96, 99, 110, 117, 173; critique of, 16, 92, 93, 95
#alllivesmatter, 94, 103, 104
alt text, 293
Amazon, 129
Amazon Web Services (AWS), 129
American Indian. *See* Indigenous. *See also* specific tribe names and languages
#amplifywomen, 131
androgyny, 54
animal, 19, 35, 38, 168, 172; abject sacrifices of, 180, 182, 193; death of, 9, 186, 188–89, 195; in game play, 234–35, 252; Oneida words for, 231
Anishinaabe creation story, 38
Anthropy, Anna, 209, 216, 244, 248, 255
Antinori, Fabio, 155
Apple, 36, 37, 186. *See also* iPhone
Application Programming Interface (API), 94, 101, 131, 188, 191
Arakawa, Shusaku, 47–48
Arch Linux distribution, 15
Arduino, 13, 19, 21, 130, 154, 160, 172, 286, 295, 297, 298; as inexpensive, 204; with instruments, 30, 41; microcontroller, 9, 30, 160; starter kit, 160; sensors, 212, 216, 296; Touch Board, 154, 156
Arellano, Sonia Christine, 51
argumentation, 110, 122
Aristotle, 35, 149, 167
Arse Elektronika conference, 216
art, 42, 48, 53, 120, 157, 226; augmented reality and, 185; concept art, 227, 236–37; eversion and, 159; folk, 39; glitch, 29; as guerilla act, 68; interactive, 155; mass produced, 52; minimalist, 56; as a network, 55; new media, 215; sound-based, 30; websites for, 248
art establishment, 54, 68
art object, 54, 57
artists' books, 51, 56–58, 60
Assassin's Creed III (video game), 224, 225, 237

Association of Teachers of Technical Writing conference, 80
ATLAS (programming language), 185
audience, 8, 14, 20, 100, 104, 170, 181, 185, 186, 223, 229, 236, 289; of art, 52, 54, 58, 68, 157, 159; of datasets, 95; disciplinary, 4, 14; and genre, 116; of multimodal texts, 18; post-PC, 145, 146, 147–48, 156, 162; practitioner, 4; and sensations, 147, 151, 154, 216; students' consideration of, 21, 42, 67, 268, 270, 272, 277, 279–81; unanticipated, 253
audio. *See* sound
augmented reality (AR), 19, 147, 180–81, 185–87, 194–95
autonomous sensory meridian response (ASMR), 212–13
avant-garde literature, 47
Azhar, Hamden, 100
Azure, 129

backyardchickens.com, 110–11
Bagnall, Gregory L., 212
Bailey, Richard W., 171–72, 173
Banks, Adam J., 201, 204, 214
Barad, Karen, 37
Bare Conductive, 148*f*, 154–56, 157, 160
Basho, 173
Benjamin, Walter, 166, 167
Bennett, William Ralph, Jr., 170
Benton-Banai, Eddie, 38
Bergson, Henri, 163n1
Bernstein, Charles, 47
Beveridge, Aaron, 17, 77, 94, 130
BioElectronicAudiospan, 40
Biber, Douglas, 115
bibliometric data, 80, 83, 84–85. *See also* citation
big data, 17, 81, 111, 129, 132, 133
BIPOC (Black, Indigenous, and people of color), 11, 80, 82, 88. *See also* people of color
bisexual, 86
Björgvinsson, Erling, 226
Blackboard, 266, 269
black bodies, 98, 100
Black identity, 54, 57, 58, 78, 88n1, 88n5
Black Lives Matter, 11, 22, 98, 100
black skin, 76
#BlackintheIvory, 88n1
#blacklivesmatter, 9, 16–17, 93, 94, 98, 101, 102–4, 131, 139
Black Mirror: Bandersnatch, 260n4
Blair, Carole, 50
blog, 96, 155, 210, 211, 246, 258; student-created, 268, 271–75, 280–81

Bluetooth onboard diagnostic scanner, 19, 180
bodies, 29–30, 36, 41–42, 49, 51, 79, 156, 216; disability and, 30, 31, 33, 35, 41, 212; deviant, 15, 32, 36, 43n4; eversion and, 151, 159, 163; gesture and, 153; as parts of circuit, 39–40, 154; racist violence against, 98, 100; touch and, 145, 147, 149–51, 154, 162–63; video games and, 208, 209, 213, 244, 251, 257; writing and, 176. *See also* Body without Organs; embodiment; mind/body distinction
Body without Organs (BwO), 150
Boggs, J. S. G., 55, 68
Bogost, Ian, 77, 244, 248, 287–88
Boler, Megan, 4, 97, 195
Borroff, Marie, 175
#BPoilspill, 187, 188, 191, 192
Bradford, Richard, 62
Breaux, Chet, 203, 204
bricolage, 10, 97
British Museum, 113, 122–23
British Petroleum (BP), 181, 186, 189; oil spill, 19, 154, 180, 181, 182, 183–84, 188–89, 195
Brock, Kevin, 95
Brown, Bob, 55
Brown, Gilian, 117–18
Brown, James J., Jr., 165, 166–67, 168, 172, 246, 251, 252–53, 287, 289, 290, 291
Bryant, Levi, 37, 167
Buckenham, George, 172
Burch, Mark, 168–69
Burek, Amy, 9
Burgess, Helen J., 5, 7, 8, 10, 11
Burke, Kenneth, 245, 246
Bush, Stuart, 30

C# (programming language), 19, 180, 187, 190
Camnitzer, Luis, 48, 52, 54, 56, 60
Campbell, Karlyn Kohrs, 116
capacitive-touch sensor, 146–47, 148*f*, 152, 154, 155, 161*f*, 185
capta, 177
Cárdenas, Micha, 215–16
carpentry, 21, 286, 287–90, 291–92, 295, 298, 299; rhetorical, 9, 10
Carter, Joyce Locke, 8
Casio keyboard, 39
Center for Biological Diversity, 189
Chang, Edmond Y., 176, 177
Charity Heartscape. *See* Porpentine
Charny, Daniel, 202
Chávez, Karma R., 79, 87

Cheap Bots, Done Quick, 172
Cheney-Lippold, John, 76
chora, 6, 163
Chun, Wendy Hui Kyong, 95
Cicero, 169
circuit-bending, 7, 39
circuits, 5, 9, 15, 17, 21, 39–40, 48, 154, 175, 271, 275, 278, 298. *See also* bodies, as parts of circuit; circuit-bending; paper-electronic circuits; sound, circuits and
circulation, 49, 51–55, 59, 59, 60–61, 66–68, 69, 216, 217
cisgender, 9, 201, 208, 214
citation: analysis, 78; data, 82, 83; mapping, 80, 81, 82–85, 87, 123; practices, 9, 16, 78–79, 83, 88, 105; politics of, 86
civic rhetoric, 54
civil disobedience, 4
civil rights, 78
Civil Rights Memorial, 50
Civil Rights Movement, 97
Clare, Eli, 32
Clark, Ian, 110
Clark, Naomi, 209
class. *See* socioeconomic class
cloud computing, 129, 131, 186
cloud server, 130, 132, 133
coalition, 77, 78, 79–81, 82, 84, 85*f*, 87, 88, 203
code, 7, 13, 19, 55–56, 93, 101–3, 147, 159, 176, 188, 190, 194, 195, 247, 296, 298; as rhetorical, 81
co-design, 29, 30, 32, 40, 41–42, 43n1, 226
coding, 5, 7, 11, 16, 81, 94, 208, 215, 227, 247, 248, 296; for augmented reality, 19; data, 129; literacy, 195; as material process, 95; qualitative, 108, 109
Cohen, Cathy, 203
Cohen, Mary Ann, 115
Colegio Atalaya, 157, 158*f*, 160
collaboration, 67, 69, 139, 151, 227, 237; between rhetor and language, 174; interdisciplinary, 140, 172; in making, 11, 17, 131, 201, 203, 214, 215, 229; by students, 29–30, 160, 266–69, 271, 277, 280, 286; by technical communicators, 77
Collens, R. J., 169, 171, 172, 174
Collier, John, 168–69
Collins, Bethany, 58–59
colonialism/colonization, 9, 10, 11, 12, 37, 40, 80, 81, 105, 146, 227, 239n2, 239n5. *See also* decolonialism/decolonization
Colton, Jared S., 97
comics, 215
#communicationsowhite, 88n1

composition instruments, 29, 36, 42
composition studies, 29; place of rhetoric in, 175–76. *See also* rhetoric and composition studies; writing studies
Compton, Kate, 172
compulsory able-bodiedness, 35
computational analysis, 92–93, 117. *See also* data, analysis
computational methods, 96, 108, 115, 123
computational rhetoric, 108–9, 123
computer-generated poetry, 19, 165–66, 167, 168, 169–72, 173, 174–75, 176
computer programming, 95, 160, 165, 166, 169–70, 172, 175, 177
computer science, 7, 93, 140, 169–70
Computers and Writing (C&W) conference, 45, 46, 47*f*
conductive paint. *See* Bare Conductive
Conference on College Communication and Composition, 8
Connor, Ulla, 115
Consensual Torture Simulator (video game), 250–51
consent: research, 281, 282; sexual, 250, 251
Cook Inlet Tribal Council (CITC), 227
coronavirus pandemic, 21–22, 95
corpus linguistics methods, 92
counter-histories, 15
counterpublics, 4
counterstory, 50
Crable, Bryan, 260n3
craft, 4, 21, 203, 206, 215, 267, 269, 273, 278, 287, 293; of methods, 123; stores, 61. *See also* craftivism
craftivism, 96, 215
Craig, Alan, 181
Creative Commons, 15
Crimson Hexagon software suite, 93
crip theory, 35
crip time, 206
Critical Art Ensemble (CAE), 4
critical making, 3, 22, 77, 84, 205, 216; actionable media and, 195; alternative genealogies of, 9, 11; assignments, 266, 267, 269, 272, 281; augmented reality and, 19, 180, 181, 182, 185, 194, 195; bot-making as, 177; carpentry and, 27; community-based, 223, 228; computational methods and, 108, 109, 131; creative writing and, 169; criticality of, 4, 11, 13; data-driving research and, 138; definition of, 3–4; digital, 94; DIY citizenship and, 4, 97; ethical aspects of, 8, 14, 15, 22, 30, 95, 104; eversion and, 18; exclusivity of, 8–9, 10, 222; Indig-

304 INDEX

enous approaches to, 10, 16, 20, 222, 226, 228; invention and, 15; lineages of, 4–5; material or nondigital, 49, 61, 68–69; networks of, 56; participatory design and, 222, 225, 226, 227, 229, 237, 238; play and, 19–20; pedagogy, 265–67, 268–69, 276–77, 279, 280; politics and, 96; as problem solving, 87; products of, 227–28, 236; queering, 201, 206–7, 217; reflection and, 195; representation and, 222; rhetoric and, 166, 177, 245; rhetoric and composition and, 5–7, 12–13, 14, 15, 22, 181, 201, 203, 205; for social futures, 22; for social justice, 76, 86, 88; students and, 21; survival and, 202; technology and, 133, 165; text mining and, 16–18, 93, 95, 96, 131; theory/practice divide and, 13, 15; as theory making, 259; video game design as, 246; zines as, 215. *See also* critical text mining; reparative making
critical race theory, 78
critical text mining, 9, 14, 16–17, 92–93, 94–96, 97, 98–99, 100, 105, 131, 139
critical thinking, 4, 56, 95, 205, 271
Crowley, Sharon, 259
cultural rhetorics, 15, 37
Cultural Rhetorics Conference, 10
cultural studies, 21
cyber security, 140
cyberspace, 18, 145–46

Dames Making Games (DMG), 214–15
Danforth, Courtney S., 43n3
Darwin, Charles, 168
data, 19, 31, 39, 81, 94, 146, 152, 155, 177, 180, 195; analysis, 9, 14, 16, 17, 92, 100, 129, 133; archiving and storing, 17, 131, 133; collection, 14, 18, 100, 130, 131, 132, 133–36, 137, 138; curation, 9, 18; diagnostic, 187; loss, 139, 140; management, 140; testing, 111; training, 110–11; visualization, 9, 16, 117, 290. *See also* bibliometric data; big data; citation, data; data-driven methods and research; data science; datasets; US census
data-driven methods and research, 77, 129, 130, 131, 132–34, 138; intersection with critical, 138, 139, 140
data science, 93, 103, 129. *See also* data-driven methods and research
datasets, 17, 19, 83, 94, 95, 96, 98, 99, 101, 103, 111, 129, 130, 131, 132, 133, 138, 139, 140, 173, 176
Davis, Diane, 20, 245–46, 259

Davis, Matthew, 50
De Copia (Erasmus), 165, 166–67, 176
Debord, Guy, 53
decolonialism/decolonization, 9, 13, 14, 16, 20, 105, 131. *See also* colonialism/colonization
decomposition, 217
Deepwater Horizon. *See* British Petroleum
Deleuze, Gilles, 150, 163n1
Delgado, Richard, 78, 79, 88
delivery, 6, 14, 292
Demosthenes, 6
depression, 206, 210, 211, 243, 244
Depression Quest (video game), 20, 242–44
design, 7, 10, 16, 45–46, 63–66, 152, 180, 298; affective, 244; of circuits, 154, 275; circulation of, 60–61; of computational programs, 108, 116, 133, 175; document, 293; graphic, 190; haptic, 18; instructional/pedagogical, 21, 29, 30, 40, 265, 266, 268, 270, 281, 298, 290; interactive, 279, 283; for social futures, 22, 202; sound intrument, 29–30, 37, 39, 40, 41; syllabus, 286, 292–93, 296; of touch-interactive interface, 147, 157, 160; "white as default," 287. *See also* adversarial design; co-design; Indigenous, game design; participatory design; speculative design; user-centered design; video game, design; visual design
design goals, 51, 56, 60
diagnosis (medical), 30, 31–32, 35, 40, 49
digital cutter, 45, 46, 48, 49, 59, 60–61, 68
digital humanities, 5, 10, 16, 17, 18, 86, 92, 95, 109, 123, 129, 145, 176, 177
Digital Humanities Summer Institute (DHSI), 86, 172
digital rhetoric, 5–6, 7, 10, 14, 19, 50–51, 93, 100, 101, 129–30, 146, 154, 159–60, 162, 165, 166, 169
digital writing. *See* writing, digital
dirt, 40–41
dirty new media, 29
disability, 15, 20, 30, 32, 35, 43n2, 85, 208, 214, 215, 293; deficit model of, 32, 35. *See also* ability/disability system; accessibility; adaptive technologies; compulsory able-bodiedness; crip theory
disability studies, 35, 51. *See also* ability/disability system; compulsory able-bodiedness; crip theory
discourse community, 110
Disney, 227
dissensus, 93, 97–98
distant reading, 92, 110

distributive justice, 97, 98
DIY (do-it-yourself), 17, 77, 203; citizenship, 4, 97, 98; ethos, 226
Documenting the Now (DocNow), 98, 103, 131, 139
Double Union, 10
Douglas, Mary, 40–41
Doyle, Richard, 165, 166, 167–68, 169, 171, 176
Drucker, Johanna, 177
Dryer, Dylan B., 100, 104
Dunbar-Hester, Christina, 8, 204

ecodelics, 166, 167, 171
Ede, Lisa, 13
Edbauer, Jenny. *See* Rice, Jenny (Edbauer)
efficiency, 53, 130, 131, 132, 133–34, 137, 138, 204
Ehn, Pelle, 226
electracy, 19, 185
electronic monument. *See* MEmorial
Electronic Numerical Integrator and Computer (ENIAC), 95
Elgin Marbles, 112, 122
E-Line Media, 227
eloquence, 19, 165–69, 170, 173, 174–75, 176, 177
eloquence adjuncts, 19, 151, 156, 165–66, 172, 174, 175, 176
eloquence machines, 166–67
embodiment, 14, 18, 50–51, 59, 61–62, 209, 215; critical making and, 95, 201, 205, 245; invention and, 6, 185; of reading, 55; touch and, 150, 151, 153; video games and, 212, 244, 245. *See also* bodies
emoji, 16, 93–94, 98, 99, 100, 101–4, 139
Emojination, 99
empathy, 29, 32, 210, 211, 245, 249
Empedocles, 149
Endangered Languages Project, 230
enthymeme, 77, 95
equality, 101; active, 97–99; Rancièrian, 93, 97–99; racial, 9, 100. *See also* inequality
Erasmus, Desiderius, 165, 166–67, 176
estrangement, 207, 210, 212, 213, 216
ethics, 8, 68; bodies and, 30; code and, 10; critical making and, 7, 8, 10, 11, 12–14, 30, 95, 104; of identification, 245–46, 247, 258, 259; of invention, 20; of making, 15, 16, 204; of mass produced art, 52; object-oriented ontology and, 11; play and, 19; Rancière and, 93, 97–99, 104; of representation, 86; of technology development, 109, 133, 138, 139, 140; text mining and, 16–17, 93, 96, 97,

98, 100, 101; in video game journalism, 242
ethnocentricism, 11
ethos, 54, 94, 123, 195, 226
eureka moment, 174, 185
Even Cowgirls Bleed (video game), 251–52
eversion, 18, 19, 145–46, 147, 149, 151, 153–55, 157, 159, 160, 162, 163, 185, 194
Evoking moves, 109–10, 111, 113, 114, 115
e-waste, 140
exclusion, 9–10, 35, 40, 41, 203–4, 212
executability, 5, 10, 13, 15, 17
executable culture, 7
experimentation, 6, 7, 14, 22, 45, 96, 108, 155, 159, 165, 171, 172–75, 209, 222, 269, 271, 276, 279, 280, 283, 287, 295

Faber, Jennie, 215
facial recognition software, 76, 105
facilitation research, 108–10, 113, 114*f*, 115–18. *See also* Evoking moves; Inviting moves; Staging moves
Faciloscope, 17, 94, 95, 108–11, 113, 115, 116–17, 119, 120, 123, 124, 124n1, 124n2, 124n3, 130
Faigley, Lester, 50
failure, 6, 39, 41, 108, 171, 210; bodily, 35; of identification, 246, 247, 259
Fairey, Shepard, 139. *See also Obama Hope* image
Fallout 4 (video game), 251
FANDSO (free assemblable nonfunctional disposable serial object), 52
Faris, Michael J., 6–7, 19–20, 43n3, 222, 226, 244
Farrell, Thomas B., 115–16, 117, 118
FedEx logo, 46
feminism, 9, 14, 203, 204, 209, 214, 215; critical making and, 16, 20; disability theory and, 35; GamerGate and, 242; intersectional, 212; new materialism and, 286; women of color feminism, 79
feminist hacker space, 9
financial cost, 17, 129
First Nations. *See* Indigenous. *See also specific tribe names and languages*
first year composition, 20, 52, 238, 265, 266, 268, 277, 283
Fisher, Stephanie, 214
Fitzgerald, Scott, 296, 299n3
Flanagan, Mary, 247
Flint water crisis, 194
Flower, Linda, 48
Floyd, George, 98, 100
font, 57
Fornaciari, Charles, 291–92, 293, 294

Fox, Sarah, 222, 230
Fragnito, Skawennati, 228
Frankfurt School, 205
Freedan, Amy, 227
Freeman, John Craig, 180–81, 194
Frentz, Thomas S., 115–16, 117, 118
Friedhoff, Jane, 247–48
Fulkerson, Matthew, 147, 149, 150, 151, 162
Furby, 39
futurity, 22, 88, 176, 202, 239n5, 269

Gaertner, David, 86
Gallagher, John, 105
GamerGate, 20, 242, 260n1
game play, 20, 206, 207, 209, 210, 212, 213, 215, 217, 234, 257, 259
games. *See* video game
Garland-Thomson, Rosemarie, 35
gender, 10, 20, 22, 35, 54, 99, 202, 203, 204, 208, 209, 214–15, 257, 258
genderqueer, 214
genre, 17, 52, 95, 108, 110, 115–17, 123, 181, 185, 195, 227, 235, 248, 287, 288, 289, 295. *See also* rhetorical genre theory (RGT)
Gerdes, Kendall, 20, 207, 209, 211, 217–18, 226
German media theory, 147, 151
gerrymandering, 68
Gershenfeld, Alan, 227
Ghazala, Q. Reed, 39–40
Gibson, William S., 18, 145–46
Gioia, Dana, 63
GitHub, 98, 102, 139
glitch, 29, 34, 40, 41
Goggin, Maureen Daly, 289, 294
Gollihue, Krystin Nicole, 10
Gomringer, Eugen, 46, 47*f*
Good Mind, 230, 234–36, 237
Google, 45, 102, 130
Google Chromebook, 134
Google page rank, 95
Google Scholar, 96
Google Trends, 135
Gorgias, 259
GPS, 186, 187, 189, 190, 191*f*
Grabill, Jeff, 124n1
Graff, Gerald, 92
gramophone, 152
Gray, Kellie M., 9, 16–17, 93, 108, 131, 138, 139
Greco-Roman rhetoric, 6
Greenpeace, 186
Grounds, Richard A., 225
Guattari, Félix, 150, 163n1

Haas, Angela, 9, 50
hackathon, 9, 10
hackerspace, 9, 214
hacking, 4, 61, 77, 81, 85, 94, 123, 124, 181
hacktivism, 5
haiku, 173, 174, 175, 176
Halberstam, J. Jack, 210
Hallowell, A. Irving, 39
Halm, Matthew, 18, 166, 185, 194, 195
Halperin, David M., 217
Hammer, Steven, 7, 9, 15, 16, 43n3, 49, 201
haptics, 18, 51, 152–53
haptic subject, 152
Haraway, Donna J., 37
Harman, Graham, 11, 290
Hart-Davidson, William (Bill), 17, 110
Hartley, John, 97
Harvey, Allison, 214
hashish, 166, 167
hashtag. *See* Twitter, hashtag; *specific hashtags*
Hatch, Mark, 265, 267
Haudenosaunee, 234, 239n3; Thanksgiving Address, 229, 230–33, 234. *See also* Iroquois; Mohawk; Oneida; Seneca
Hawhee, Debra, 174
Hawisher, Gail E., 37, 203
Hawkins, Jeffery, 225
Hayes, John R., 48
Hayles, N. Katherine, 50, 54, 92
Hearst, Marti A., 117, 118
Hedge-O-Matic, 130
Heidegger, Martin, 11, 205
Heideggerian quietism, 11
Heilker, Paul, 100
Hendren, Sara, 213
Hermes, Mary, 38
Hertz, Garnet, 4, 8, 9, 10, 101
heterosexuality, 9, 35, 201, 208
Hewlett Packard (HP), 76
hexis, 101
Hillgred, Per-Anders, 226
Hirst, Graeme, 118, 124n5
historical materialism, 286
Hockema, Stephen, 94, 95
Holmes, Steve, 9, 16–17, 93, 97, 105, 108, 131, 138, 139, 244
Holzer, Jenny, 52–53, 54, 55, 60, 68
@HomeworkCopia, 172
Horner, Bruce, 13
Howe, Daniel C., 173
human-computer interaction, 131
humanities, 5, 16, 57, 93, 104, 131, 132, 140, 160, 165, 169–70, 177, 195, 205, 287
hypertext, 50, 242, 245, 246, 248, 255

identification, 51, 152, 211, 242, 244, 245–46, 249, 252, 254, 258, 260n2, 260n3; failure of, 247, 259; violence of, 242, 245
identity, 4, 10, 30, 38, 53, 64, 78, 85–86, 89n4, 97, 118, 150–51, 201–203, 214, 217, 224, 228, 237, 239n2; authorial, 268; brand, 36; city, 51; gamer, 242; maker, 204
Igoe, Tom, 296, 299n3
immigration, 180–81
inclusion, 8, 29, 78, 79, 80, 88, 88n1, 99, 104–5, 203, 205, 209, 213–14, 216, 225, 292, 294
Indian. *See* Indigenous. *See also specific tribe names and languages*
Indigenous, 9, 78, 80, 86, 121, 237, 239n2; game design, 9, 20, 209, 222–23, 225–30, 237, 238–39, 239n2; intellectual traditions, 11, 37–38, 286; making practices, 15, 16. *See also* Native American, representation; *specific tribe names and languages*
inequality, 97, 99, 100, 214, 287. *See also* equality
@InfiniteQuintilian, 19, 173
information architecture, 166, 281
information theory, 34
Institutes of Oratory (Quintilian), 173
Intel, 134, 137
interdisciplinary, 57, 129, 130, 133, 138
interface, 17, 34, 77, 82*f*, 84, 85*f*, 123, 176, 184–85, 190, 191*f*, 192, 194, 217, 252; augmented reality, 19, 181; co-design of, 40; physical, 20, 209, 212–13, 216; politics of, 37, 80–81; touch, 147, 152, 153, 155, 158*f*, 159–60
Internet of Things, 18, 291
intersectionality, 13, 14, 99, 212, 214, 217
Inuit, 79
Iñupiaq, 227–28
invention, 6, 7, 8, 14, 15, 19, 20, 146, 151, 163, 174, 181, 185, 194, 201, 202, 208, 249, 259, 286, 287–89, 299
Inviting moves, 109, 110, 111, 112, 113, 114, 115, 118–19, 121
iPhone, 152–53, 248
Iroquois: as a derogatory name, 239n3; language group, 232. *See also* Haudenosaunee
Itchuaqiyaq, Cana Uluak, 9, 16, 17, 104–5, 123, 131, 138

Jagoda, Patrick, 10, 247
Jalbert, Kirk, 226
Jamieson, Kathleen Hall, 116

Java (programming language), 160
JavaScript, 100, 178n4, 247
Jones, John, 21, 265, 268, 281
Jones, Natasha N., 78–79, 80, 84–85, 86, 87, 89n4, 89n5
Jones, Steven E., 18, 145

Kaggle Institutes Dataset, 111
kairos, 6, 11, 19, 22, 95, 174, 167
Kant, Immanuel, 98
Keeling, Kara, 203
Keep Austin Weird stickers, 51–52, 54
Keogh, Brendan, 210
Khaled, Rilla, 228
Khan, M.Bawar, 19, 147, 153, 160
Kimes, Jason, 182, 184
Kimmerer, Robin Wall, 290–91
King, Kaki, 49
Kittler, Friedrich, 152
Klimas, Chris, 247
Knight, Aimée, 7
knitting, 21, 96
Konkol, Margaret, 172, 178n2
kopas, merritt, 208, 216, 250, 251
Kostelanetz, Richard, 64

Lancaster, Zak, 92, 104
language action paradigm, 115–16
Lanham, Richard, 165, 176
LaPensée, Elizabeth, 20, 226, 228, 239n2
Latour, Bruno, 37, 39, 43n5, 123, 205
Laufer, Susan B., 47
Law, John, 124n10
LCD screen, 295, 296, 297*f*
Le Dantec, Christopher, 222, 230
learning management system (LMS), 292, 293, 294. *See also* Blackboard
Leavitt, Peter, 224
Lee, Jennifer 8., 99
Leitch, Amy, 215
Lenovo ThinkPad, 134
Leskovec, Jure, 111
Lewis, Jason E., 228
lexical chains, 118
Li, Xin, 111
liberal arts education, 165
liberalism, 97, 217
lissencephaly, 31, 49
literacy, 203, 206, 217; coding, 195; control, 212; digital, 291; materiality of, 50
littleBits, 6
Lorde, Audrey, 96
Louisiana Gulf Coast. *See* British Petroleum
Love, Christine, 251
Lund Dean, Kathy, 291–92, 293, 294

machine learning, 99, 109, 110
Machine Room, 10
machine-oriented ontology, 167
Madej, Krystina, 57
Madsen, Deborah Lea, 227
MAK Museum, 155, 156*f*
Make magazine, 8
maker approach, 77, 129, 130
maker community, 135, 280
maker cultures, 3, 6, 7, 8, 10; criticality of, 4; ethos of, 94; feminist critique of, 204; exclusions within, 9, 201, 203–4
maker identity, 204, 280
maker labs, 5, 130
maker mentality, 5, 7, 8, 11, 245, 280, 283
maker movement, 5, 20, 86, 177, 201, 202, 203–4, 205, 216–17, 269
maker ontologies, 9
maker orientation, 104, 130, 131, 132
maker pedagogy, 203, 218
maker rhetoric, 216
maker turn, 16
"Maker's Bill of Rights" (Hertz), 8, 9, 10
makerspace composition, 45
makerspaces, 7, 8, 9, 11, 45, 61, 131, 201, 204, 214, 265, 267, 268, 269
Makey Makey, 13
making, 21, 286, 287, 288, 298; bodies and, 41; digital and nondigital, 95; Indigenous, 9; networked, 56; privileged, 22; politics of, 22. *See also* critical making
maktivism, 5, 77
manifesto, 4, 11, 40, 202, 238, 246, 248
Manito, Gitchie, 38
Mann, Steve, 77
Manning, Erin, 13, 147, 150–51, 156, 162
mapping, 37, 75–77, 78, 79, 80–81, 82–88, 123, 162, 176, 182, 194, 234, 248, 249*f*; obfuscated, 252, 253
Marcotte, Jess, 212–13
Markov chains, 173
Martin, Trayvon, 93
MassMine, 18, 130, 131–32, 134, 135–37, 140
Massumi, Brian, 37
material composition, 4, 6–7, 51, 53, 59, 60, 66
material form, 49, 51, 52–53, 56, 60–61, 205, 245
materiality, 7, 8, 57, 66; of coding, 95; of colonization, 12; of communication, 286; of critical making, 4, 94, 95, 97, 131, 177, 201, 215, 245; of interactions, 146, 147, 153, 162, 195, 190; of literacy and writing, 49–50; of making, 18, 19; of objects, 6, 10, 16; of practice, 6, 55; of processes, 56, 69; of programs, 12;

16; of theory, 11, 13; of work, 56; of world, 288, 289
material rhetoric or rhetorical materialism, 5, 6, 14, 15, 49–50, 201
Mattern, Shannon, 287, 288
May, Todd, 97
McAllister, Michael, 30
McAuley, Julian, 111
McGarrigle, Conor, 180
McGregor, Bree, 20–21, 288, 298
McLuhan, Marshall, 36, 270
McRuer, Robert, 35
Merhmand, Elle, 215–16
memes, 168, 291
MEmorial, 19, 154, 180, 181, 182, 184–85, 194, 195
Menkman, Rosa, 34
mentorship, 203, 214, 222, 229, 236–37
metamedium, 147, 180, 181, 195
Michael V. Drake Institute for Teaching and Learning, 299n1
Michel, Anthony J., 6
Michelangelo, 152
Microsoft, 129; Visual Studio, 190; Word, 37
Milberry, Kate, 94
Miller, Carolyn R., 115, 116–17
mind/body distinction, 34, 150
minimalism, 46, 49, 54, 56, 60, 61
misogyny, 35, 242, 260n1. *See also* sexism
Moby Dick (Melville), 173, 177
modding, 10
Mohawk, 224, 228, 231, 232. *See also* Haudenosaunee
Moore, Kristen R., 78–79, 86, 87, 89n4, 89n5
Morey, Sean, 19, 147, 153, 160
Morris, Jane, 118, 124n5
Mount Rushmore, 184
move analysis. *See* Evoking moves; faciltiation research; Inviting moves; Staging moves
MPR 121 sensor, 147, 148*f*, 154, 160, 161*f*
multimodal composition, 5, 13
multimodality, 5–7, 11, 14, 18, 50, 146, 160, 162, 265, 270, 287, 288, 293, 298
multiply marginalized and underrepresented (MMU) scholars, 77, 78, 79, 80, 84–86, 87, 88, 88n1, 89n4, 89n5, 104, 123
Muñoz, José Esteban, 202
music therapy, 36
myth of noiselessness, 32–35, 36, 40, 41, 49

Nakamura, Lisa, 209
National Endowment for the Humanities, 130, 236

Native American: representation, 209, 222–26, 228, 236, 237, 238–39, 239n1. *See also* Indigenous peoples; *specific tribe names and languages*
Native American Graves Protection and Repatriation Act (NAGPRA), 120
Native Self Sufficiency Center, 232
natural language processing, 92, 108, 109, 123, 124n3
Natural Language Toolkit, 124n6
needlework, 288–89, 293, 294, 298
neoliberalism, 16–17
Neon, Lydia, 254
Netflix, 260n4
networked art or composition, 15, 55
networks, 32, 49, 56, 60, 61, 68, 79, 83, 84, 130, 138, 139, 162, 188, 204, 248, 268, 276, 291; of agents, 34, 95; global, 246; petroleum, 184, 186
new materialism, 29, 37, 286
new materialist rhetorics, 15
new media, 29, 50, 202, 217, 288; art, 215; rhetoric, 6; studies, 9
New York Graphic Workshop, 52
Nguyen, Minh-Tam, 110
No World Dreamers: Sticky Zeitgeist (video game), 209–11
Noble, Safiya U., 81, 95
noise, 30, 31, 33–34, 35–36, 40–41, 42, 43n3, 49, 119, 168. *See also* myth of noiselessness
Noisebridge, 10
nonbinary, 214
nonhuman, 12, 195, 224; adjuncts, 167; aliveness, 39; bodies, 32, 40; co-design with humans, 41; instruments, 39; relations with humans, 38. *See also* agency, nonhuman
nontextual rhetoric, 286, 287, 291, 298
nontraditional students, 265–66

O'Neill, Gloria, 227
Obama Hope image, 139
object-oriented ontology (OOO), 10–11, 21, 37
object-oriented rhetoric (OOR), 10–11, 15, 21
#occupy, 98
Ojibwe, 38–39
Omizo, Ryan M., 17, 94
Oneida, 20, 229–31, 234–5, 236–37. *See also* Haudenosaunee
online education, 20–21, 22, 265–66, 267–68, 277, 281
online forums, 102, 109, 110, 111–12, 114, 115, 117, 119, 124n4, 248, 294

ontology, 9, 10–11, 15, 37, 42, 167
open source, 18, 94, 100, 102, 135, 147, 159, 207, 242

pandemic. *See* coronavirus pandemic
paper writing device (PWD), 15, 48, 51, 56, 60, 61, 63, 69
paper-electronic circuits, 20–21, 267, 268–69, 270–73, 274–76, 278
Parasite (video game), 207–8
Parisi, David, 152–53
Parker, Elizabeth, 288–89, 293
participatory design, 20, 222, 225, 226–27, 228, 229, 230, 236, 237, 238
partitions of the sensible, 97, 98, 99, 101
pedagogy, 5, 203, 265, 287, 288, 298; carpentry and, 21; digital, 170; game design, 238; humanities computing, 177; language acquisition, 238; public rhetoric, 6. *See also* maker pedagogy; writing, pedagogy
Penn Treebank, 173
people of color, 10, 215, 286. *See also* BIPOC
Peterka-Bonetta, Jessica, 93, 94, 102, 103
Petersen, Emily January, 78
Pew Research Center, 93
phenomenology, 11
physical computing, 5, 7, 15, 18, 20, 30, 32, 94, 130, 145–46, 159, 162, 181, 186, 212, 215
Pilsch, Andrew, 19, 151, 156, 185
Piper, Adrian, 53–54, 60, 68
planned obsolescence, 36
Plato, 6
play, 19, 176–77, 201, 289; critical, 20, 211, 249
Player 2 (video game), 254–55
poetry: aleatory, 165; concrete, 46; materiality of, 56–57; remixed, 213; visual aspects of, 63–64. *See also* computer-generated poetry
police, 98, 99, 100, 103, 104
Porpentine (Charity Heartscape), 207–8, 209–11, 246, 247–48, 252–53, 256, 258
Porter, Tom, 231, 239n5
portfolio, 275, 280
post-PC era, 145–46, 157, 160, 162
posthumanism, 151, 162, 287
Potawatomi, 290
Potter, Harry, 159
Powell, Malea, 9, 105
Pozo, Teddy, 211
Pristash, Heather, 289
procedural rhetoric, 244, 245, 247, 251, 259

process: composing or writing, 48, 172, 175, 265, 267, 270, 272, 278, 279, 280, 281
Processing (programming language), 159, 160, 162, 178n4
processing time, 137, 138
processing power, 17
professional communication. *See* technical and professional communication; technical communication
prognosis, 31–32
prototyping, 7, 46*f*, 47*f*, 56, 180, 182, 212, 226, 231, 237, 288; interactive, 286, 292, 294, 295, 296, 298
public/private distinction, 34
public rhetoric, 6, 14, 16, 19, 52, 100, 101
public sphere, 4, 104
pynx, 6
Python, 99, 100
Pytlewska, Alicja, 155

queer, 11, 14, 86, 205, 215, 218, 256; bodies, 35; games and game designers, 10, 20, 202, 207, 208–9, 210–11, 212, 216, 217, 242, 244, 247, 248, 253; futurity, 202, 216; making practices, 201, 202, 203, 212; as rupture or excessiveness, 217; worldmaking, 19, 201, 206
Queer OS, 203
queer theory, 205
queer rhetorics, 80
Queers in Love at the End of the World (video game), 209, 255–56
Quinn, Zoë, 20, 242–43
Quintilian, 173
quipus, 50
Quispe-Agnoli, Rocío, 50

R (programming language), 17, 93, 94, 99, 100, 101–2
race, 35, 86, 99, 202, 203, 204, 214, 215; nonwhite, 8, 85; social construction of, 58
racial logics, 76
racism, 81, 88n1, 95, 99, 105, 113, 206, 225; structural, 22; national, 59
Ramsay, Stephen, 92
Rancière, Jacques, 93, 97–99, 100, 104
Raspberry Pi, 9, 17–18, 94, 130, 132, 133, 134, 135–38, 139, 140, 204
Rath, Richard Cullen, 51
Ratto, Matt, 4, 5, 13, 17, 19, 20, 56, 81, 94–95, 131, 177, 201, 205, 206–7, 222, 226–27; *DIY Citizenship* (Ratto and Boler), 4, 97, 195
Ravelry, 96
Ravenscroft, Alison, 286, 287

R-Bloggers, 102
Reddit, 130
Reid, Alex, 163n1
reflection, 13, 19, 81, 177, 195, 202, 205, 228, 268, 271, 273, 274
relational event, 38
relational network, 162
relationality, 8, 11, 12, 15, 29, 30, 31, 32, 37–39, 40–41, 42, 51, 57, 62, 80, 147, 167, 207, 218, 229, 231, 246, 255, 258; of touch and bodies, 145, 157, 150–51, 159. *See also* relations
relations: colonial, 10; emerging, 93; ethical, 10, 246, 254, 259; material, 287, 290–91, 194, 198; political, 22. *See also* relationality
remix, 10, 213
reparative making, 19–20, 201–2, 205, 206–7, 208, 209–10, 211, 213–14, 215–17, 218, 244
reparative reading, 206
representation. *See* Native American, representation
reprogramming, 3, 5, 7–8, 9, 10, 11, 12, 15, 16, 17, 18, 21, 22, 42, 48, 94, 102, 109, 111, 120, 123, 166
retrofit, 41
rhetoric and composition studies, 3, 4, 9, 11, 18, 21, 22, 77, 80, 92, 93, 100, 104, 105, 131, 132, 175; making and, 5–7, 12–13, 14, 15, 22, 181, 201, 203, 205, 216, 217. *See also* composition studies; writing studies
rhetoric and writing studies. *See* rhetoric and composition studies; writing studies
rhetoric-aesthetic work, 55, 56, 59
rhetorical ecology, 51
rhetorical genre theory (RGT), 108, 110, 115, 123. *See also* genre
rhetorical situation, 80, 100, 116
Rhode Island School of Design, 5
Rhodes, Jacqueline, 217, 287
rhythmic entrainment, 168–69, 172
Rice, Jenny (Edbauer), 51
Ridolfo, Jim, 6
Rieder, David M., 5, 18, 157, 159, 166, 181, 185, 186, 194, 195, 289–90
Riley-Mukavetz, Andrea, 10
RiTa library, 173
Ritual of the Moon, 206
Rivers, Nathaniel, 287, 289, 290, 291
Roadkill Tollbooth Actionable Monument (RTAM), 182–86, 189–90, 194
Rober, Mark, 21
Robinson, David, 101

Rogers, Melissa, 214
Ronald Reagan Memorial Turnpike, 182, 187
Ronell, Avital, 20, 245, 260n2
Roth, Dan, 111
Ruberg, Bo, 244
Rudman, Joseph, 169–70, 172, 178n1
Rustle Your Leaves to Me Softly (video game), 212–13

sadomasochism, 250
Sample, Mark, 176
Sanchez, Tony R., 225
Sanders, Elizabeth B.-N., 43n1
Saper, Craig, 15, 55–56
Sayer, Robert, 57
Sayers, Jentery, 9 201
Sayre, Henry, 47–48, 53
Schaechterle, Inez, 289
Science Buzz discussion forum, 11–12, 113, 119, 124n4
Scratch, 238
Scrivener, 247, 260n4
Seattle Attic, 10
Second Life, 216
Sedgwick, Eve Kosofsky, 19, 201, 205–6
Selfe, Cynthia L., 37, 81, 203
Selfe, Richard J., Jr., 37, 81
Seminole, 225
Semiosis, 48, 60, 62
Seneca, 232. *See also* Haudenosaunee
Sengers, Phoebe, 204
SenseLab, 13
sensor, 18, 41, 42, 131, 145, 146–47, 148*f*, 154–55, 159, 160, 185, 186, 187, 188, 194, 195, 212, 215–16, 290, 296, 298, 299n3
sentiment analysis, 94, 99, 103
sewing, 21, 42, 146
sex, 208, 216; violence and, 242, 249, 250, 251
sexism, 81. *See also* misogyny
sexuality, 20, 25, 202, 203, 205, 209, 214, 250, 251. *See also* bisexual; heterosexuality; queer
Shannon, Claude, 33–34
Shepherd, Dawn, 95
Sheridan, David M., 6–7, 15, 42, 245, 265, 267, 268
Shiloh, Michael, 296, 299n3
Shipka, Jody, 5
Sicart, Miguel, 212
Sierra, Wendi, 9, 20, 209, 246
signal-to-noise ratio, 34
Silicon Valley, 186
Silge, Julia, 101

Skins 1.0 workshop, 228, 229, 237
Skwarek, Mark, 180, 181, 185, 186, 194
Sky Woman story, 234–35
Sligh, Clarissa T., 57–58, 59
Smith, Cherise, 53–54, 68
Snow, C. P., 195
social justice, 14, 57, 76, 78–79, 86, 214
social media, 88n1, 98, 100, 101, 104, 131, 138, 188, 195, 200, 295
socioeconomic class, 11, 41, 99, 203, 204, 217, 224
Söderlund, Lars, 290
solidarity, 215, 245, 246, 247, 249, 253–54, 259
solidarity machines, 20, 246, 247, 259
sonic rhetoric. *See* sound
sound, 5, 33, 34, 42, 146, 152, 156–57, 216, 247; accessibility and, 297–98; in augmented reality, 181, 195; circuits and, 39, 42; materialist approach to, 37; music and, 29, 30, 36, 41, 45; processing, 134; recording, 230. *See also* noise; myth of noiselessness
soundscape, 155–157, 213
spaCy, 119, 120, 121*f*, 124n3
SparkFun Electronics, 146
Speak & Spell, 39
speculative design, 7
speculative realism, 37
speech act, 116, 117
Sperrazza, Whitney, 96
Spinoza, Baruch, 163n1
Spinuzzi, Clay, 226
Squinkifer, Dietrich "Squinky," 212
Srauy, Sam, 225
Staging moves, 109, 110, 111, 113, 115
Staley, David, 16
Stappers, Pieter Jan, 43n1
Steam (video game platform), 243
Stedman, Kyle D., 43n3
Steele, Claude M., 88n2
Stein, Gertrude, 55
Stone, Kara, 206, 218
Strong Fire, A (video game), 20, 222, 229, 234, 238
style, 166–67, 175, 235, 246
subjectivity, 150, 151, 163, 185
suffrage, 289
supercomputer, 132, 133
supervised machine learning, 109, 110
survivance, 226, 227, 239n2
Swales, John, 110
Swalesean move analysis, 108, 110, 115
Syllabot, 286, 295–98, 299n2
syllabus, 21, 266, 286, 291–96, 298

312   INDEX

tactical media, 4, 5, 7
tactile, 60, 162, 226; craft and, 267; engagement, 45, 49, 57, 59, 62, 68; literacies, 50; objects, 16, 57; practices, 21; processes, 270; rhetoric, 49, 51; visual-tactile experiences, 57
Tate Modern, 52–53
techne, 86
technical and professional communication (TPC), 78, 83, 86, 89n1. *See also* technical communication
technical communication, 16, 76–78, 80, 84, 85, 86, 87, 88, 123. *See also* technical and professional communication
technical writing. *See* technical and professional communication; technical communication
Technoculture, Arts and Games Research Centre, 212
technological rhetoric, 37
technology: apolitical or neutral view of, 14, 76, 95
technotext, 54
term weighting, 117
terministic screens, 14
text mining, 11, 16–17, 92, 93, 94, 95–96, 98, 99, 100, 101, 103, 104, 105, 139. *See also* critical text mining
text segmentation, 117
TextMeshPro, 188
TextTiling algorithm, 117
TF-IDF, 117, 119, 120, 124n7
Thanksgiving Address. *See* Haudenosaunee, Thanksgiving Address
theory/practice divide, 9, 13, 15, 19, 95, 176, 259
Thiel, Tamiko, 186, 194
things, 10, 40, 287, 288, 290–91, 298; making things, 5, 7, 11, 13, 21, 195, 207, 214, 215, 248, 286, 287
three-dimensional objects, 5, 45, 57, 58, 276f. *See also* paper writing device (PWD)
3D printing, 6, 11, 291
TileBars text analysis, 117
tinkering, 7, 8, 22, 88, 159, 181, 194, 201, 204, 265, 267, 279, 280, 283
Tinnell, John, 184, 185, 194, 195
Tinney, Craig, 188
tiny computer, 130, 131, 132
topic modeling, 115
topics: for analysis, 108, 117–18, 119
topoi, 92, 175
touch, 30, 59–60, 66, 146–47, 149, 152–53, 162–63, 212–13; circulation of, 49, 60, 66, 68; interactivity and, 18, 147, 149, 154–62; as relational, 145, 147, 150–51; rhetorical, 51, 61, 147, 149, 151. *See also* Arduino, Touch Board; bodies, touch and; capacitive-touch sensor; interface, touch; touchscreens
touchscreens, 145, 147, 153, 185, 212
TRACE Innovation Initiative, 180
Tracery library, 172
Track the Recovery website, 95
tranarchafeminist, 207
#trans, 98
transduction, 18, 146, 186–87, 195
transgender, 20, 78, 86, 99, 207, 209, 210, 211, 214, 215, 247, 252
Transocean. *See* British Petroleum
Trump, Donald, 96
Tsing, Anna Lowenhaupt, 123
Tumblr, 130, 248, 258f
Tuscone Women Techmakers Hackathon, 10
tweets, 9, 17, 100, 101, 102, 104, 139–40, 172, 187, 188, 191–92. *See also* Twitter
Twine, accessibility of, 208–9, 244, 248; for course assignments, 238; games, 20, 207, 208, 209, 242, 244, 245–49, 252–53, 259
Twitter, 19, 100, 101, 104, 130, 131–32, 135, 139, 187; API, 94, 101, 131, 188, 191; development portal, 139; hashtags, 93, 94, 98, 99, 102–4, 190, 191–92; script, 188. *See also specific hashtags*; tweets; Twitter bot; TwitteR package; webscraping
Twitter bot, 19, 172–73, 176, 177
TwitteR package, 17, 100, 101
typeface. *See* font
typewriter, 152, 175
Tyvek, 156

Ulmer, Gregory L., 6, 19, 174, 181, 182, 185, 194, 195
Ultra Business Tycoon III (video game), 252–53
Unicode, 99
Unity (software), 19, 180, 187, 188, 189–92, 194
Upton, Thomas A., 115
US census, 223
usability, 130, 134, 290
user-centered design, 226

Van Horn, Nicholas, 17, 77, 94
Van Kuppevelt, Jan, 118, 121
Vandenberg, Peter, 100
Vasalou, Asimina, 228
vector machine learning libraries, 109
vector space models, 115
vectorization. *See* TF-IDF

Vee, Annette, 195
video, 5, 156, 162, 277, 282; accessibility of, 298; assets, 19, 180; processing, 134; student-created, 271, 273, 274, 275; tutorials, 265, 269
video game: accessible software for the creation of, 209, 242, 247–48; cultures, 213–15, 242–43; 260n1; design, 9, 10, 20, 202, 206, 207, 208, 209, 210, 212–13, 214–15, 216, 222–23, 225–29, 231, 236–37, 238, 239n2, 246, 247–48; embodiment and, 213, 242; and identification, 242, 247, 259; Indigenous, 222, 226–31, 234–39; and inclusion, 214–15, 225; interface, 30, 212–13; making, 206, 207, 214, 228–31; mainstream or traditional, 207, 210, 243, 249, 251–52; and Native American characters, 223–225; queer indie, 20, 202, 208, 209, 216, 217; physical controllers, 212–13; text-based, 246; and violence, 250–52; weird, 217. *See also specific video game titles*; GamerGate; Native American, representation; procedural rhetoric; Twine
Village Voice, 54
Vimeo, 156
violence, 57, 98, 100, 139, 177, 207, 224, 242–43, 245, 247, 249–53, 258
virtual, 18, 37, 145, 145–47, 156, 157, 163n1, 186, 188, 224, 259; monuments, 181, 182, 184–85
visual artists, 160
visual aspects of text, 62, 208, 247. *See also* font
visual design, 279, 283
visual forms, 6
visual learning, 278
visual material, 46
visual mediation, 48
visual objects, 16
visual representation, 83, 86, 267. *See also* data, visualization
visual rhetoric, 288, 298
visual rhythms, 53, 58
visual-tactile reading experiences, 75, 60
Vitanza, Victor J., 175–76
Vizenor, Gerald Robert, 239n1
volvelle, 57, 63–64, 68
Vonnegut, Kurt, 99
VOSviewer, 81, 82f, 83–85, 86
Voyant, 100
Vuforia AR SDK, 19

Wallace, David Foster, 243
Walters, Shannon, 51, 145, 147, 149, 150, 151, 162

Walton, Rebecca, 78–79, 86, 87, 89n4, 89n5
wampum belts, 50–51
Warner, Michael, 202, 205
wearable rhetorics, 18
wearables, 18, 94
webscraping, 9, 17, 18, 93, 94, 95, 98, 99, 100–102, 103
Weimer, Maryellen, 295–96
Western exceptionalism, 113
white supremacy, 57, 76
whiteness, 8, 76, 78, 97, 100, 217; of academia, 8, 79; of digital humanities methods; 86; of maker cultures, 9, 201, 204
Whitson, Roger, 5, 7, 8, 10, 11
Wikipedia, 33f, 130
Wilkens, Matt, 96
Willard, Frances, 289
Willard Dress, 289, 293
Williams, Dmitri, 223, 239n1
Williams, Miriam, 80
Williams, Paul, 233
Williams, William Carlos, 53, 55, 62–63
Wisniewski, Nicholas, 224
With Those We Love Alive (video game), 256–58
Womack, Anne-Marie, 292, 293
Women's Studies Multimedia Studio (University of Maryland), 214
Wood, Sue Carter, 289
Woolgar, Steve, 123
wonder, 147, 156, 162, 163
word frequency counts, 115
WordPress, 268, 269, 270, 271, 273, 274
working class. *See* socioeconomic class
Works Progress Administration, 230
World of Warcraft (video game), 224
worldmaking, 22, 207, 215, 254, 259; queer, 19, 201, 206
writing, 11, 14, 92, 134, 167, 174, 209, 217, 245, 248, 286; and agency, 175–76; creative, 169; digital, 5, 13, 146, 154, 194, 203, 275; literary, 169; multimodal, 288; pedagogy, 6, 37, 194, 203, 247, 265, 266, 269–70, 271, 272, 275, 280, 287, 288, 295, 298; portfolios, 275, 280; process, 265, 267, 270, 280, 281; reflective, 271, 278; scholarly conventions of, 246
writing studies, 7, 14, 130, 146, 154, 180, 181, 217, 286, 287, 288, 299. *See also* composition studies; rhetoric and composition studies
Writing Studies Tree, 130
Wylie, Sara Ann, 226
WYSIWYG, 190, 207, 208

Wysocki, Anne Frances, 50
Wysocki, Rick, 11, 12

Xbox, 248

Yale Computer Center, 175
yarn-bombing activism, 4
Yancey, Kathleen Blake, 50, 60
Yergeau, M. Remi, 41

YORICK (poetry robot), 169, 171, 172
YouTube, 21, 130, 156
Yuchi, 225
Yuen, Elaine, 30
Yule, George, 117–18

Zimmerman, George, 93
zine, 4, 9, 202, 215
Zuray, M., 64*f*

www.ingramcontent.com/pod-product-compliance
Lightning Source LLC
Chambersburg PA
CBHW071230070526
44583CB00017B/2116